Digital Immunoassays

Drawing from extensive laboratory experience and international research, this groundbreaking book is the first to systematically explore digital immunoassays, covering everything from basic concepts to advanced applications.

The book is organized into seven chapters. It begins with an introduction to immunoassay fundamentals and then provides detailed coverage of digital immunoassay principles. Next, it explores discretization, the key feature of digital immunoassays, before examining the implementation strategies, technical principles, and performance of heterogeneous and homogeneous digital immunoassays. The final chapter addresses practical applications in areas such as tumor marker, neurodegenerative disease, and virus detection.

With its comprehensive approach, in-depth content, and broad scope, this book is ideal for researchers in the life sciences, medical professionals, and chemical metrologists. It also serves as an invaluable reference for educators, graduate students, and professionals involved in research and development or management within the *in vitro* diagnostic industry.

Hongwei Gai is a Professor of Analytical Chemistry at the Jiangsu Normal University, China. His current research interests are focused on single molecules/nanoparticles optical imaging, digital immunoassay, and their applications in medical laboratory tests.

Qingquan Zhang is a Professor of Analytical Chemistry at Jiangsu Normal University, China. His current research activities include the development and application of digital immunoassay techniques and droplet microfluidics.

Xiaojun Liu is a Professor at Jiangsu Normal University, China. Her current research focuses on the development and application of new digital immunoassays for biomarker detection.

Chenghua Zong is a Associate Professor at Jiangsu Normal University, China. Her current research activities include the development and application of new analytical methods for biosensing and analyzing.

Digital Immunoassays

Hongwei Gai
Qingquan Zhang
Xiaojun Liu
Chenghua Zong

CRC Press
Taylor & Francis Group
Boca Raton London New York

CRC Press is an imprint of the
Taylor & Francis Group, an **informa** business

SCIENCE PRESS

The cover image is created by Hongwei Gai.

This book is published with financial support from the Natural Science Foundation of China (NSFC grant 22174056) and "343" Industry Development Project of Xuzhou (gx2023013).

First edition published 2026
by CRC Press
2385 NW Executive Center Drive, Suite 320, Boca Raton FL 33431

and by CRC Press
4 Park Square, Milton Park, Abingdon, Oxon, OX14 4RN

CRC Press is an imprint of Taylor & Francis Group, LLC

Translated by Xiaojun Liu

Published by CRC Press, Taylor & Francis Group LLC, an informa Company, under exclusive license granted by China Science Publishing & Media Ltd. for English language throughout the world excluding Mainland China, and with non-exclusive license for electronic versions in Mainland China.

ISBN: 978-1-041-21362-8 (hbk)
ISBN: 978-1-041-21828-9 (pbk)
ISBN: 978-1-003-72921-1 (ebk)

DOI: 10.1201/9781003729211

Typeset in Minion
by SPi Technologies India Pvt Ltd (Straive)

Contents

Figures

Tables

Preface

Severe diseases such as viral infections, malignant tumors, cardiovascular diseases, and neurodegenerative disorders significantly impact health and quality of life, while also placing a heavy burden on society's economy. Early screening and diagnosis are critical for combating these diseases and improving patients' survival rates and life quality. To achieve early detection, the first challenge is improving the sensitivity of measurement methods. In the early stages of disease, antigen concentrations in the blood are often at the femtomolar (fM) level or even lower, whereas conventional immunoassays can only detect antigens in the nanomolar (nM) to picomolar (pM) range. The second challenge is advancing the widespread adoption of personalized medicine. Current diagnostic standards use population averages as the threshold, considering values above the threshold as positive and those below as negative. However, it is not appropriate to use population averages as a threshold for individual early screening and diagnosis, as biomarker expression varies greatly between individuals. For individuals with low biomarker expression, even a 1,000-fold increase in expression may still fall within the "normal" range, thus missing the opportunity for early diagnosis and screening.

Digital immunoassay is a technology used to quantify targets by counting *individual immunocomplex molecules*. When the detected object is a single molecule, it is referred to as single-molecule counting immunoassay. This technology demonstrates ultrahigh sensitivity, enabling the detection of abnormal biomarker expressions in the early stages of severe diseases, even revealing changes below the "normal" range that current methods cannot identify. Digital immunoassays are believed to address these challenges effectively. Portable and automated instruments designed for digital immunoassays can track changes in an individual's biomarker levels over time, establish correlations between biomarker levels and disease progression, and assess the likelihood of disease onset. This represents a shift from a reactive approach to disease management, focusing on prevention before the disease occurs rather than responding after its onset. It is anticipated that digital immunoassay technology will play a crucial and irreplaceable role in fundamental research on natural phenomena and laws, biomarker screening and validation, and clinical diagnosis and monitoring of severe diseases.

Digital immunoassay is widely regarded as the next generation of immunoassay technology and has emerged as a key area of investment within the *in vitro* diagnostics (IVD) industry in recent years. Leading international companies in this field have secured substantial capital investment and gained widespread recognition within the IVD sector, establishing monopolies in the instrument market. To develop digital immunoassay instruments with independent intellectual property in China, it is essential to pursue a differentiated path. Simply following the same technical approaches as existing platforms will not overcome the commercial and technological barriers, nor enable competition at the same level.

This book is based on decades of single-molecule detection research conducted by Professor Hongwei Gai's team at Hunan University, along with over ten years of digital immunoassay research at Jiangsu Normal University. It provides a comprehensive overview of the fundamental principles, key technologies, classifications, and recent advancements in digital immunoassay applications. The book also explores the current landscape and future trends in the development and industrialization of instruments for digital immunoassay. Additionally, it discusses the challenges faced in instrument development and shares the research group's insights into this emerging technology.

This book was structured by Dr. Hongwei Gai, who developed the overall framework and authored Chapters 1, 2, 5, and 6. Chapters 3 and 4 were written by Dr. Qingquan Zhang. Dr. Xiaojun Liu contributed to the writing of Chapter 5 and translated the entire book into English, while Dr. Chenghua Zong was responsible for proofreading the full manuscript. Dr. Linfen Yu and Yang Wei from Shenzhen Borui Biotechnology Co., Ltd. provided valuable input for the section on microdroplet-based discretization in Chapter 3. Several graduate students from the research group assisted in collecting documents for Chapter 6 and organizing the book's references. These students include Yuting Su, Yue Han, Wenwen Chai,

Zihan Xu, Jiajia Hu, Yafang Wang, Chaonan Huang, Ye Rong, Xiaoxiao Yang, and Yameng Wang.

We would like to express our sincere gratitude to the National Natural Science Foundation of China, the Jiangsu Provincial Department of Science and Technology, the Jiangsu Provincial Department of Education, and Jiangsu Normal University for their strong support of our research. Special thanks are extended to the graduate students who have been part of our research group, including those from Hunan University – their contributions have been a vital foundation for the completion of this book. Hongwei Gai is also deeply grateful to Professor Bingcheng Lin of the Dalian Institute of Chemical Physics, Chinese Academy of Sciences, who served as his doctoral supervisor. Professor Lin has provided consistent support throughout Gai's doctoral studies and continuing into his career as an independent principal investigator. This book would not have been possible without his guidance and encouragement.

Acknowledgments

THE AUTHORS WOULD LIKE to thank the following funding sources:

National Natural Science Foundation of China (22174056, 20705007, 21075033, 21575053, 21775057, 21405064, 21505058, 21804060)

Jiangsu Provincial Department of Science and Technology Natural Science Fund (BK20140233, BK20150228)

Jiangsu Provincial Department of Education Excellent Science and Technology Innovation Team Project, Major Basic Research Project for Universities, "Qinglan Project" Young and Middle-Aged Academic Leaders Project, Jiangsu Province "Six Talent Peaks" High-Level Talent, and "333 High-Level Talent Training Project".

"343" Industry Development Project of Xuzhou (gx2023013).

Shenzhen Borui Biotechnology Co., Ltd.

Dalian Institute of Chemical Physics, Chinese Academy of Sciences:
Professor Bingcheng Lin

Jiangsu University:
Professor Yewang Zhang

Hunan University Graduate Students:
Qun Li, Nan Li, Huaping Chen, Rui Han, Xingbo Shi, Xiaobing Bu, Lichun Sun, Xinliang Liu, and Haixia Ding.

Jiangsu Normal University Graduate Students:
Yang Tu, Minmin Li, Suli Dong, Meng Xu, Min Wang, Shuzhan Wang, Conghui Huang, Hanlin Li, Tian Lei, Bo Li, Mimi Sun, Xuebing Zhang, Yusu Zhang, Jing Wang, Wentao Tian, Jiaqiu Chen, Zhangjian Wu, Yuanyuan Sun, Shi Lü, Jiajia Li, Xiaoting Jin, Mengyi Ge, Hongwei Ding, Weiwei Zheng, Zhen Wang, Xinyi Lin, Ziyin Zhong, Yuting Su, Yue Han, Wenwen Chai, Zihan Xu, Jiajia Hu, Yafang Wang, Chaonan Huang, Ye Rong, Xiaoxiao Yang, Yameng Wang, Baiyuan Zhao, Dandan Zhao, Yue Liu, Qian Li, Silu Zang, Shuangshuang Ma, Haiwei Lin, Shan Wang, and Xishuang Yao.

CHAPTER 1

Introduction to Immunoassays

OVERVIEW OF IMMUNOREACTION

Every day, the human body encounters countless foreign substances that invade through various systems, including the skin, digestive system, and respiratory system. Fortunately, most of these substances do not cause significant harm, thanks to the immune system. The immune system plays a critical role in defending the body against damage from these invaders. Immunity refers to the body's ability to distinguish between self and non-self-substances, triggering an appropriate response to the latter. This response often involves the production of specific antibodies and the activation of particular lymphocytes. The immune response – also called an immunoreaction – is a complex series of processes initiated by the immune system to combat foreign invaders. Immunity is a well-established field, extensively covered in immunology literature.[1–3] To help beginners quickly understand the basics, we will provide an overview of both immunity and immunoassays.

Antigen

An antigen is a substance that triggers an immune response and can specifically bind to the products of that response. While many substances in nature can act as antigens, not all foreign substances qualify. Antigens have two key properties: immunogenicity and immunoreactivity. Substances

DOI: 10.1201/9781003729211-1

that have both properties are called complete antigens. On the other hand, substances that only have immunoreactivity are known as haptens or incomplete antigens.

Immunogenicity refers to an antigen's ability to trigger an immune response, either by producing antibodies or activating specific lymphocytes. This is a key characteristic that defines an antigen. For an antigen to be immunogenic, it must be recognized as foreign by the immune system. "Foreignness" includes substances that are not encoded by the organism's own genes, as well as those the immune system hasn't encountered during its development. Under normal circumstances, the immune system doesn't attack the body's own cells or tissues. But when the immune system becomes dysfunctional or the body faces harmful triggers, it can mistakenly identify its own substances as foreign, leading to an autoimmune response. Certain characteristics – such as the antigen's size, structure, and chemical stability – are important for its immunogenicity. Generally, larger antigens, especially those with complex structures and stable chemical properties, are more likely to provoke a strong immune response. Most antigens are proteins, often with molecular weights greater than 10 kDa.

Immunoreactivity is the ability of an antigen to bind to the specific antibodies or lymphocytes produced in response to that antigen. This binding serves as a typical sign of immunoreactivity and is highly specific – an antigen can only interact with the antibodies it triggers. The specificity of an antigen is determined by its antigenic determinants, also known as epitopes. An epitope consists of the chemical groups and corresponding structures on the antigen's surface. Peptide epitopes typically contain 5 to 15 amino acids, while polysaccharide ones are made up of 5 to 7 monosaccharide units. Nucleic acid hapten epitopes usually consist of six to eight nucleotide residues. The total number of epitopes that an antigen can present is known as its binding valence. Some haptens have only one epitope, which means they can bind to just one antibody and are classified as monovalent antigens. On the other hand, natural antigens often have multiple identical or different epitopes, making them multivalent antigens. Different antigens can share the same antigenic epitope, called common antigenic epitopes. Antibodies can not only bind to their specific antigens but also react with other antigens containing these common epitopes – a phenomenon known as cross-reactivity. While specificity is crucial for immunoassays, cross-reactivity can affect the accuracy of these tests.

Antibody

When B lymphocytes are activated by antigens, they proliferate and differentiate into plasma cells. These plasma cells then secrete antibodies, which are a type of immunoglobulin (Ig). While all antibodies are Igs, not all Igs function as antibodies – some have a similar chemical structure but don't have the same antigen-binding activity. Active antibodies specifically recognize and bind to their matching antigens. In the blood and other tissue fluids, antibodies form immune complexes when they interact with antigens. This process can block the pathways that allow antigens to damage cells, inactivating them through mechanisms like neutralization, agglutination, precipitation, and complement activation and fixation.

As shown in Figure 1.1, an Ig monomer consists of two identical heavy chains (marked as H in the figure) and two identical light chains (marked as L in the figure). The light chains are made up of about 220 amino acids, with a molecular weight of around 25 kDa, while the heavy chains contain between 450 and 550 amino acids and have a molecular weight of about 50 to 70 kDa. Each heavy chain is linked to its corresponding light chain by disulfide bonds, forming a "Y"-shaped structure. At the N-terminus of each chain, there's a sequence of over 100 amino acids that shows significant variability, which is why this region is called the variable region (V region). This V region typically serves as the antigen-binding site, with the variable regions of the heavy and light chains known as V_H and V_L, respectively. Within the V_H and V_L domains, there are three hypervariable

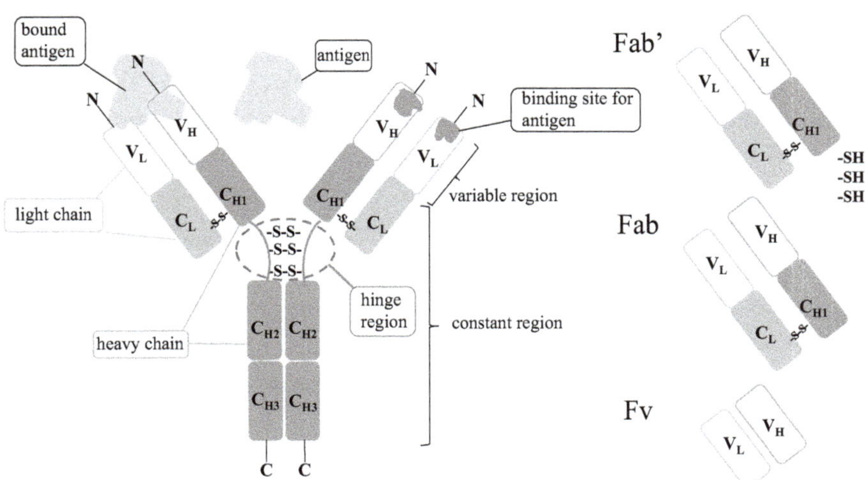

FIGURE 1.1 Structural diagram of an immunoglobulin monomer.

regions (HVRs), characterized by a high frequency of amino acid sequences. These HVRs are crucial for the precise interaction of the Ig with antigen epitopes and are also referred to as complementarity-determining regions (CDRs). In contrast, the amino acid sequences outside the variable regions are relatively constant and are called the constant region (C region). The C region of the heavy chain is further divided into three to four structural domains, such as C_{H1}, C_{H2}, and C_{H3}, while the constant region of the light chain has only one structural domain. The segment between the C_{H1} and C_{H2} domains is known as the hinge region. It's rich in proline and provides flexibility, allowing the immunoglobulin to change its conformation. This flexibility helps expose binding sites and enables the antibody to accommodate antigens at different distances. The hinge regions of the two heavy chains are connected by disulfide bonds. The part of the antibody responsible for antigen binding is called the Fab fragment, and the variable region within the Fab fragment is referred to as the F_v region.

Antibodies are crucial in clinical diagnosis, immunoassays, and disease treatment. The effectiveness of immunoassays largely depends on the quality of the antibodies used. To produce large quantities of antibodies, artificial preparation methods are typically employed. Depending on how they are made, antibodies can be classified into three main types: polyclonal antibodies, monoclonal antibodies, and genetically engineered antibodies.

Polyclonal Antibody

When an antigen is injected into an animal, it activates multiple B cells, which then produce antibodies that enter the bloodstream. The antigen usually contains a mix of different antigen components and multiple epitopes. Even a pure protein antigen can have several antigenic epitopes. The antibodies produced in response to these multi-epitope antigens are diverse and are called polyclonal antibodies. In addition, antibodies made by different B lymphocytes against the same antigenic epitope are also considered polyclonal. Serum collected from immunized animals, convalescent patients, or immunized populations that contains these polyclonal antibodies is known as immune serum or antiserum.

Monoclonal Antibody

Monoclonal antibodies are produced by plasma cells that originate from a single B lymphocyte in response to a single antigenic epitope. These antibodies are identical in properties such as specificity, affinity, biological characteristics, and molecular structure. Because plasma cells have a limited lifespan and are difficult to culture, Köhler and Milstein developed a groundbreaking method in 1975. They fused mouse B cells with myeloma cells to create hybridomas. These hybridomas combine the myeloma cells' ability to grow indefinitely with the B cells' ability to produce specific antibodies. The antibodies produced by these hybridomas are known as monoclonal antibodies. This innovation has had a huge impact on immunology and earned the Nobel Prize in Physiology or Medicine in 1984.

Genetically Engineered Antibody

Most monoclonal antibodies today are derived from mice, which can sometimes cause side reactions in humans. However, new antibody molecules that are partially or fully humanized can now be created using advanced genetic engineering techniques. These include human-mouse chimeric antibodies, bispecific antibodies, and fully humanized antibodies. Genetically engineered antibodies retain the high uniformity and strong specificity of monoclonal antibodies while overcoming the issues often seen with mouse-derived antibodies. As a result, these engineered antibodies are already being used in a wider range of applications and are expected to play an even bigger role in the future, especially as fully human antibodies become more common in antibody-based treatments.

Antigen-Antibody Reaction

The binding of antigens to antibodies is a specific and reversible reaction, and it can be described using the law of mass action:

$$[Ag]+[Ab]\underset{k_d}{\overset{k_a}{\rightleftharpoons}}[AgAb], \tag{1.1}$$

where [*Ag*], [*Ab*], and [*AgAb*] represents the concentrations of the antigen, the antibody, the antigen-antibody immunocomplex, respectively. k_a is the association rate constant in units of L/(mol·s), also referred to as k_{on}. k_d is the dissociation rate constant in units of s^{-1}, also known as k_{Off}. In common biological systems, k_A typically ranges from 10^3 to 10^7 L/(mol·s) while k_d ranges from 10^{-6} to 10^{-1} L/(mol·s). The rate constants can be measured using surface plasmon resonance (SPR).

When the reaction reaches equilibrium, the following equations are derived:

$$k_a[Ag][Ab] = k_d[AgAb], \tag{1.2}$$

$$K_{eq} = \frac{k_a}{k_d} = \frac{[AgAb]}{[Ag][Ab]}. \tag{1.3}$$

Here, K_{eq} represents the equilibrium constant in units of L/mol, also known as the association constant or affinity constant (K_a). Its reciprocal is the dissociation constant (K_d) in units of mol/L. Both the association and dissociation constants serve as indicators of binding strength: a larger association constant or a smaller dissociation constant signifies a stronger affinity and greater stability in the antigen-antibody complex. The dissociation constants for most antibodies range from 10^{-9}–10^{-6} mol/L. Antibodies with a dissociation constant of 10^{-9} L/mol are classified as high-affinity antibodies, while those with a dissociation constant of 10^{-12} mol/L are considered ultrahigh-affinity antibodies. High-affinity antibodies facilitate rapid reactions, require lower consumption, and exhibit higher sensitivity in clinical testing. The affinity and dissociation constants can be calculated from the rate constants or determined using techniques such as capillary electrophoresis and immunoassays.

OVERVIEW OF IMMUNOASSAYS

An immunoassay is a chemical measurement technique used to capture target molecules in a system through the specific interaction between antigens and antibodies. The signal generated by these captured molecules, along with the reagents used in the process, can then be quantified. The target molecules can be either antigens or antibodies. Since other substances in the system typically don't interfere with this specific immunoreaction, immunoassays offer high specificity and sensitivity. As a result, they are widely used in fields such as medical testing, biopharmaceuticals, and environmental chemistry.

Immunoassays were first developed by American scientists Berson and Yalow in 1959.[4] To quantify insulin in blood, they labeled insulin standards with radioactive iodine (I^{131}) and added the labeled standards and insulin antibodies into the samples. The I^{131}-labeled insulin competed with the endogenous insulin in the samples for binding to the antibodies. The insulin bound to the antibodies was separated from the unbound one using paper chromatography, and the insulin levels were quantified based on the radioactivity intensities of the two fractions. For this pioneering work, Yalow was awarded the Nobel Prize in Physiology or Medicine in 1977. In 1968, Miles and Hales marked radioactive elements on antibodies rather than on antigens and conducted single epitope-based immunoassays.[5] In 1971, Swedish scientist Perlmann published the first paper on enzyme-linked immunosorbent assays (ELISA) where alkaline phosphatase was used as a reporting probe instead of radioactive elements to detect IgG.[6] In the same year, Dutch scientist Schuurs employed horseradish peroxidase as a reporting tag to quantify chorionic gonadotropin in urine, thus inventing enzyme immunoassay (EIA) technology.[7] These two studies are regarded as the origins of enzyme-labeled immunoassays. In 1973, Belanger first reported a sandwich-structured immunocomplex-based immunoassay for the alpha-fetoprotein detection,[8] which subsequently gained widespread adoption. The advancement of immunoassays was significantly accelerated in 1975 when Köhler and Milstein established hybridoma technology,[9] enabling the large-scale production of monoclonal antibodies. They were awarded the Nobel Prize in Physiology or Medicine in 1984 for their contributions. In 1976, chemiluminescence was used in immunoassays as a detection technique.[10] To this day, it remains the most widely commercialized and commonly used method in hospital laboratories. With these advancements in place, immunoassays rapidly evolved, finding widespread application in clinical testing. In recent years, immunoassays have expanded into various applications, and immunological technologies have advanced toward greater sensitivity and the capability for multicomponent analysis within a single assay. In 1992, Sano et al. introduced polymerase chain reaction (PCR) into immunoassays by utilizing antibodies labeled with amplifiable DNA. This innovation transformed protein detection into PCR product detection, achieving detection levels as low as 10^{-18} mol in serum.[11] In 2010, Quanterix, an American company, developed single-molecule ELISA technology,[12] allowing for the detection of protein antigens at sub-femtomolar concentrations. This innovation heralded the advent of digital immunoassays. A more comprehensive history of immunoassay development can be found in the relevant literature.[13–15]

IMMUNOASSAY CATEGORY

Over the past 60 years, immunoassays have significantly evolved into various analytical methods, driven by the continuous development of innovative signal amplification and detection strategies. Immunoassays can be classified from multiple perspectives. For example, they may be categorized into heterogeneous and homogeneous immunoassays, depending on whether a separation step is required; single-target and multicomponent immunoassays, based on the number of target analytes; and labeled versus nonlabeled immunoassays, determined by the presence of a reporting tag. Labeled immunoassays are further divided into several types, including radioimmunoassays, fluorescent immunoassays, enzyme immunoassays, chemiluminescent immunoassays, electrochemical immunoassays, colorimetric immunoassays, Raman immunoassays, plasmon resonance immunoassays, and liposome-amplified immunoassays, among others. Immunoassays can also be classified by the type of labeling used: antigen-labeled immunoassays (also called competitive immunoassays) and antibody-labeled immunoassays (known as immunometric assays). In terms of immunocomplex configuration, they are classified as single-binding site or dual-binding site immunoassays, with the latter commonly referred to as sandwich immunoassays. The platform used for the immunoassay also plays a role in classification, including capillary electrophoresis immunoassays, microfluidic chip immunoassays, lateral flow immunoassays, flow injection immunoassays, and mass spectrometry immunoassays. Additionally, immunoassays can be divided into ensemble immunoassays and single-molecule (or digital) immunoassays, depending on whether the signal is derived from single-target molecules. These categories often overlap and integrate, such as in multicomponent homogeneous Raman immunoassays using microfluidic chips or quantum dot (QD)-labeled multicomponent homogeneous digital immunoassays. Next, we highlight some of the key and most commonly used categories of immunoassays.

Nonlabeled Immunoassays and Labeled Immunoassays

Nonlabeled immunoassays are traditional immunotechniques that detect antigens or antibodies by measuring changes in physicochemical properties resulting from antigen-antibody interactions. These properties primarily include precipitation, agglutination, immune diffusion, and immunoturbidity. While these immunoassays are relatively straightforward and were established in the early days of immunoassay technology, some continue to be utilized in clinical testing today. However, a common limitation of these methods is their low sensitivity, rendering them unsuitable for quantitative detection. Consequently, labeled immunoassay techniques were developed. Labeled immunoassays involve the attachment of detectable signals – such as radioactive, optical, or electrical markers – to

TABLE 1.1 Classification of Commonly Used Labeled Immunoassays

Labeling Species	Analyte	Assay Category	Number of Binding Sites
Antigen	Antigen (Figure 1.2)	Competitive assay	1
Antibody	Antibody (Figure 1.3)		
	Antigen (Figure 1.4)		
	Antigen (Figure 1.5)	Sandwich-structure assay	2

antigens or antibodies, a process known as labeling. These signaling substances are referred to as tags, or reporters, or labels, while the antibody or antigen labeled with these tags is termed a probe. The quantification of the antibody or antigen is based on the signal strength of the probe. Labeled immunoassays significantly enhance detection sensitivity and dynamic range, thereby broadening the methodologies and applicable contexts for immunoassays.

Labeled immunoassays represent the predominant techniques in immunoassays. In these assays, labels are attached to either antigens or antibodies. When labels are linked to antigens, the assays typically function in a competitive format, primarily for antigen detection. In contrast, labeled antibodies are more commonly used, often resulting in a sandwich assay that utilizes dual-binding sites to form a stable immunocomplex. Furthermore, labeled antibody-based immunoassays can also be configured as single-binding site competitive assays, enabling the detection of both antigens and antibodies. Table 1.1 summarizes these labeled immunoassays.

Competitive Immunoassays

Antigen Detection Using Labeled Antigen In competitive immunoassays utilizing labeled antigens, it is essential to prelabel the antigen with signal tags. This labeled antigen must be the same as the antigen present in the sample. The typical analytical process is illustrated in Figure 1.2. A fixed amount of antibody is immobilized on a solid phase, to which both the labeled antigen and the unlabeled antigen from the sample are introduced. These two antigens compete for binding to the antibody. After a designated incubation period, the unbound antigens are separated and removed; therefore, this allows for the detection of the labeled antigen on the solid phase. A stronger signal indicates that a greater number of binding sites on the antibodies are occupied by the labeled antigen, suggesting

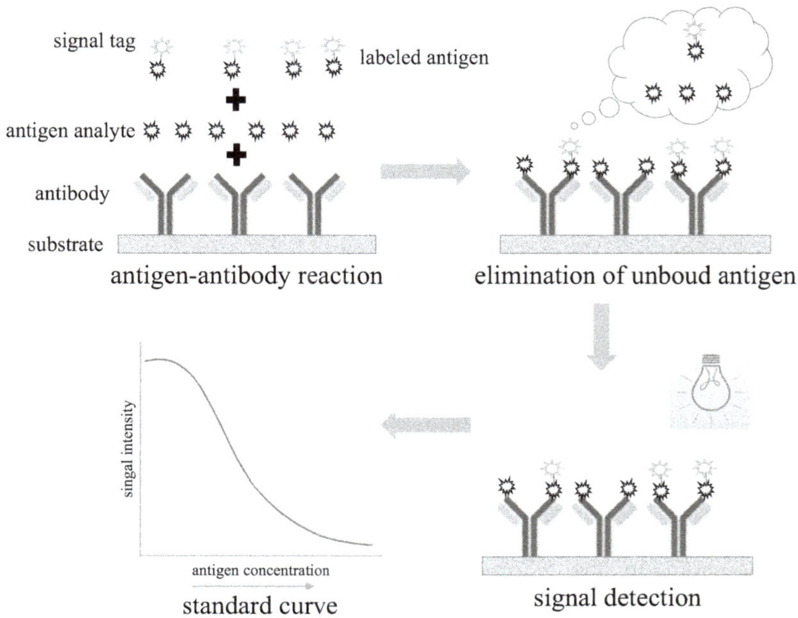

FIGURE 1.2 Competitive immunoassay process: antigen detection using labeled antigen.

a lower concentration of unlabeled antigen in the sample. Conversely, a weaker signal reflects a higher concentration of antigens in the sample. The concentration of the unknown sample is then quantified using a pre-established standard curve. From the perspective of occupancy theory, competitive immunoassays represent an indirect detection method since it detects the unoccupied binding sites rather than analyte of interest. These assays require only one type of antibody and are particularly suited for the detection of small molecules with low molecular weights, such as drugs and hormones. In competitive immunoassays, it is crucial to precisely determine the amounts of both the fixed antibody and the labeled antigen, as both quantities must be optimized to prevent excess. If the fixed antibody is in excess, both labeled and unlabeled antigens will bind completely, resulting in no observable signal difference. Conversely, if the labeled antigen is in excess, it will dominate the competition, and the unlabeled antigen will fail to produce a significant signal difference. Therefore, competitive immunoassays are often referred to as reagent-limited analyses. Additionally, competitive immunoassays that use labeled antigens are limited by the affinity constant between the antigen

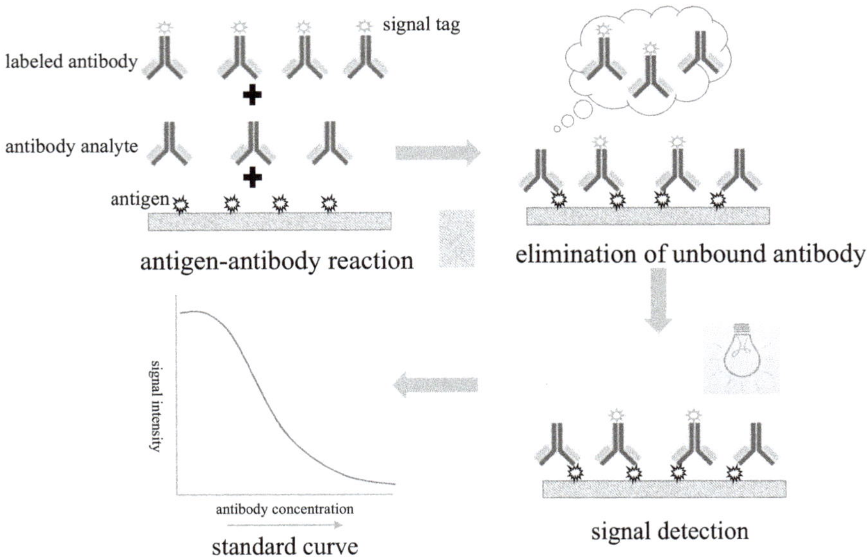

FIGURE 1.3 Competitive immunoassay process: antibody detection using labeled antibody.

and antibody. Analytes with low affinity for the antibody may not be effectively detected. Moreover, labeling small antigens can sometimes interfere with their recognition by the antibody, potentially affecting the performance of the assay.

Antibody Detection Using Labeled Antibody The detection of antibodies using labeled antibodies mirrors the methodology employed in antigen detection using labeled antigens, with the roles of antigens and antibodies reversed. In this approach, the antibody is labeled with signal tag, while the antigen is immobilized on a substrate. The labeled antibody competes with the endogenous antibody present in the sample for binding to the fixed antigen. A strong signal from the labeled antibody indicates that it occupies numerous binding sites, suggesting a low concentration of the unlabeled antibody in the sample. Conversely, a weak signal indicates that the labeled antibody occupies fewer sites, implying a high concentration of the target antibody in the sample. This process is illustrated in Figure 1.3.

Antigen Detection Using Labeled Antibody The process of antigen detection using labeled antibodies is illustrated in Figure 1.4. The antigen is immobilized onto a solid substrate. A labeled antibody probe, along with the

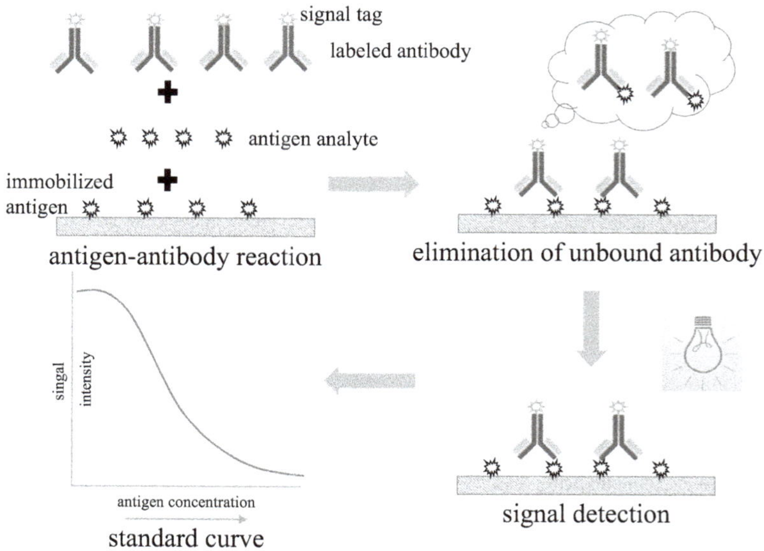

FIGURE 1.4 Competitive immunoassay process: antigen detection using labeled antibody.

antigen present in the sample, is introduced to the fixed antigen. The fixed antigen competes with the added target antigen for binding to the labeled antibody. Following the reaction, unbound antibodies are washed away, and the signal intensity of the antibodies on the solid substrate is measured. A lower concentration of the added target antigen results in a greater number of antibodies binding to the solid substrate, yielding a stronger signal. Conversely, a higher concentration of the target antigen leads to fewer antibodies binding to the solid substrate, resulting in a weaker signal.

Immunometric Assays

Immunometric analysis refers to a type of immunoassay that utilizes labeled antibodies. For antigen detection, it primarily employs a sandwich-structured immunocomplex. For antibody detection, the analysis typically operates through a competitive format (as shown in Figure 1.4) or via a so-called second antibody detection (as illustrated in Figure 1.7).

Antigen Detection Through Sandwich-Structured Immuncomplex The sandwich-structure-based immunoassay, also referred to as a noncompetitive immunoassay, requires two antibodies: a capture antibody and a detection antibody. The capture antibody, which can be either fixed on

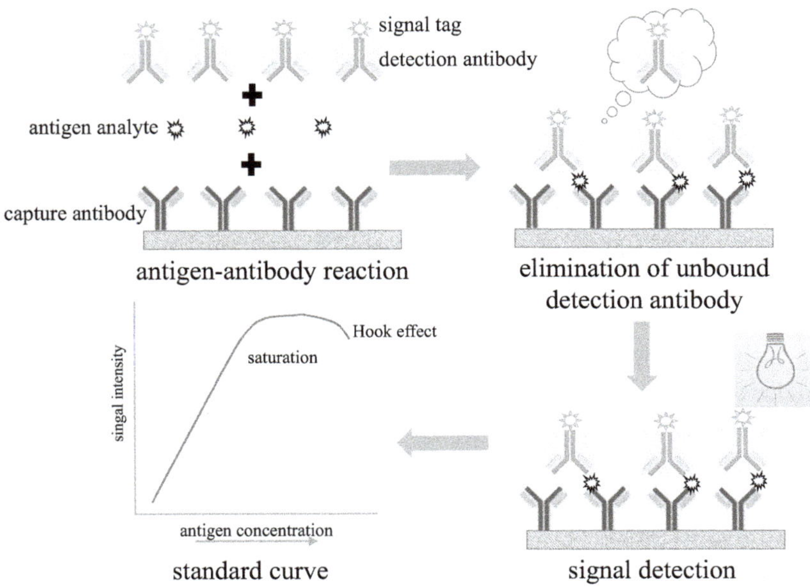

FIGURE 1.5 Sandwich-structured immunoassay process.

a surface (as illustrated in Figure 1.5) or not, is designed to specifically capture antigen present in the sample. The detection antibody, which carries a labeled tag, binds to a different epitope on the antigen. Together, the capture antibody, antigen, and detection antibody form a sandwich-structured immunocomplex. Since both the capture and detection antibodies are used in excess to maximize the capture of the antigen, it is commonly referred to as the reagent excess method. Once the antigen and antibody have reacted completely, the unbound detection antibodies are removed. The signal intensity of the bound detection antibody is then utilized for quantification. As the concentration of the antigen increases, a greater number of immunocomplexes are formed, resulting in a stronger signal. The concentration of the sample is quantified based on a pre-established standard curve that correlates the signal intensity of the detection antibody with the antigen concentration. As shown in Figure 1.5, the relationship between signal intensity and concentration initially follows a linear trend, with signal intensity increasing proportionally to antigen concentration. This portion of the curve represents the calibration range for quantification. Once all available binding sites on the solid phase become saturated, further increases in antigen concentration no longer enhance the signal, entering the saturation phase. If the antigen concentration continues to rise beyond this point, a decrease in signal intensity may occur, a phenomenon known as the Hook effect. This effect arises from the excess antigen in the liquid phase, which competes with the antigen for binding to the labeled detection antibody. As a result, the labeled antibody cannot effectively bind to the capture antibody, preventing the formation of labeled immunocomplexes on the solid substrate and leading to a reduction in signal intensity. Compared to competitive immunoassays, the sandwich-structured immunoassay offers superior specificity and detection sensitivity. The sandwich immunoassay for antigen detection is one of the most widely used and well-established techniques, representing a major approach in digital immunoassay methodologies.

A novel variation of the sandwich-structured immunoassay, known as the open sandwich-structured immunoassay, has been developed. In this method, the variable regions (V_L and V_H) of an antibody, instead of two complete antibodies, are utilized to form the sandwich-structured immunocomplex, as illustrated in Figure 1.6. Traditional sandwich-structured immunoassays are well-suited for detecting proteins that possess multiple antigenic epitopes, but they are less effective for detecting small molecules with only a single epitope. In contrast, the open sandwich-structured immunoassay is not limited by the number of epitopes and can detect small molecules with molecular weights below 1,000 Da, such as pollutants and hormones.[16]

Antibody Detection Using a Labeled Secondary Antibody (Indirect Immunoassay) As shown in Figure 1.7, the analyte antibody is added to the antigen fixed on a solid-phase surface, where it binds specifically

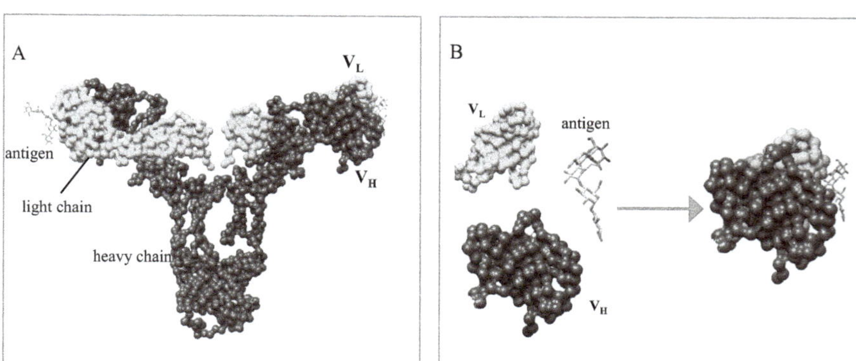

FIGURE 1.6 Structures of antibody (A) and open-sandwiched immunocomplex (B).

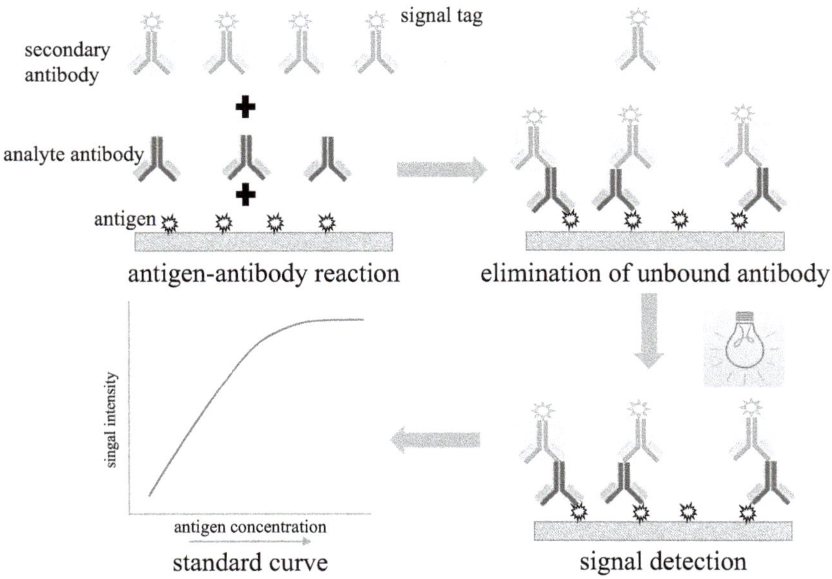

FIGURE 1.7 Indirect immunoassay process for antibody detection.

to the antigen. As previously discussed, antibodies can also serve as antigens and have corresponding specific antibodies. These antibodies, which can specifically bind to either the antibody or the antigen-antibody complex, are referred to as secondary antibodies or anti-antibodies. The excess secondary antibody labeled with signal tags is then introduced into the immunoreaction solution. After incubation, unbound secondary antibodies are removed, and the signal intensity of the secondary antibody remaining on the solid substrate is measured. A stronger signal intensity corresponds to a higher concentration of the target antibody in the sample.

Compared to competitive immunoassays, sandwich-structured immunoassays offer higher sensitivity. This improved sensitivity can be attributed to several factors. First, detection specificity is enhanced due to the use of dual antibodies (or both V_H and V_L domains), which increases binding precision. Second, in sandwich-structured assays, the signal intensity is directly proportional to the concentration of the target molecule. This means a low background signal when the target is present at low concentrations, thereby improving detection accuracy. In contrast, competitive immunoassays exhibit high signal intensity at low target concentrations, and it is more challenging to differentiate between variations in high-intensity signals.

Heterogeneous Immunoassays

The general process of heterogeneous immunoassays begins with the immobilization of capture antibodies onto the surface of a solid substrate, such as microtiter plates or microspheres. This is followed by the binding of the antigen to the attached capture antibody, the binding of detection antibodies to the captured antigen, and the labeling of the sandwich-structured immunocomplex with optical, electrical, or magnetic signals. The antigen concentration is then determined based on the intensity of the signal. This process demands multiple washing steps to remove unbound reactants, including antibodies, antigens, and labeled tags, which is why the assay is referred to as "heterogeneous". The ELISA is a typical example of a heterogeneous immunoassay, as illustrated in Figure 1.8.

The first step in a heterogeneous immunoassay is to immobilize capture antibodies onto the surface of a solid substrate. The density and orientation of the immobilized antibodies, as well as the strength of their binding to the substrate, play a critical role in the immunoassay's performance. The surface of the solid substrate, where the antibodies are attached, must contain active groups for antibody attachment and exhibit good biocompatibility and stability to minimize antibody activity loss during immobilization. Commonly used solid substrates include polydimethylsiloxane (PDMS), glass, polystyrene (PS), nitrocellulose membranes, and polyvinylidene fluoride (PVDF). The chemical and physical properties of these materials are different significantly, and as a result, their binding ability and the methods used to immobilize antibodies must be adapted accordingly. Each material has its own advantages and disadvantages, so

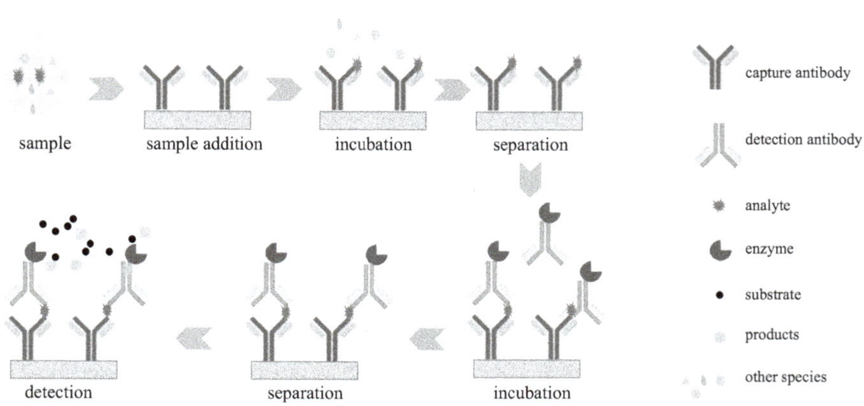

FIGURE 1.8 ELISA process.

the selection of an appropriate solid substrate should be based on the target analyte and the specific requirements of the immunoassay. Depending on the orientation of the antibody's Fab region, immobilization techniques can be classified into random immobilization and oriented immobilization.

Random Immobilization of Antibodies

When antibodies are randomly immobilized onto the surface of a solid substrate, they often adopt irregular spatial orientations. As a result, the Fab regions are distributed randomly and are not always as far from the surface as possible. This irregular orientation of the Fab regions can alter the conformation of the antibody, increase steric hindrance, impair its immunological activity, and reduce antigen-binding efficiency. Although random immobilization results in a loss of antibody functionality, the method itself is versatile, and the associated technologies are well established, making it the most widely used technique for antibody immobilization. These methods are further categorized based on the immobilization mechanism into physical adsorption, covalent conjugation, and affinity binding.

Physical Adsorption In this method, noncovalent interactions such as ionic bonds, hydrophobic interactions, electrostatic forces, and van der Waals forces are employed to bind the antibody to the solid substrate. This technique is simple and does not require coupling reagents or antibody modification. However, the binding strength is highly sensitive to various factors, such as ionic strength, pH, and temperature. More importantly, due to the relatively weak nature of the binding, antibodies may desorb during the immunoassay process, which compromises antigen capture efficiency and affects the assay reproducibility.

Physical adsorption is commonly used to immobilize antibodies on polymer such as PS and PVDF membranes. Commercial ELISA microplates are typical examples of PS-based physical adsorption. In this process, antibody diluted in carbonate buffer (pH 9.6) is added to the wells and incubated, allowing the antibody to adsorb onto the surface. However, the active sites on the microplate are limited, resulting in low antibody coupling efficiency. To enhance this efficiency, the PS surface can be modified with polyethylenimine (PEI), making it positively charged. Gold nanoparticles (GNPs) are then deposited at the bottom of the wells to form a GNP-PEI film for antibody immobilization. The capture antibody sits in the GNP-PEI well and absorbs onto the surface. Using this PEI-GNP platform, carcinoembryonic antigen (CEA) immunoassays are performed without blocking. Additionally, a thin PS film is created by spin-coating a PS solution is spin-coated onto silicon wafers, which increases surface roughness and the number of active sites. Compared to traditional immunoassays conducted in microplates, those performed on PS thin films demonstrate improved sensitivity to high-concentration antigens, thereby expanding the dynamic range.

Siloxane polymers synthesized by the sol-gel method are commonly used as substrates for protein adsorption. Jia et al. synthesized a liquid hybrid polysiloxane by hydrolyzing three organic silicon precursors.[17] The C=C double bond from one of the precursors was retained in the final structure. When the sol was spin-coated onto a silicon wafer and exposed to ultraviolet (UV) light, the C=C double bond underwent a photochemical reaction. After thermal removal of the solvent, a transparent, thin polysiloxane polymer film was formed. The resulting polymer film exhibited good transparency and moderate hydrophobicity, making it suitable for protein adsorption. Three fluorescently labeled proteins were used to assess the adsorption performance of the polysiloxane polymer film. After immersing the film in the protein solution, each protein formed a stable, dense, and uniform adsorption layer on the film. The adsorption efficiency was significantly higher than that observed on bare silicon wafers, providing a stable and reliable interface for biosensor applications or immunoassays.

PDMS is a commonly used substrate in microfluidics, known for its superior hydrophobicity, flexibility, transparency, and biocompatibility. In the early stage of microfluidics, protein molecules were generally immobilized on PDMS surfaces via physical adsorption. Silicon-based microstructures were fabricated using contact photolithography and reactive ion etching, followed by hydrophilic modifications of oxygen plasma treatment. PDMS blocks were then reversibly sealed onto these silicon microstructures, forming microfluidic channels. Fluorescent-labeled antigen solutions were introduced into the microchannels and adsorbed onto the bottom surface of the PDMS, resulting in antigen strips. After blocking, new silicon-based microstructures were orthogonally sealed onto the antigen strips. Fluorescent-labeled antibody molecules were then introduced into the microchannel, where they reacted with the antigen, forming a fluorescent microarray. This approach enables the easy implementation of multicomponent immunoassays at various concentrations.

Covalent Conjugation In the process of covalent conjugation, an antibody is chemically bonded to a substrate through its active groups and the functional groups present on the substrate surface. The stable covalent bonds formed during this process provide enhanced immobilization stability, resistance to detachment, high coupling efficiency, and uniform distribution of the protein. Active groups in the natural amino acids that can undergo covalent bonding include primary amines, carboxyl groups, and thiol groups, among others.

Primary amines are the major functional groups used for covalent conjugation. Carboxyl groups, present at the C-terminus or in the exposed residues of glutamic acid and aspartic acid within proteins, can rapidly react with primary amines to form peptide bonds under carbodiimide activation. Commonly used carbodiimide reagents include 1-(3-dimethylaminopropyl)-3-ethylcarbodiimide (EDC) and 1-cyclohexy l-3-(2-(4-morpholinyl) ethyl) carbodiimide (CMC). These reagents enable the direct coupling of antibodies to carboxyl or amine groups on a solid substrate surface in a one-step reaction. Additionally, EDC in combination with N-hydroxysuccinimide (NHS) activates the carboxyl groups on the substrate surface. The activated carboxyl groups then react with the amine groups on the antibodies to form stable peptide bonds. The use of EDC and NHS together enhances activation efficiency and minimizes antibody self-crosslinking.

The mechanism of EDC/NHS activation of carboxyl groups is as follows: EDC reacts with the carboxyl group on the substrate surface to form an unstable O-acyl urea intermediate. This intermediate can undergo three possible reaction pathways: (1) It reacts with NHS to form a relatively stable succinimidyl ester and urea (Step 2 in Figure 1.9). (2) It reacts with a nearby carboxyl group, grafting an anhydride onto the surface and releasing urea (Step 3 in Figure 1.9). (3) It undergoes an intramolecular acyl migration rearrangement to form N-acyl urea (Step 4 in Figure 1.9). In principle, the O-acyl urea intermediate can directly react with the primary amine of an antibody to form an amide bond, but this reaction occurs at a relatively slow rate. The N-acyl urea formed in the third pathway is not reactive. NHS in this process serves to form a more stable succinimidyl ester product, which competes with the anhydride pathway and accelerates the reaction rate. The anhydride can further react with NHS to generate an activated ester and carboxylic acid. Under more acidic conditions, the anhydride can also serve as a secondary pathway for succinimidyl ester formation, although surface activation efficiency is relatively low.

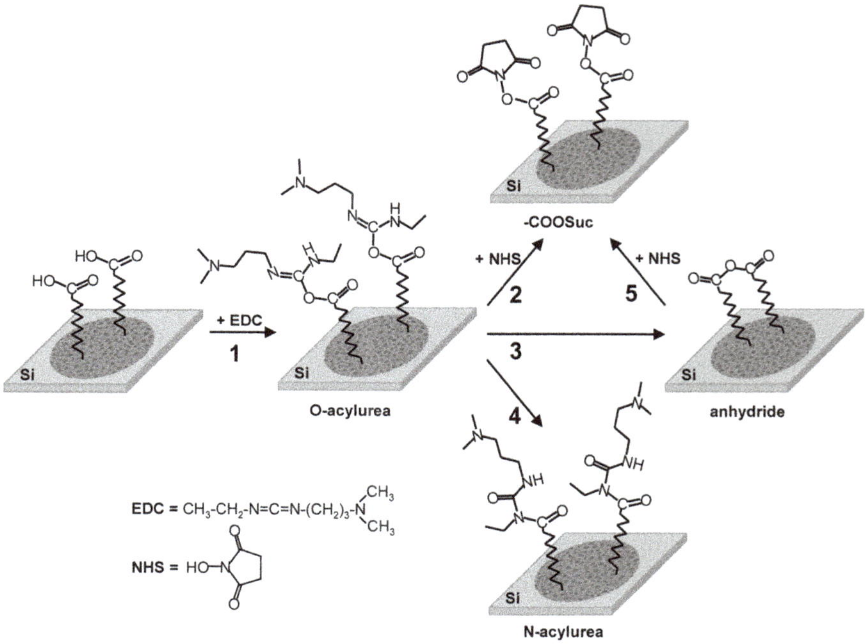

FIGURE 1.9 Activation of carboxylic groups using EDC/NHS.

Source: Reproduced from Sam, S.; Touahir, L.; Andresa, J. S. Semiquantitative study of the EDC/NHS activation of acid terminal groups at modified porous silicon surfaces. *Langmuir*, 2010, 26, 809–814.[18] Copyright 2010 American Chemical Society.

Glutaraldehyde is another commonly employed coupling agent for immobilizing antibodies onto solid substrates. The typical procedure for antibody immobilization on a glass substrate using glutaraldehyde is outlined as follows: Initially, the glass substrate is treated by soaking it in chromic acid overnight, followed by ultrasonic cleaning in deionized water for two 10-minute cycles. The substrate is then dried with nitrogen gas. Subsequently, the clean glass is immersed in a 1% (v/v) 3-aminopropyltriethoxysilane (APTES) in acetone for 30 minutes. The substrate is then washed three times with acetone and dried with nitrogen gas, resulting in an amine-modified surface. The amine-modified glass is then treated with a 10% (v/v) glutaraldehyde at 37°C for 1 hour. Excess glutaraldehyde is removed by washing the substrate with deionized water and drying it with nitrogen gas. The substrate is then placed in a 0.7 mg/mL antibody prepared in 10 mM phosphate-buffered saline (PBS, pH 7.4) overnight at 4°C. PBS is used to remove any nonspecifically bound antibodies. To block remaining active sites, the glass substrate is incubated in a 4% bovine serum albumin (BSA) for 30 minutes. Following this blocking step, the covalent coupling of the antibody to the glass surface is complete. Epoxy groups can be introduced onto the glass substrate by replacing APTES with 3-Glycidyloxypropyltrimethoxysilane. These epoxy groups can then react with various functional groups such as carboxyl groups, primary amines, thiols, and hydroxyl groups, forming esters, secondary amines, thioethers, and alkyl ethers, respectively. As a result, they can be utilized for the covalent immobilization of antibodies. Additionally, 1,4-butanediol diglycidyl ether is often employed to convert hydroxyl groups into epoxy groups on the surface of solid substrates.

Affinity Binding The affinity between biotin and avidin is frequently utilized in affinity binding methodologies. Biotin, with a molecular weight of approximately 244 Da, is commonly extracted from egg yolk and liver tissue. Avidin, a glycoprotein with a molecular weight of about 66 kDa, consists of four identical subunits. Avidin is primarily derived from Streptomyces species, therefore also known as streptavidin. A single avidin molecule can bind up to four biotin molecules, with an affinity constant ranging from approximately 10^{-15} to 10^{-13} mol/L. The biotin-avidin interaction is highly specific and stable, which underpins its widespread use in immunoassays.

Ouerghi et al. synthesized biotinylated pyrrole and subsequently electro-polymerized it onto a clean gold electrode to form a polypyrrole film.[19] The biotin groups exposed on the film bound to streptavidin, thereby immobilizing avidin on the electrode surface. Biotinylated antihuman IgG was then immobilized onto the electrode through its binding to the remaining available sites on avidin, allowing for target detection. The linear range for this method spanned from 10 to 80 ng/mL, with a detection limit (LOD) as low as 10 pg/mL.

The surface of PDMS is primarily composed of silicon-hydroxy groups. Due to the low reactivity of silicon-hydroxy group, the efficiency of covalent conjugation is usually poor. Additionally, the high hydrophobicity of PDMS leads to significant nonspecific adsorption of proteins. To address these challenges, Cheng et al. self-assembled a phospholipid bilayer on PDMS using a mixed solution of biotinylated phosphoethanolamine and phosphatidylcholine.[20] This bilayer not only significantly reduced nonspecific protein adsorption by a factor of 100 to 1,000 but also provided biotin functional groups for subsequent antibody immobilization. In their approach, streptavidin was introduced into the PDMS channel, where it bound to the biotin groups, and biotinylated antibodies were then attached to the streptavidin sites, effectively immobilizing the antibodies on the PDMS surface. This method was employed for the detection of foodborne Staphylococcus aureus enterotoxin B, achieving a LOD as low as 0.5 ng/mL.

Oriented Immobilization of Antibodies

Oriented immobilization preserves the natural conformation of the Fc fragment of the antibody after it is attached to a solid substrate. This ensures that there is no steric hindrance that could interfere with antigen binding, thereby maintaining the antibody's activity to the greatest extent. Oriented immobilization is widely regarded as one of the most effective protein coupling methods for immunoassays. Using Immunoglobulin G (IgG) as an example, as shown in Figure 1.10, IgG consists of 2 heavy chains (gray/blue) and 2 light chains (green). Antigens bind to the variable regions V_L and V_H. The C_{H2} domain bears a carbohydrate moiety (orange hexagon). scFv refers to a single-chain variable fragment, consisting of the variable light V_L and V_H while Fab refers to an antigen-binding fragment

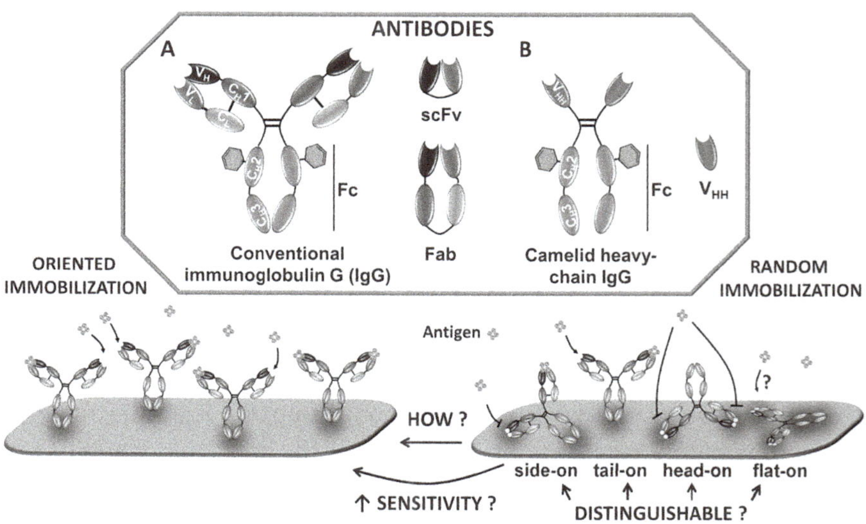

FIGURE 1.10 Scheme of oriented immobilization of antibodies: typical structure of immunoglobulin G (IgG) (A) and camelid heavy-chain antibody (B).

Source: Reproduced from Trilling, A.K.; Beekwilder, J.; Zuilhof, H. *Analyst*, 2013, 138, 1619–1627 (ref 21), with permission of the Royal Society of Chemistry.

of an immunoglobulin (Ig), containing both the variable and constant regions necessary for antigen recognition. Camelid heavy-chain antibody contains only two heavy chains, which are composed of one variable (V_{HH}; antigen-binding unit) and two constant (C_{H2} and C_{H3}) domains. Antibodies can be immobilized in an oriented or random fashion as shown in Figure 1.10.

Oriented Immobilization Using Glycosylation Sites A glycoprotein contains one or more oligosaccharide units that are covalently linked to a protein. These molecules function in various roles, including as enzymes, antibodies, hormones, and receptors. The oligosaccharide chains attached to glycoproteins play key roles in the body, such as cell recognition, immunogenicity, and maintaining structural stability. Outside the body, these oligosaccharide chains can serve as reactive sites for the oriented immobilization of proteins. A common approach is to oxidize neighbor diols in the glycan moiety to reactive aldehyde groups using periodate. These aldehyde groups then react with primary amines (or hydrazines) on a solid substrate to form Schiff bases (or stable hydrazone bonds). Schiff bases are easily cleaved under acidic conditions, so $NaBH_4$ or NaBH3CN is used to reduce Schiff bases to secondary amines, thereby stabilizing the protein. This strategy is also applicable to the biotin-avidin system, where avidin, a typical glycoprotein, serves as an oriented immobilization site. Additionally, lectins and boronic acid derivatives are commonly employed for glycoprotein-based oriented immobilization.

Glycosylation sites are commonly located on the Fc fragment of antibodies. When antibodies are immobilized by oxidizing the glycosyl groups in the Fc region, the conjugation site is located far from the antigen-binding sites. This minimizes steric hindrance, allowing better interaction between the antibody and antigen, and typically enables binding of more antigens. However, the immobilization process involves antibody oxidation, which can be cumbersome and may lead to partial loss of activity due to the antibody structural damage.

Immobilization Using Protein A or Protein G Protein A, a cell wall protein derived from Staphylococcus aureus, specifically binds to the Fc region of human and mammalian IgG antibodies. The natural Protein A consists of five IgG-binding domains, along with additional non-Fc-binding domains of unknown function, and has a molecular weight of approximately 42 kDa. Using genetic engineering techniques, amino acids of Protein A can be conveniently modified to create reactive sites. For instance, a histidine tag can be incorporated at the C-terminus of Protein A. The histidine-tagged Protein A can then be immobilized in an oriented manner on solid surfaces modified with iminodiacetic acid. The Fc region of IgG binds to Protein A immobilized on a solid substrate surface. The resulting IgG-Protein A complex can be further crosslinked with dimethyl pimelimidate (DMP) to enhance antibody stability. Protein G, a cell wall protein derived from Streptococcus species, has a molecular weight of approximately 25 kDa. Compared to Protein A, Protein G exhibits stronger affinity for mammalian IgG and is particularly suitable for immobilizing monoclonal and polyclonal IgG from mammals with weaker binding to Protein A. Recombinant Protein G, however, typically does not bind to human IgM, IgD, or IgA. For example, when immobilizing rabbit antihuman serum albumin antibodies on a solid surface with aldehyde groups, the amount of antibody immobilized through direct conjugation of antibodies to aldehyde groups is 3.5 times greater than when Protein A is used as a bridging agent. However, the antigen-binding capacity of antibodies immobilized via Protein A is twice as high as that of antibodies directly conjugated to aldehyde groups. This demonstrates that Protein A-based immobilization enhances antibody orientation, thereby improving antigen-binding performance.

Cysteine-Mediated Oriented Immobilization Specific binding sites are introduced into native proteins using modern molecular biology techniques. For instance, site-directed mutagenesis allows the introduction of specific amino acid residues, while gene fusion techniques enable the incorporation of specialized amino acid sequences. A widely used strategy for oriented immobilization is to introduce free cysteine residues at specific positions within the protein. The thiol group of introduced cysteine serves as a reactive site for oriented immobilization. Cysteine residues are able to form covalent bonds with functional groups on solid substrate, such as iodoacetyl groups, disulfide bonds, p-chloromercuribenzoic acid amide, or maleimide, resulting in effective and oriented protein immobilization. This approach is suitable for antibodies that lack natural cysteine residues, such as small molecule-engineered antibodies, as well as those containing native cysteine. For antibodies with natural cysteine residues, it is crucial to assess whether the presence of cysteine affects the antibody's activity or stability. If no adverse effects are observed, native cysteine residues can be mutated to other amino acids, and new cysteine residues can be introduced at positions optimized for oriented immobilization.

Homogeneous Immunoassays

In heterogeneous immunoassays, multiple incubation steps are required to conduct sequential reactions. For example, in ELISA, multiple incubation steps are needed to conduct the reaction between the target molecule in the sample and the immobilized antibody, the interaction between the captured target molecule and the detection antibody, and the reaction between the detection antibody and the enzyme. Multiple washing steps are used to remove excess reagents and interfering molecules in the sample. As the number of target species increases, more cycles of incubation, binding, and washing are required. Consequently, experimental conditions become more complex, the errors accumulate, and it is increasingly challenging to ensure measurement accuracy. In contrast, homogeneous immunoassays require only the mixing of the sample with a capture antibody and a detection antibody labeled with a signal tag, eliminating the need to remove excess reagents. Compared to heterogeneous immunoassays, homogeneous immunoassays offer several advantages: (1) Rapid Analysis: The reaction occurs entirely in solution rather than at the interface between a solid surface and the solution, resulting in shorter diffusion distances and reduced incubation times. Additionally, the simplified operational steps significantly shorten the overall measurement period. (2) Broad Application Scope: Homogeneous immunoassays can detect antigens with moderate or weak affinity for antibodies, as no immunocomplex is lost during the process. (3) Improved Measurement Accuracy: Since only a single mixing step is required, the accumulation of procedural errors is minimized. Furthermore, solution-phase reactions reduce the likelihood of conformational changes in antibodies, thereby enhancing measurement accuracy. (4) Automation Potential: Homogeneous immunoassays are well-suited for automation in detection systems, offering significant potential for the development of personalized or bedside diagnostic devices. (5) Accelerated conversion: Homogeneous immunoassays offer significant advantages in expediting biomarker screening, particularly in facilitating the transition from laboratory research to clinical practice. Theoretically, homogeneous immunoassays would exhibit superior sensitivity compared to heterogeneous immunoassays. This is because in heterogeneous immunoassays, the separation and washing steps disrupt the reaction equilibrium, favoring dissociation and reducing the formation of immunocomplexes, thereby lowering sensitivity. However, in practice, homogeneous immunoassays are hindered by higher background noise due to the absence of separation steps, which ultimately compromises sensitivity. As a result, the sensitivity of current homogeneous immunoassays is typically one to two orders of magnitude lower than that of heterogeneous immunoassays.

The key to achieving homogeneous immunoassays lies in establishing detectable differences between the immunocomplex and the free tag-labeled antibody. In the reaction solution, both the free tag-labeled antibody and the immunocomplex contain signal tags, making it a significant challenge to distinguish between them without a separation step. The following sections provide a summary of homogeneous immunoassays, focusing on how to distinguish between the immunocomplex and the free tag-labeled antibody.

Signal Transfer

Fluorescence Resonance Energy Transfer-Based Immunoassays

Fluorescence resonance energy transfer (FRET) refers to a nonradiative energy transfer between a donor fluorophore and an acceptor molecule under specific conditions. The acceptor, which can be another fluorophore or a nanomaterial such as graphene, absorbs the transferred energy. The conditions required for FRET include the following: (1) Distance: The distance between the donor and acceptor must fall within an appropriate range for energy transfer to occur. When both the donor and acceptor are fluorophores, this distance typically cannot exceed 10 nm. FRET efficiency decreases as the distance between the donor and acceptor increases. (2) Spectral Overlap: The emission spectrum of the donor must effectively overlap with the absorption spectrum of the acceptor. (3) Dipole Orientation: The dipole moments of the donor and acceptor must be suitably aligned for energy transfer. The outcomes of FRET include the quenching of the donor's fluorescence and the emission of light by the acceptor. Figure 1.11A illustrates the fundamental principle of FRET-based immunoassays, employing a fluorophore emitting green fluorescence as the donor and a fluorophore emitting red fluorescence as the acceptor. In the absence of the antigen, no immunocomplex is formed, and the donor fluorophore emits green fluorescence upon excitation by blue light, while the acceptor remains nonemissive. Upon the addition of antigen, a sandwich-structured immunocomplex is formed. When the donor is excited, energy transfer occurs, exciting the acceptor and resulting in red fluorescence. The intensity of the red fluorescence correlates with the antigen concentration, enabling antigen detection without the

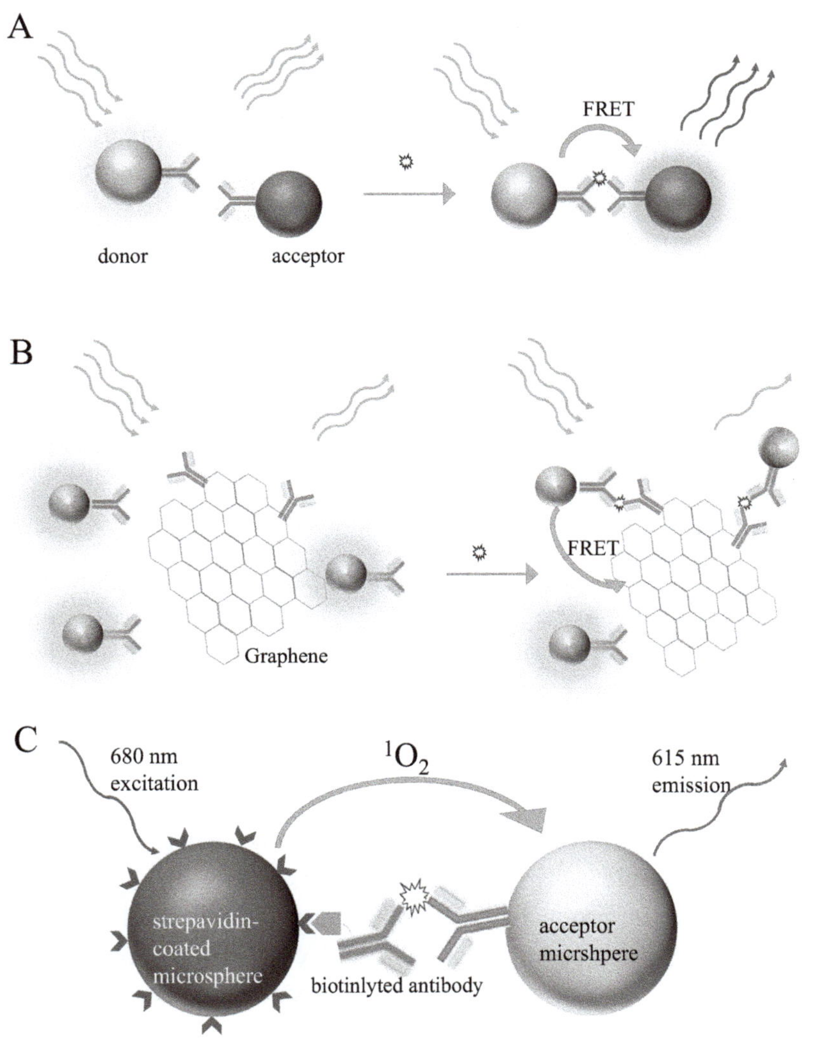

FIGURE 1.11 Principles of FRET-based homogeneous immunoassays: (A) sandwich-structured immunoassay and (B) graphene quenching QD-based immunoassay.

Source of Figure 1.11B: Reproduced from Liu, M.; Zhao, H.; Quan. X. et al. *Chemical Communications*, 2010, 46, 7909–7911 (ref 22), with permission of the Royal Society of Chemistry.

need for separation steps. The LODs for FRET-based immunoassays are typically in the picomolar (pM) range.

FRET in an open sandwich-structured immunoassay offers distinct advantages. The V_H and V_L regions of the antibodies are labeled with donor and acceptor fluorophores, respectively. Compared to the immunocomplex formed by two full antibodies, the open immunocomplex exhibits stronger FRET signals due to the significantly smaller distance between the V_H and V_L regions. This configuration enhances the efficiency of energy transfer, making the open sandwich-structured immunoassay more sensitive and effective in detecting the target analyte.

Figure 1.11B illustrates another mode of FRET in which QDs are used as the donor and graphene serves as the acceptor. Upon antigen binding, the QDs and graphene form an immunocomplex, and energy from the excited QDs is transferred to the graphene, resulting in a reduction of the QD's fluorescence intensity.[22] The maximum distance at which FRET can occur is as high as 22 nm in this method, thereby extending the dynamic range. This enhanced distance allows for greater flexibility in the design of sensitive immunoassays.

Luminescent Oxygen-Channeling Immunoassays The principle of luminescent oxygen-channeling immunoassay (LOCI) is commercially known as Alpha LISA. In this assay, a photosensitive agent, benzothiazine blue, is coated on the donor microspheres. Upon exposure to 680 nm laser light, benzothiazine blue decomposes oxygen to produce singlet oxygen. The acceptor microspheres are coated with a luminescent agent, a dimethylthienyl derivative chelated with europium, which emits light at 615 nm with a half-life of 0.3 s upon excitation by singlet oxygen.[23] When the donor and acceptor microspheres are linked together by an antigen, the singlet oxygen produced from the donor microspheres diffuses to the acceptor microspheres, exciting them and causing light emission. The diffusion distance of singlet oxygen is approximately 200 nm, effectively overcoming the distance limitations inherent in FRET-based methods.

Signal Recovery

Enzyme Multiplied Immunoassays

Enzyme Multiplied Immunoassay Technique (EMIT®) is based on the principle that the binding of an enzyme-labeled antigen to its corresponding antibody modulates the enzyme activity. Specifically, the enzyme activity is damaged when an immunocomplex forms, such as when steric hindrance or conformational changes in the enzyme occur upon antigen binding to the antibody. This in turn decreases the detectable signal. The antigen in a sample to be tested competes with the enzyme-labeled antigen for the antibody binding sites, thereby restoring enzyme activity and increasing the etectable signal (see Figure 1.12A). Initially, EMIT was used in drug abuse monitoring, where lysozyme was employed to detect drugs in urine. At that time, EMIT was not suitable for detecting the target in serum. However, by replacing lysozyme with malate dehydrogenase (MDH) or glucose-6-phosphate dehydrogenase (G6PDH), EMIT was successfully adapted for detecting drugs in serum. Detecting proteins using EMIT is challenging, as the interaction between the antibody and the protein has minimal impact on enzyme activity. Although a few studies had reported

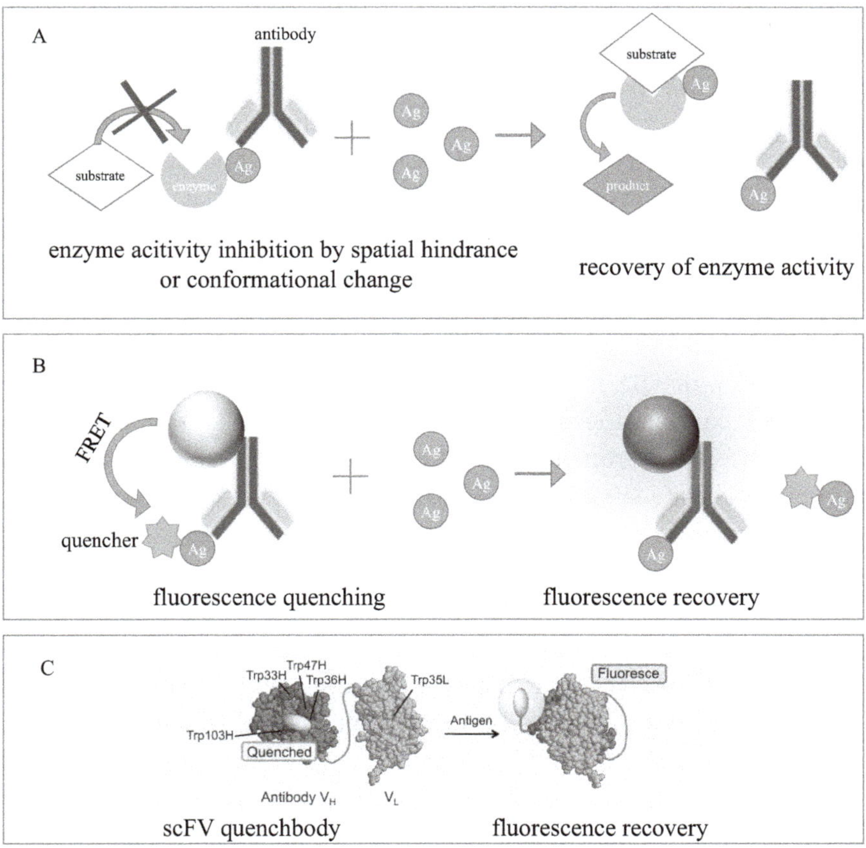

FIGURE 1.12 Principles of the enzyme multiplied immunoassay (A), fluorescence quenching-recovery-based immunoassays (B), and quenchbody-based immunoassays (C).

Source of Figure 1.12C: Reproduced from Abe. R.; Ohashi, H.; Iijima, I. et al. "Quenchbodie": quench-based antibody probes that show antigen-dependent fluorescence, *Journal of the American Chemical Society*, 2011, 133, 17386–17394 (ref 26). Copyright 2011 American Chemical Society.

the use of EMIT for protein detection, these attempts had limited success and did not significantly advance the application of EMIT for protein detection.

Fluorescence Quenching-Recovery-Based Immunoassays The antigen labeled with a fluorescence quencher functions as the receptor, while the antibody labeled with a fluorescent emitter acts as the donor. Upon binding, FRET causes quenching of the donor's fluorescence. When the sample is introduced, the target antigen competes with the labeled antigen for binding to the antibody. If the target antigen successfully displaces the labeled antigen, the quencher is released from the antibody, resulting in fluorescence recovery (see Figure 1.12B). The recovered fluorescence intensity is positively correlated with the concentration of the target antigen in the sample. Using this method, Mattoussi et al. detected 2,4,6-trinitrotoluene (TNT) in water by conjugating TNT and an anti-TNT antibody to a quencher and QDs, respectively.[24] When the quencher-labeled TNT bound to the QD-labeled antibody, the fluorescence of the QDs was quenched. In the presence of a sample, TNT competed with the labeled TNT for binding to the antibody, leading to the recovery of QD fluorescence. The fluorescence intensity increased with TNT concentrations. In a similar approach, Kokko et al. detected estradiol (E2) in serum using a quencher (QSY21)-labeled E2 and an europium chelate-labeled anti-E2 antibody.[25] Upon addition of E2, it displaced the QSY21-labeled E2 within the immunocomplex. This release of QSY21-labeled E2 eliminated FRET and allowed the fluorescence of the europium chelate to recover. The LODs reached 18 pM in buffer and 64 pM in serum. These methods require careful consideration of the donor-quencher matching and the potential impact of quencher-labeled antigens on antigen-antibody binding. They are not universally applicable. In 2011, Ueda et al. discovered the phenomenon of fluorescence quenching and recovery in single-chain variable fragment (scFv) antibodies, leading to the development of the quenchbody-based immunoassay (Figure 1.12C).[26] The core of quenchbody was an scFv, where the light and heavy variable chains were connected by a peptide. A fluorescent dye was linked to the N-terminal of the scFv via a flexible short peptide (4–25 amino acids). Due to hydrophobic interactions, the dye entered the variable region and was quenched by tryptophan residues, which transferred electrons to the dye in its excited state. When the quenchbody bound to the antigen, the dye was expelled from the variable region as the Fv structure stabilized.

This movement of the dye away from the tryptophan residues prevented electron transfer, resulting in fluorescence recovery. The fluorescence intensity increased as antigen concentration increased. This method offers advantages of simplicity and speed and has been applied in drug, protein, and cell imaging with promising prospects. However, the construction of quenchbodies is still relatively complex, and the method requires further optimization.

Size Change

After antigen-antibody binding, the physical size of immunocomplex changes, which in turn affects the rotational speed and diffusion coefficient. Quantitative immunoassays are performed by detecting these parameters that can reflect these changes.

Fluorescence Polarization-Based Immunoassays The polarization direction of light refers to the orientation of the electric field vector in an electromagnetic wave. Fluorescence polarization, or fluorescence anisotropy, describes the degree of emitted fluorescence polarization. When a fluorescent molecule is excited by polarized light, the emitted fluorescence typically maintains a certain degree of polarization. This polarization is influenced by the rotational motion of the molecule. If the fluorescent molecule is stationary and excited by light with a specific polarization, the emitted fluorescence aligns with the polarization of the excitation light. However, if the molecule undergoes rotational motion, the emitted light becomes partially depolarized as the molecule's orientation changes during the emission process. The extent of this depolarization is influenced by the size, shape, and dynamics of the molecule. Larger molecules generally rotate more slowly than smaller ones, resulting in less depolarization of the emitted fluorescence. This phenomenon can be used to distinguish between larger and smaller molecules, such as differentiating an antigen-antibody complex from free antigen or antibody in solution. Fluorescence polarization immunoassay (FPIA) is typically designed in a competitive format for detecting small molecules, as illustrated in Figure 1.13A. An antigen is labeled with a fluorescent tag and binds to an antibody, forming an immunocomplex. When a sample is introduced, the antigen in the sample competes with the labeled antigen for binding to the antibody, causing the labeled antigen to be displaced and released into the solution. As the labeled antigen's rotational speed increases, its fluorescence polarization value decreases. The change in polarization is

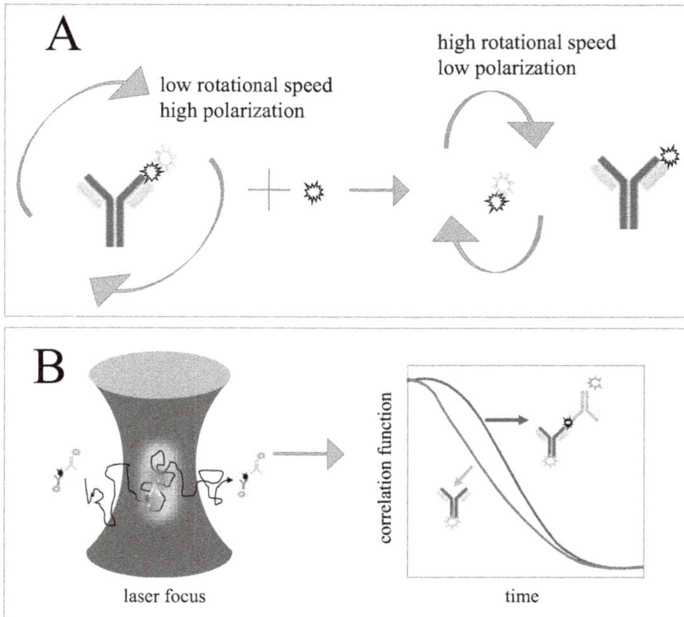

FIGURE 1.13 Principles of two size-dependent homogeneous immunoassays: (A) competitive fluorescence polarization immunoassay and (B) fluorescence correlation spectroscopy-based immunoassay.

directly related to the concentration of the antigen in the sample. FPIA offers advantages such as simplicity and reliability. However, its dynamic range and minimum detectable concentration are generally inferior to those of other methods. It is most suitable for samples with concentrations in the mg/mL range. Additionally, when high-affinity antibodies are used, the competitive reaction can take longer, which may limit the assay's utility for rapid detection. FPIA can also be designed as a noncompetitive format for detecting macromolecules. In this approach, the antibody is labeled with a fluorescent dye, and the polarization difference between the labeled antibody-antigen complex and the free labeled antibody is measured. However, a major challenge in this type of FPIA is insufficient resolution. For example, the polarization difference between a 30 kDa monomer and a 60 kDa dimer is 0.047, which is detectable. However, the polarization difference between a 50 kDa monomer and a 100 kDa dimer decreases to 0.033, making it more difficult to detect. Similarly, when a 150 kDa anti-IgG antibody binds to a 50 kDa IgG, the resulting polarization change is only 10%, leading to limited detection sensitivity. To improve sensitivity, one potential solution is to reduce the molecular weight of the antibody.

For instance, Fukuyama increased the polarization change to 60% by replacing the full antibody with a smaller variable heavy chain (15 kDa), thereby enhancing the assay's sensitivity.[27]

Fluorescence Correlation Spectroscopy-Based Immunoassays Another technique that leverages changes in molecular size to develop homogeneous immunoassays is fluorescence correlation spectroscopy (FCS). FCS utilizes laser confocal technology to excite fluorescent molecules. Due to Brownian motion, these molecules move in and out of the laser focus zone, causing fluctuations in the fluorescence signal. By analyzing the correlation function of these fluorescence fluctuations, important parameters such as molecular diffusion time, diffusion coefficient, and concentration can be determined. The left side of Figure 1.13B illustrates the green laser beam, with the glowing elliptical area representing the laser beam focus zone. The right side of Figure 1.13B shows a correlation curve derived from the correlation function. This curve provides the characteristic diffusion time, and the molecular diffusion coefficient can be calculated by combining the characteristic diffusion time with the volume of the focus zone. The characteristic diffusion time varies with the size of the molecules, making it possible to distinguish between free antibodies and antibody-antigen complexes. FCS encounters a similar challenge as FPIA – namely, limited sensitivity and dynamic range when detecting large molecular targets, as their hydrodynamic changes are relatively small. To enhance the molecular weight difference between the antigen and the immunocomplex, Ren Jicun et al. employed gold nanoparticles as tags and developed a scattering correlation spectroscopy-based homogeneous immunoassay.[28]

Single-Molecule Homogeneous Immunoassays

In labeled immunoassays, the antibody is conjugated with a signal tag, and as a result, the immunocomplex formed between the antibody and antigen is also tagged with the same label. In homogeneous immunoassays, there is no operation to separate the immunocomplex from the free antibody. Consequently, both the labeled antibody and the labeled immunocomplex coexist in the solution. The primary challenge in homogeneous immunoassays is distinguishing between these two components, as they share the same signal tag.

When the signal properties remain unchanged, it's challenging to distinguish between the signals from the antibody and the immunocomplex at the ensemble level. However, it might be possible to differentiate them at the single-molecule level. Our research group has been engaged in developing a single-molecule spectral imaging techniques based on transmission grating. We discovered that some of their optical properties of QDs and GNPs are suitable for homogeneous immunoassay. For example, each QD in a QD aggregate exhibits asynchronous blinking, spectral blue shift, and photobleaching. Using these properties, we have achieved breakthroughs in diffraction-limited measurements, ultrasensitive detection of mercury ions, and fM-level homogeneous immunoassays. Using single-molecule spectral imaging technology, we can observe the spectral differences between immunocomplexes containing multiple QDs and free antibodies with a single QD. This enables the identification and quantification of immunocomplexes and thereby precise quantitative measurement. Single-molecule immunoassays are a subset of digital immunoassays (DIs), and the principles underlying these techniques are discussed in more detail in the following sections.

Homogeneous immunoassays offer several advantages, including simplicity, ease of automation, and are seen as a key direction for the future development of immunoassays. However, a major challenge remains that their sensitivities have not yet reached the levels of heterogeneous immunoassays. Single-molecule homogeneous immunoassays have already achieved sensitivities comparable to those of traditional heterogeneous immunoassays, although they require instrument automation. Table 1.2

TABLE 1.2 Comparison between Representative Homogeneous Immunoassays

Homogeneous Immunoassay Characteristics	Analyte	LOD	Dynamic Range	Ref.
Noncompetitive fluorescence polarization	Rabbit IgG	NA	0.45–41 nM	27
Competitive fluorescence polarization	mycotoxin	~1 nM	~1.5–46 nM	29
Fluorescence correlation spectroscopy	Alpha-fetoprotein	20 pM	20 pM–5 nM	30
Resonance light scattering correlation spectroscopy	Alpha-fetoprotein	100 pM	100 pM–10 nM	31
Quenching-recovery	estradiol	18 pM	18 pM–0.4 nM	25
FRET	Alpha-fetoprotein	6 pM	0–13.9 nM	32
Quenchbody	Human epidermal growth factor receptor 2	20 pM	EC50 = 0.3 nM	33a
Single-molecule	Alpha-fetoprotein	3.4 fM	43.7–1399.3 pM	33b

provides a comparison of the performance of representative homogeneous immunoassays.

Single- and Multicomponent Immunoassays

Immunoassays that detect only one target in a single run are called single-component immunoassays, while multicomponent immunoassays can simultaneously quantify two or more analytes in one run. Unlike multi-indicator immunoassays, which conduct serial or parallel runs of single-component assays, multicomponent immunoassays emphasize the simultaneous determination of multiple analytes within a single test. This approach ensures consistency in experimental conditions, and thereby reduces measurement errors that may arise from variations between serial or parallel tests. Multicomponent immunoassays for tumor biomarkers, for example, can significantly reduce the likelihood of false positives and false negatives, improve diagnostic accuracy for malignant tumors, and help to avoid overtreatment. A panel of biomarkers typically provides greater accuracy than a single marker for the early diagnosis of cancer. For instance, the FDA-approved OVA1 tumor diagnostic biomarker panel, which consists of five proteins, achieves 91.4% accuracy in predicting ovarian cancer, compared to just 65.7% accuracy with the individual marker CA125.[34] In multicomponent immunoassays, different antibodies are often required, with each antibody labeled with a distinct tag (a commonly used labeling strategy).

A typical example of a commercial multicomponent immunoassay is the multicolor microsphere and flow cytometry developed by Luminex Corporation, also known as liquid chip technology (Figure 1.14). PSs are coated with capture antibodies, and detection antibodies are labeled with phycoerythrin (PE). The target analyte links the PSs and detection antibodies, forming a sandwich-structured immunocomplex. As the immunocomplex passes through a laser beam, red and green lasers excite the PSs and PE, respectively. Fluorescence intensity in the green channel correlates with the amount of antigen, while fluorescence in the red channel identifies antigens. The PSs contain two or three different dyes in specific ratios, which enables modulation of the fluorescence spectra to generate hundreds of distinct signals. Consequently, up to 500 targets can be detected in a single multicomponent immunoassay.[35] In 2018, Cohen et al. used the CancerSEEK technology to detect circulating tumor DNA and 39 protein biomarkers in blood.[36] By comparing protein level differences between healthy individuals and patients diagnosed with eight common

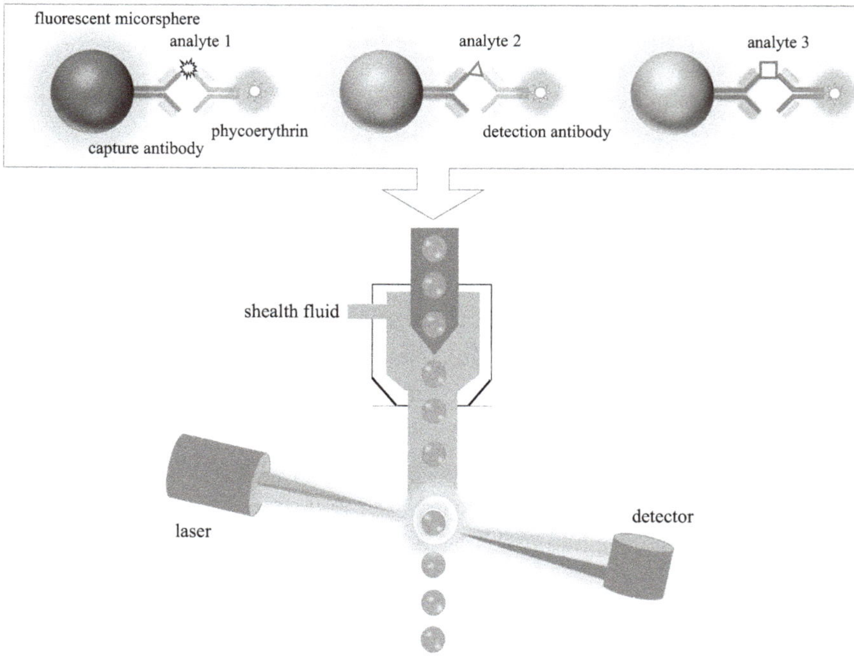

FIGURE 1.14 Principle of liquid chip-based multicomponent immunoassay.

cancers, they identified that the levels of nine proteins were below or reached the LODs of current immunoassays, leading to the identification of eight biomarker candidates.

Multicomponent immunoassays can be classified into heterogeneous and homogeneous types, depending on whether separation steps are required. Multicomponent heterogeneous immunoassays require multiple separation steps to remove excess detection antibodies and detection tags. As the number of targets increases, more cycles of incubation, binding, and washing are needed, making the process more complex. This complexity also increases the risk of error accumulation, which can complicate accurate measurements. In contrast, multicomponent homogeneous immunoassays only require the sample to be mixed with labeled capture antibodies and detection antibodies, without the need for separating unbound antibodies. As a result, multicomponent homogeneous immunoassays offer simpler measurement operations and more precise results, indicating they offer significant advantages in terms of efficiency and accuracy.

The key to developing a successful multicomponent homogeneous immunoassay lies in the design of a sensing strategy that ensures significant differentiation between the signal from the immunocomplex and that from the free antibody and identification of multiple targets in complex medium. In recent years, several strategies have been developed to construct multicomponent homogeneous immunoassays, including fluorescence polarization, resonance energy transfer, Raman spectroscopy, luminescent oxygen channel, single-molecule counting, DNA proximity immunoassays, and liquid chip technologies. Among these, the test capacity of the first three strategies is generally limited to two or three targets, mainly due to constraints in tag selection and encoding capabilities. The latter four strategies can test four or more targets, but each comes with its own set of limitations. These limitations include the difficulty of miniaturizing instruments (e.g., the need for four laser beams), low detection efficiency resulting from complex and laborious processes (e.g., 300 sampling steps), and prolonged measurement times (e.g., up to 24 hours). Moreover, the LODs of current multicomponent homogeneous immunoassays typically range from the nM to pM scale, which is insufficient to meet the precision requirements for complex samples. The accuracy of signal recognition and the ability to resist interference are critical factors in determining the sensitivity and reliability of immunoassays, particularly for complex sample measurements. Therefore, the development of highly sensitive multicomponent homogeneous immunoassays is essential for the accurate detection of tumor biomarkers, advancing precision medicine, and addressing the future challenges in immunoassay technology.

COMMON LABELED IMMUNOASSAYS

Labeling refers to the chemical attachment of a tag to an antigen or antibody, creating a conjugate that can be used to detect immunocomplexes. This is an indirect detection strategy and remains one of the most widely employed methods in immunoassays. This section reviews these common labeled immunoassays, categorizing them based on the type of signal generated by the tags.

Radioactive Isotope Labeling

Radioimmunoassay (RIA) is a cornerstone technique in modern immunoassays. RIAs utilize radioactive isotopes as tags that replace corresponding atoms in molecules to form radioactive tracers. The atomic nuclei of these isotopes are inherently unstable and undergo spontaneous decay to achieve a more stable configuration. During this decay process, energy is emitted in the form of particles (such as α or β particles) or electromagnetic radiation (including X-rays or γ rays). The emitted radiation is detected by a scintillation counter, which consists of scintillators and a photomultiplier tube. The scintillator absorbs the radiation and emits faint scintillation light, which is subsequently converted into an electrical signal by the photomultiplier tube. The intensity of this electrical signal is directly proportional to the radiation energy emitted by the radioactive isotope. Commonly used isotopes in RIAs include iodine-125 (^{125}I) and tritium (^{3}H). γ rays emitted by ^{125}I possess strong penetrating power, allowing for contactless detection without the need for close proximity to the scintillation counter. In contrast, beta particles emitted by ^{3}H have lower penetrating strength, requiring direct contact with the scintillator for detection. Additionally, isotopes that emit gamma rays generally exhibit much higher specific activities than those emitting beta particles, meaning that fewer isotopes are needed for detection. In terms of detectable strength, one atom of ^{125}I is equivalent to 51.6 atoms of ^{3}H. RIA is primarily conducted as a competitive immunoassay, where the labeled isotope (either ^{125}I or ^{3}H) is attached to the antigen and competes with the unlabeled target antigen for the limited binding sites on the antibody. The higher the concentration of the target antigen, the less labeled antigen binds to the antibody, resulting in a lower signal intensity. Initially, RIA was performed using iodine-131 (^{131}I) to quantify insulin levels in serum; however, due to the radiation hazards associated with ^{131}I, it was later replaced by ^{125}I, which emits lower levels of γradiation. Maintaining the immunological activity of the labeled molecules is crucial for accurate results. For tritium labeling, there is typically no significant difference in the immunological activity between the labeled and unlabeled molecules, as the chemical structure remains unchanged. However, for iodine labeling, a change in immunological activity is often observed. This is because the hydrogen atom on the tyrosine phenyl group is typically replaced by radioactive iodine during the iodination reaction. To achieve high radioactivity, multiple iodine atoms may be attached to a single protein molecule. However, the more iodine atoms that are labeled, the greater the likelihood that immunological activity will be altered. Typically, one to two iodine atoms are used for labeling to minimize this effect.

In its early stages, RIA was regarded as an ultrasensitive technique capable of detecting proteins at the ng/mL level and small molecules at the pg/mL level, with the ability to measure nearly a thousand different targets. RIA significantly advanced the fields of medicine and life sciences, providing essential diagnostic tools for the detection and treatment of various diseases. The technique experienced rapid growth until the 1990s, but after 2005, it entered a period of stagnation. Today, while RIA still holds a place in clinical immunoassays, its use is gradually declining. In 2018, A national survey on nuclear medicine in China highlighted this trend. The number of departments conducting RIA tests decreased from 337 in 2016 to 216 in 2018.[37] The number of RIA samples also dropped by 20%, from 12.16 million in 2016 to 9.73 million in 2018. In contrast, the number of non-radioactive immunoassay samples increased by nearly 50%, from 58.65 million to 87.73 million. This shift is driven by growing concerns over the health risks associated with radiation exposure and the environmental impact of nuclear waste. Consequently, alternative non-radioactive labeling methods – such as enzyme labeling, fluorescent labeling, and scattering labeling – have rapidly developed and gained popularity.

Enzyme Labeling

In enzyme-labeled immunoassays, an enzyme is conjugated to an antigen or antibody. After immunoreaction, the enzyme is attached to the immunocomplex. The detection object shifts from the immunocomplex to the product generated by the enzyme-catalyzed conversion of the substrate. Enzymes possess high catalytic activity, with each enzyme molecule capable of converting 10^3 to 10^7 substrate molecules per minute. This actually significantly amplifies and enhances the signal, allowing for highly sensitive detection of immunocomplexes. Even at very low concentrations, large quantities of detectable products can be generated. Common enzymes used in enzyme-labeled immunoassays include horseradish peroxidase (HRP), alkaline phosphatase (ALP), β-galactosidase (β-Gal), glucose oxidase, glucoamylase, carbonic anhydrase, acetylcholinesterase, urease, pyrophosphatase, and β-lactamase. Among these, HRP, ALP, and β-Gal are the most widely used, accounting for approximately 80%, 20%, and less than 1% of total usage, respectively. Detection techniques based on these enzymes include colorimetric, fluorescent, chemiluminescent, electrochemiluminescent, and electrochemical methods, depending on the substrate, enzyme, and product. Given the current advances in DIs, the following section focuses on the mechanisms of HRP, ALP, and β-Gal in colorimetric, fluorescent, and chemiluminescent detection techniques. For a more comprehensive overview of immunoassays, readers are encouraged to consult the relevant literature and specialized books.

HRP

HRP is extracted from the roots of the horseradish plant and primarily consists of seven isoenzymes. Isoenzyme C (HRPC) is the most commonly used in analytical chemistry, accounting for more than 50% of total HRP applications. When HRP is mentioned, it typically refers to HRPC. HRPC is a holoenzyme, with heme serving as its cofactor. Its apoenzyme is a glycoprotein with molecular weight of approximately 42,100 Da. The glycoprotein contains 308 amino acid residues and 8 sugar chains, along with two calcium-binding sites. Calcium ions play a crucial role in stabilizing the HRPC structure, as demonstrated by the loss of structural stability in the enzyme's active site upon removal of calcium ions. The carboxyl groups of glutamic acid and aspartic acid residues, the amino groups of lysine, and the thiol groups of cysteine are available for conjugating HRPC to matrix surfaces or antibodies. HRPC has an isoelectric point of pH 8.8, with an optimal pH range for activity between 5.5 and 9. When stored as lyophilized powder at 4°C, HRPC can remain stable for years. In solution, it can be stored for several months at 4°C, provided its concentration exceeds 0.1 mg/mL. Due to its sensitivity to light, HRPC should be stored in darkness. The purity of HRPC is assessed by the absorbance ratio at 403 nm (heme) and 280 nm (protein). A lower ratio indicates a higher level of impurities in the preparation. A ratio between 3.0 and 3.4 is considered indicative of good purity. However, it is important to note that high purity does not necessarily correlate with high enzymatic activity.

HRPC catalyzes the oxidation of hydrogen peroxide or other peroxides. The general reaction is as follows (reactions 1.4):

$$H_2O_2 + 2AH_2 \xrightarrow{\text{HRPC}} 2 \cdot AH - H_2O, \tag{1.4}$$

where, AH_2 represents the reducing substrate (hydrogen donor), and ·AH is the free radical. The free radical can either combine to form a stable product or continue to react.

Reactions 1.4 are further refined as follows (1.5, 1.6, and 1.7):

$$HRPC + H_2O_2 \rightarrow HRPC - I + H_2O \tag{1.5}$$

$$HRPC - I + AH_2 \rightarrow HRPC - II + \cdot AH \tag{1.6}$$

$$HRPC - II + AH_2 \rightarrow HRPC + \cdot AH + H_2O, \tag{1.7}$$

where, HRPC-I and HRPC-II are intermediates in the enzyme's catalytic reaction.

In HRPC-labeled immunoassays, the choice of detection technique – such as colorimetry, chemiluminescence, or electrochemical methods – depends on the substrate used. Given that DIs primarily focus on luminescence, the following section will concentrate on colorimetric and luminescence-based immunoassays.

Colorimetry-Based Immunoassays Colorimetry-based immunoassays use the color change of a solution as both a qualitative and quantitative measure before and after enzyme catalysis. In the early stage, the common used substrates mainly included guaiacol, o-tolidine (OT), 5-amino salicylate, pyrogallol, o-phenylenediamine (OPD), o-dianisidine. The officially approved method by the Chinese government for detecting HRP (referred to as GB standards) involves oxidizing 2-methoxyphenol into brown 4-methoxyphenol by hydrogen peroxide with the aid of HRP (GB/T 32131-2015). The absorbance of the product is measured at 436 nm. However, due to the carcinogenicity of these substrates, their usage has gradually decreased. Commonly used substrates now include 3,3′,5,5′-tetramethylbenzidine (TMB) and 2,2′-azinobis(3-ethylbenzothiazoline-6-sulfonic acid) (ABTS). TMB is considered a safer substrate and is widely used in diagnostic tests. Performance comparisons between various HRP colorimetric substrates have confirmed that TMB shows the highest sensitivity in competitive immunoassays, even comparable to fluorescent substrates. HRP catalyzes the oxidation of TMB by hydrogen peroxide, resulting in the formation of a blue product. When both TMB and hydrogen peroxide concentrations are kept constant, an increase in enzyme concentration leads to a more intense color change. Kasetsirikul et al. developed a paper-based ELISA for detecting antibodies against the novel coronavirus, as shown in Figure 1.15a.[38] The chromatography filter paper is coated with the nucleocapsid antigen of the novel coronavirus. The sample is added sequentially, followed by the HRP-labeled detection antibody and TMB. After allowing the immunoreaction to occur, the paper is imaged using a camera. If antibodies against the novel coronavirus are present in the sample, they will bind to the coated antigen to form an immunocomplex. The HRP-labeled detection antibody binds to the immunocomplex, while unbound detection antibodies and free sample components are washed away. Only the bound immunocomplex remains on the paper. When TMB and hydrogen peroxide are added, hydrogen peroxide oxidizes TMB in the presence of HRP, resulting in a blue-green color on the paper. The intensity of the color is directly proportional to the concentration of viral

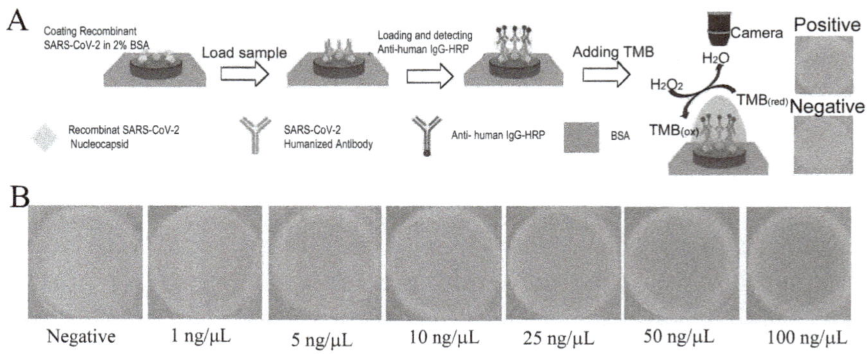

FIGURE 1.15 Test process and quantification performance of a colorimetric, paper-based device for detecting humanized antibodies against SARS-CoV-2.

Source: Reproduced from Kasetsirikul, S.; Umer, M.; Soda, N. et al. *Analyst*, 2020, 145, 7680–7686 (ref 38), with permission of the Royal Society of Chemistry.

antibodies: the higher the antibody concentration, the darker the color (Figure 1.15b). The results can be visually assessed within 30 minutes, with a minimum detectable concentration of 10 ng/μL.

Chemiluminescence-Based Immunoassays Chemiluminescence refers to the emission of light resulting from a chemical reaction. This phenomenon occurs when electrons in a molecular bond absorb chemical energy, causing the molecules to transition to an excited state. As the electrons return to a lower energy state, light is emitted. In HRP-catalyzed chemiluminescence assays, the most commonly used substrate is luminol and its derivatives, collectively known as the HRP-H_2O_2-luminol system. Luminol serves as a hydrogen donor, as illustrated in reaction 1.4. The resulting product exhibits a maximum emission peak at 425 nm. The LOD for free prostate specific antigen (PSA) in this assay is 0.03 ng/mL.[39] However, the chemiluminescence intensity is relatively low, with a low quantum yield (0.01) and short emission duration. To enhance sensitivity, the use of enhancers is required. Typically, enhancers increase the detection sensitivity by three to five orders of magnitude. Phenol iodide (PIP) is one of the most commonly used enhancers. The development of novel enhancers, including both single and synergistic types, is a key research focus in enzyme-catalyzed chemiluminescence immunoassays. With the application of these enhancers, the LODs of enhanced chemiluminescence-based immunoassays reach as low as 10^{-18} to 10^{-15} mol/L. In addition to the usage of enhancers, increasing the amount of enzyme also improves sensitivity. For example, by using multiple HRP-labeled GNPs, each immunocomplex contains several HRP molecules, resulting in an eightfold increase in the chemiluminescence signal. This approach allows for the simultaneous detection of tumor markers using spatially resolved chemiluminescence imaging.[40] It is important to note that detection performance is optimized when hydrogen peroxide is generated in situ, rather than added externally, as hydrogen peroxide is unstable.

Fluorescence-Based Immunoassays Fluorescence is the emission of light that occurs when a molecule absorbs photons and subsequently re-emits them as visible light. A key feature of fluorescence is that the emission occurs almost immediately (on the order of nanoseconds) after the absorption of light, and the emitted light has a longer wavelength (lower energy) than the absorbed light. Fluorescence is a form of photoluminescence and is widely used in biological and chemical assays. Common fluorescent substrates include 3-(p-hydroxyphenyl)propionic acid (HPPA) and 10-acetyl-3,7-dihydroxyphenoxazine (Amplex Red). HPPA, when reacted with hydrogen peroxide at pH 8.5 in the presence of HRP, forms a dimer that emits fluorescence at 404 nm when excited at 320 nm. Discovered by Zaitsu et al. in 1980, HPPA increases the sensitivity of immunoassays by 5 to 50 times compared to OPD-based colorimetric substrates.[41] While HPPA offers only a modest improvement in sensitivity over TMB, its dynamic range is significantly broader. Amplex Red is a commercially available fluorescent substrate that exhibits no fluorescence. When catalyzed by HRP, it reacts with hydrogen peroxide to generate a strongly fluorescent molecule with an emission wavelength of 587 nm and an excitation wavelength of 563 nm. Although fluorescence-based detection is generally more sensitive than colorimetry-based detection, when considering sensitivity, cost, and testing requirements, the HRP-TMB colorimetric kit appears to be a viable and effective option.

Alkaline Phosphatase

ALP is another commonly used enzyme tag in ELISA and chemiluminescence-based immunoassays. As a nonspecific phosphoesterase, ALP has a molecular weight ranging from 80 to 100 kDa, depending on its source. The optimal pH for ALP activity lies between 8 and 9.6. Inorganic phosphate can inhibit ALP activity, making PBS unsuitable as a buffer. Commercial ALP is typically derived from calf intestinal mucosa, and it is more expensive than HRP. The challenges associated with its purification and the higher cost limit its widespread use. ALP is widely present in the human body, with significant concentrations in organs such as the liver, bones, intestines, and kidneys. It plays essential roles in cell proliferation, apoptosis, and signal transduction. Abnormal levels of ALP in serum are associated with various diseases, including diabetes, prostate cancer, and bone disorders, making it a potential diagnostic biomarker. Although methods for detecting ALP can be adapted for use in ALP-labeled ELISA, its higher cost, lower stability (compared to HRP), and difficulties in purification make it less commonly employed in enzyme-labeled immunoassays. The following section provides a brief overview of the common approaches used for detecting ALP in enzyme-labeled immunoassays.

Colorimetry-Based Immunoassays In traditional ELISA, p-nitrophenyl phosphate (pNPP) is commonly used as a substrate for ALP. As depicted in Figure 1.16A, ALP catalyzes the hydrolysis of pNPP to produce yellow p-nitrophenol (pNP), which has a maximum absorption at 400 nm under alkaline conditions (pH > 9.0). Antigen concentration is determined semi-quantitatively by the color change, while quantitative measurement is achieved by monitoring changes in absorbance. pNP also inactivates ALP, and sodium hydroxide is typically employed to terminate the reaction. This substrate is widely used in commercial ELISA kits. To enhance the

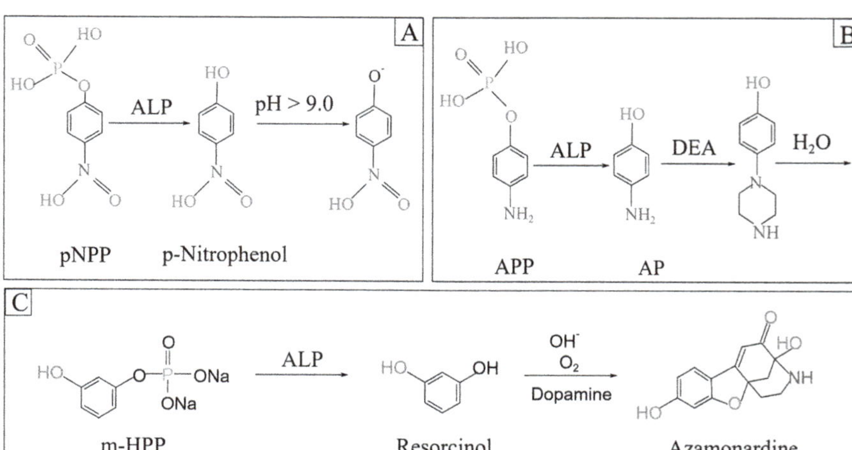

FIGURE 1.16 Processes of ALP catalysis, with (A) pNPP, (B) APP, and (C) mHPP as substrates.

sensitivity of colorimetric ELISA, new substrates have been developed. For example, 4-aminophenyl phosphate (APP) is hydrolyzed by ALP to produce 4-aminophenol (AP) (Figure 1.16B), which then reacts with ethylenediamine (DEA) to generate a product with absorption peak at 365 nm. The intensity of this peak is directly proportional to ALP activity, and this substrate was applied in the detection of AFP with a LOD of 0.1 ng/mL.[42] Inspired by the reaction between resorcinol and dopamine under alkaline conditions to form azamonardine, a dual-readout immunoassay has been developed, combining both fluorescence and colorimetric detection. mHPP is hydrolyzed by ALP to yield resorcinol, which then reacts with dopamine to form azamonardine, a compound that exhibits an absorption peak at 420 nm and fluorescence emission at 460 nm (Figure 1.16C). This dual-readout assay has demonstrated significant improvements in the detection of myocardial troponin compared to traditional colorimetric ELISA. Specifically, the LOD of the conventional pNPP-based assay was 0.15 ng/mL, while the mHPP-based colorimetric assay achieves a lower LOD of 0.04 ng/mL.[43]

Fluorescence-Based Immunoassays A commonly used fluorescent substrate for ALP is 4-methylumbelliferyl phosphate (4-MUP). Upon hydrolysis, 4-MUP is converted to 4-methylumbelliferone, which emits fluorescence at 445 nm when excited at 372 nm. This hydrolysis reaction has been successfully adapted to microfluidic chips. Due to the micron-scale dimensions of the microfluidic channels, the time required for reagent mixing and reaction is significantly reduced. Consequently, the total detection time was shortened from 920 minutes in a microplate to just 118 minutes in a microchip, and reagent consumption decreased from 500 μL to 4 μL. However, despite these improvements in efficiency, detection performance was not significantly enhanced.[44] To improve sensitivity, a promising strategy is the development of novel enzymatic fluorescent substrates. Calcium Green, a commonly used fluorescent probe, is excited at 480 nm and emits fluorescence at 512 nm. However, when it binds to cerium (Ce^{3+}), the fluorescence of calcium green is quenched. In the presence of ALP, pNPP is hydrolyzed to produce p-nitrophenol and inorganic phosphate. The inorganic phosphate strongly binds to Ce^{3+}, thereby releasing calcium green and restoring its fluorescence. When ALP is linked to antibodies and calcium green-Ce^{3+} is used as a reagent, immunoassays can be performed by measuring the degree of fluorescence recovery. The LOD for ALP activity was 0.023 U/L, while for AFP, the value was 0.041 ng/mL.[45] In another application, ALP catalyzes the hydrolysis of ascorbic acid phosphate (AA-phosphate) to produce AA. AA enhances the fluorescence of various organic dyes, including polydopamine nanoparticles, QDs, carbon dots, and gold nanoclusters. The product of the reaction between AA and o-phenylenediamine (OPD) emits fluorescence at 425 nm when excited at 360 nm. This substrate has been adapted to ELISA for AFP detection, with a LOD of 0.21 ng/mL.[46] At 70°C, ALP catalyzes the hydrolysis of N-[3-(trimethoxysilyl)propyl]ethylenediamine (DAMO) over 20 minutes, producing orange-red silicon-containing nanoparticles that emit yellow-green fluorescence. ALP catalyzes the hydrolysis of 4-aminophenyl phosphate (APP) to produce 4-aminophenol (AP). With these substrates, the LOD for ALP activity was 0.0022 U/L, while the ELISA LOD for PSA was 0.0041 ng/mL.[47]

Chemiluminescence-BasedImmunoassays 3-(2′-spironaphthyl)-4-methoxy-4-(3″-phosphoryloxy)-benz-1,2-dioxane (AMPPD), synthesized by Bronstein et al. in 1988[48], is widely employed as ALP substrate in commercial chemiluminescence-based ELISA assays. The chemiluminescent mechanism of AMPPD is illustrated in Figure 1.17. ALP catalyzes the hydrolysis of AMPPD, cleaving the phosphate group and generating an unstable intermediate, AMPD. This intermediate rapidly decomposes into adamantanone and an excited-state anion product (MOB^-). Upon relaxation of the MOB^- anion to its ground state, light is emitted at 470 nm. The LOD for thyroid-stimulating hormone (TSH) using the chemiluminescence reaction wass 0.007 mU/L, compared to 0.03 mU/L with colorimetric detection.[49] For total PSA detection, the LOD for ELISA using AMPPD was 0.05 ng/mL.[39] While chemiluminescence-based detection generally exhibits lower sensitivity than fluorescence-based one, it is preferred for its simplicity and cost-effectiveness due to the absence of a required light source in the instrument.

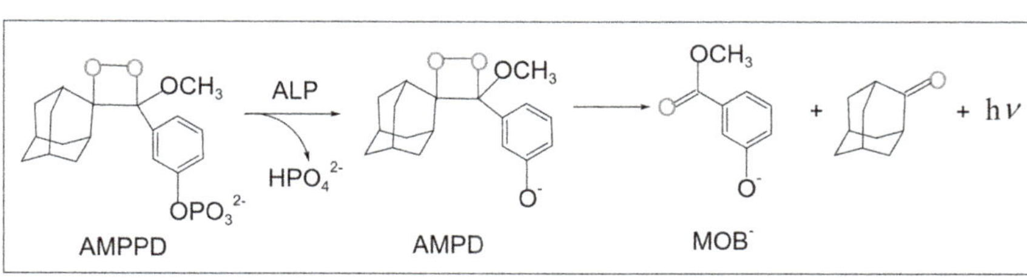

FIGURE 1.17 ALP catalysis process with AMPPD as the substrate.

β-D-Galactosidase

β-Gal, also known as β-D-galactoside galactohydrolase, is widely distributed in plants, animals, and microorganisms. Microorganisms are the primary providers of this industrial enzyme, which has a molecular weight of approximately 500 kDa and an optimal pH range of 6–8. However, its relatively large molecular size limits its broader application. β-Gal exhibits two main catalytic activities: hydrolyzing lactose and transglycosylating substrates to synthesize oligosaccharides. The enzyme-catalyzed synthesis of oligosaccharides has become a key alternative to the oligosaccharides found in human milk. Nonetheless, the transglycosylation activity of β-Gal is relatively low, and the hydrolysis reaction happening at the same time often interferes with its transglycosylation function. Both deficiency and overexpression of β-Gal have been linked to various rare diseases, especially those associated with cellular aging. The enzyme primarily catalyzes the hydrolysis of 1,4-glycosidic bonds, resulting in the release of galactose and alcohol. Most β-Gal substrates consist of galactosyl residues and signaling molecules that are linked by glycosidic bonds. The hydrolysis of these substrates by β-Gal leads to the release of the signal molecule, which is often accompanied by a detectable color change or fluorescence emission (Figure 1.18).

In colorimetric assays, o-nitrophenyl-β-D-galactoside (oNPG) serves as a substrate for β-Gal. The hydrolysis of oNPG produces o-nitrophenol (oNP), which absorbs maximally at 405 nm. The oNP is measured using a standard spectrophotometer, as well as a handheld ion mobility spectrometer, as demonstrated by Snyder et al.[50] In their assay, Bacillus subtilis was immobilized on a carrier and used as an antigen, and it captured monoclonal antibodies. β-Gal-labeled antibodies bound to the captured monoclonal antibodies, and oNPG was used as the substrate. The amount of oNP generated was then used to quantify the microorganism.

Two common fluorescent substrates for β-Gal are Fluorescein di (β-D-galactopyranoside) (FDG) and Methylumbelliferone

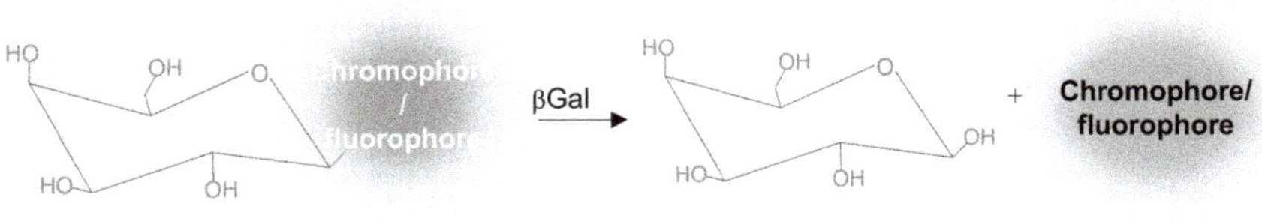

FIGURE 1.18 Universal process of β-Gal catalysis.

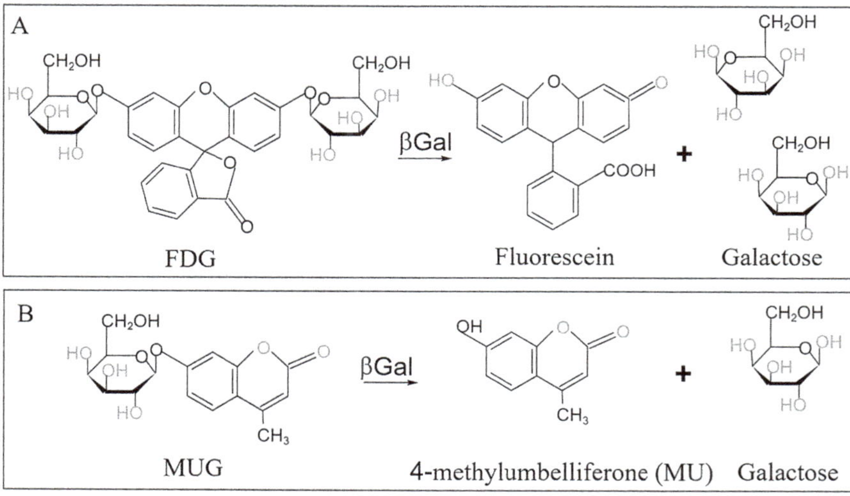

FIGURE 1.19 Processes of β-Gal hydrolyzing FDG (A) and MUG (B).

β-D-galactopyranoside (MUG). Neither FDG nor MUG emits fluorescence on its own; instead, their hydrolysis products do. Therefore, β-Gal activity is quantified by measuring fluorescence intensity. The hydrolysis of FDG produces fluorescein and two molecules of galactose (Figure 1.19A). Fluorescein can be excited at 491 nm and emits fluorescence at 514 nm. The FDG hydrolysis reaction is a two-step process. In the first step, Fluorescein monogalactoside (FMG) and one molecule of galactose are generated. In the second step, FMG is further hydrolyzed to produce fluorescein and another molecule of galactose. It is important to note that FMG itself does not emit fluorescence. The conversion rates from FDG to FMG and from FMG to fluorescein are 1.9 mmole/min·mg and 22.7 mmole/min·mg, respectively. As a result, when FDG is used as a substrate, the fluorescence intensity remains relatively low within the first ten minutes of the reaction. This suggests that a longer incubation time is required for optimal enzyme activity measurement. Rotman et al. encapsulated FDG and diluted β-Gal into droplets of approximately 14 μm in diameter.[51] According to Poisson distribution, some droplets were expected to contain a single enzyme molecule. By measuring changes in fluorescence intensity, the activity of individual enzyme molecules and their reaction constants were determined. This study marked the beginning of single-molecule enzyme detection and droplet-based enzyme activity measurements. As shown in Figure 1.19B, the hydrolysis of MUG produces 4-methylumbelliferone (MU) and galactose. MU has an excitation peak at

372 nm and an emission peak at 445 nm. Over a 60-minute reaction period, the LODs of FDG and MUG are nearly identical. However, MUG is considered more suitable for rapid detection due to its one-step hydrolysis process. It is important to note that the fluorescence of MU is easily quenched when the pH of the solution drops below its pKa of approximately 8.0.

In addition to the reactions shown in Figure 1.18, theoretically, the colorimetric and fluorescent techniques used to measure β-Gal activity can also be adapted for enzyme-linked immunoassays. Hu et al. designed a colorimetric approach for detecting β-Gal activity using 4-amino phenyl β-D-galactoside (pAPGP) as substrate. Upon hydrolysis of pAPGP by β-Gal, AP and galactose are generated. The AP then reduces Fe^{3+} to Fe^{2+} that forms a red complex with a 1,10-phenanthroline derivative (hydroxy red ferrous phenanthroline disulfonate, BPDS). This complex exhibits absorbance at 535 nm, and the intensity of this absorbance is directly proportional to β-Gal activity. When β-Gal is conjugated to antibodies, this method can be used for antigen quantification in enzyme-labeled immunoassays.[52]

Performance Comparison of Three Enzyme-Labeled Immunoassays

HRP, ALP, and β-Gal are the three most commonly used enzyme tags in immunoassays, and their performance has been compared in numerous studies. Among these, HRP has the smallest molecular weight (40 kDa), which results in the least steric hindrance and minimizes the impact of labeling on the immunoreaction. ALP, although having the highest catalytic activity, is susceptible to inhibition by phosphates and EDTA. β-Gal has the largest molecular weight (500 kDa), which may affect its suitability in certain applications. In practice, the choice of enzyme and substrate depends on the specific detection requirements and environmental conditions. The sensitivity and precision of enzyme-labeled immunoassays are influenced by various factors, including the type of immunoassay (competitive or sandwich), the method used for enzyme labeling, and the labeling efficiency (i.e., the number of enzyme molecules conjugated to each antigen molecule).

Imagawa et al. compared the advantages and disadvantages of β-Gal and HRP in sandwich-structured immunoassays.[53] MUG and HPPA were used as fluorescent substrates for β-Gal and HRP, respectively, with AFP as the target analyte. In terms of enzyme sensitivity alone, β-Gal demonstrated superior sensitivity compared to HRP. After 10 minutes and

100 minutes of reaction, the detectable product catalyzed by β-Gal was 0.2 amol and 0.02 amol, respectively, whereas for HRP, the detectable product was 5 amol and 0.5 amol, respectively. However, HRP and HRP-Fab conjugates exhibited less nonspecific adsorption than β-Gal and β-Gal-Fab conjugates, leading to higher sensitivity for HRP in sandwich immunoassays targeting AFP. Furthermore, HRP required a higher amount of enzyme (250 fmol/tube for 100 minutes) and longer incubation times, compared to β-Gal, which required only 2.9 fmol/tube and 30 minutes for optimal performance.

Porstmann et al. studied the conversion rates of three enzymes and their antibody conjugates under varying reaction times and temperatures, using both colorimetric and fluorescent methods.[54] They found that fluorescent substrates had lower LODs than colorimetric substrates, with HRP exhibiting the lowest LOD. In terms of enzyme sensitivity alone, HRP showed the highest conversion rate for ABTS. However, after conjugation to IgG, the catalytic efficiency of all three enzymes decreased to varying extents. Their study also revealed that fluorescent substrates had lower conversion efficiency than colorimetric substrates. The LOD for fluorescent substrates was lower than for colorimetric substrates, likely because the fluorescent product is more sensitive to detection.

Hosoda et al. compared the performance of competitive enzyme-labeled immunoassays using HRP, ALP, and β-Gal as tags, with hormones such as testosterone and cortisol as target analytes.[55, 56] The sensitivity of the assays, whether using fluorescence or colorimetric detection, was ranked from highest to lowest as follows: HRP, ALP, and β-Gal. The best colorimetric sensitivities for HRP and ALP reached levels comparable to those of fluorescence-based methods.

Grandke et al. conducted a comprehensive comparison of the performance of HRP and ALP in direct and indirect competitive immunoassays for detecting caffeine, evaluating sensitivity, dynamic range, accuracy, and precision.[57] Using TMB and pNPP as colorimetric substrates for HRP and ALP, respectively, and HPPA and 4-MUP as fluorescent substrates for HRP and ALP, respectively, they measured caffeine concentrations in common consumer products such as coffee, soft drinks, tea, and cosmetics. Overall, the direct competitive immunoassays outperformed the indirect competitive assays. The performance of the direct competitive immunoassays is summarized in Table 1.3. HRP-labeled immunoassays demonstrated superior performance compared to ALP-labeled assays, regardless of whether fluorescence or colorimetric detection was used. Fluorescence-based

TABLE 1.3 Comparison between Enzymes and Their Corresponding Substrates

Enzyme-Substrate	Sensitivity (Concentration Corresponding to the Median Signal of the Standard Curve, µg/L)	Measurable Range (Concentration Range with a Measurement Variance Less Than 30% on the Standard Curve, µg/L)	Accuracy (Slope Compared to That of LC-MS/MS Measurements)	Precision (CV, %)	
				Intra-plate	Inter-plate
HRP-TMB	0.095	0.033–33	0.96	1.0–9.9	0.9–18
HRP-HPPA	0.075	0.105–0.623	0.83	1.8–9.4	0.4–9.8
ALP-pNPP	0.817	0.128–33	1.05	6.7–41	1.5–29
ALP-4MUP	0.890	0.191–33	0.99	6.2–52	5.3–50

HRP-labeled immunoassays showed slightly better sensitivity and precision than colorimetric assays, which exhibited a broader dynamic range and higher accuracy. The study of colorimetric competitive immunoassays for cortisol revealed that results obtained with HRP were more sensitive to variations in sample conditions compared to ALP. Specifically, cortisol levels in serum samples subjected to multiple freeze-thaw cycles were found to be higher when HRP was used as the label.[58]

In both fluorescent and colorimetric assays, HRP is generally considered superior. It also offers distinct advantages in chemiluminescence immunoassays. For example, Yu et al. compared the performance of HRP-luminol-hydrogen peroxide and ALP-AMPPD chemiluminescence kits in competitive enzyme immunoassays for detecting chemical drugs.[59] They found that the HRP-based assay exhibited faster luminescence kinetics, reaching maximum luminescence intensity within approximately 2 minutes, whereas the ALP-based assay took around 30 minutes to reach its peak intensity. There were no significant differences in LODs, precision, or dynamic range between the two kits.

Nanozyme-Based ELISAs

Enzyme-labeled immunoassays are among the most widely used immunoassays. However, the preparation and purification of natural enzymes can be costly, and their catalytic activity is highly sensitive to environmental conditions. Additionally, they are difficult to recover and reuse, posing challenges related to cost, quality control, and long-term storage. In 2007, Chinese scientist Yan et al. first discovered that Fe_3O_4 nanoparticles exhibited peroxidase-like catalytic activity and could effectively replace natural enzymes in immunoassays.[60] Since then, various nanomaterials, including noble metals, metal oxides, metal sulfides, carbon-based materials, and metal-organic frameworks, have been found to possess enzyme-like activity. These materials, known as nanozymes or enzyme mimics, offer several advantages, including a large surface-to-volume ratio, ease of surface modification, high stability, and low cost. The potential of nanozymes as enzyme tags in immunoassays has been well demonstrated. The following section discusses several common types of nanozymes.

Peroxidase-Like Nanozyme Nanozymes with peroxidase-like activity are the most diverse, including metal nanoparticles (e.g., gold, platinum, palladium), metal oxide nanomaterials (e.g., Fe_3O_4, MnO_2, CeO_2), metal-organic frameworks (MOF; e.g., Cu-MOF, hemin-Au@MOF, Fe-MOF), and composite nanostructures (e.g., Au@Pt, Au@PtCu, gold-graphene, alloy materials, Au-Pt/SiO_2, Pd@Pt). Many of these peroxidase-like nanozymes have been successfully applied in immunoassays.[61]

ALP-Like Nanozyme Fewer nanomaterials exhibit ALP-like activity compared to peroxidase-like nanozymes. ALP-like nanozymes include CeO_2 nanoparticles, Zr(IV) cluster MOF materials, and CeO_2/ZrO_2 composite nanomaterials. They have been less commonly applied in immunoassays. For instance, PAA-modified CeO_2 nanoparticles were conjugated with anti-PSA detection antibodies. Using CDP-star as a chemiluminescent substrate, the ALP-like nanozyme-based, sandwich-structured immunoassay achieved a LOD of 53 fg/mL for PSA.[62]

Oxidase-Like Nanozyme Oxidases catalyze the oxidation of substrates in the presence of oxygen, and they are typically named after their substrates, such as glucose oxidase, alcohol oxidase, and lactate oxidase. The oxidase-like properties of nanoparticles vary depending on their composition and structure, and these properties have been applied to detect various antigens. Notable examples include Au@Pt nanoparticles for detecting interleukin-2 through HRP-like catalysis with TMB and OPD as substrates, MnO_2 nanoflowers for detecting C-reactive protein via catalysis of OPD, zeolitic imidazolate framework (ZIF) nanobands for detecting E. coli O157:H7 through TMB catalysis, and $Co(OH)_2$ nanocages for detecting ochratoxin A by catalyzing TMB. Some oxidases utilize the same substrates as peroxidases; however, unlike peroxidases, oxidases do not require hydrogen peroxide for their catalytic reactions.

In general, nanozymes exhibit higher catalytic activity, greater antibody loading capacity, enhanced stability, and reduced susceptibility to degradation compared to natural enzymes. Moreover, they are easier to produce. However, nanozymes lack specific substrate-binding sites, meaning that a single nanozyme can catalyze multiple reactions, resulting in lower selectivity and specificity. The functional diversity of nanozymes is still somewhat limited, with a primary focus on catalyzing oxidation reactions. Furthermore, because catalytic activity occurs at the surface, it can be reduced or inhibited when the surface is coated with antibodies. Therefore, the application of nanozymes in immunoassays requires further investigation.

Fluorescence Dye Labeling

Immunoassays using fluorescent tags theoretically offer higher sensitivity compared to those employing other labeling tags. However, in practice, their sensitivity is often compromised by background fluorescence, scattered light, and light quenching. For instance, the LODs of fluorescent immunoassays are typically 10–50 times higher in serum than in buffer solutions. Background fluorescence is primarily caused by serum proteins, NADH, bilirubin, and other auto-fluorescent substances, while light scattering is mainly due to proteins and colloidal particles. To address these challenges, fluorescent labels generally require long-wavelength emission (500–700 nm), a large Stokes shift (greater than 50 nm), and a long fluorescence lifetime (greater than 20 ns). Commonly used fluorescent tags include Alexa dyes, FITC, rhodamine B, rhodamine 6G, Lucifer Yellow, and Cy3. Among these, Alexa dyes are preferred for fluorescent dye-labeled immunoassays due to their superior quantum yield, photostability, pH insensitivity, and labeling efficiency, although cost may be a consideration. Moreover, fluorescent labeling enables the feasibility of homogeneous immunoassays without the need for separation, expanding the range of immunoassay modalities. The previously mentioned FRET-based, fluorescence polarization-based, and FCS-based homogeneous immunoassays all fall within the category of fluorescent immunoassays. In contrast, radioimmunoassays and enzyme immunoassays cannot be performed in a homogeneous format. It is also worth noting that time-resolved fluorescence immunoassays and infrared dye labeling-based immunoassays merit particular attention due to their unique advantages.

In time-resolved fluorescence immunoassays, long-lived fluorescent tags are used to label immunocomplexes. Pulsed light is employed as the excitation source, and detection occurs after a delayed time period. During this delay, background fluorescence and scattered light, which have short lifetimes of less than 10 ns, decay, while long-lived fluorescent tags continue to emit light. As a result, background interference is minimized, improving the signal-to-noise ratio and enhancing sensitivity. To achieve this, the fluorescence lifetime of the probe must exceed 50 ns, a requirement that is met by rare organic dyes. Currently, chelates of rare-earth ions are the most effective time-resolved fluorescent tags, with Eu, Tb, Sm, and Dy being commonly used lanthanide elements. These elements have fluorescence lifetimes ranging from 100 to 1000 μs, and multicomponent homogeneous immunoassays can be achieved by labeling different reagents with distinct lanthanide elements.[63]

To minimize background signals, near-infrared fluorescent tags present a viable alternative. These probes emit light in the range of 650–900 nm, which reduces self-fluorescence background because the matrix typically absorbs and emits light below 600 nm. Additionally, Rayleigh scattering intensity is inversely proportional to the fourth power of the wavelength, meaning that as the wavelength increases, scattering background signals are further reduced. Near-infrared light offers other advantages, including deeper tissue penetration and lower phototoxicity. Common near-infrared dyes include anthocyanins, oxazole compounds, BODIPYs, and phthalocyanine dyes. However, many of these dyes suffer from issues such as poor water solubility, limited photostability, low biocompatibility, and restricted functionalization capabilities. Future developments are likely to focus on the invention of new near-infrared tags, including near-infrared QDs, rare-earth complexes, and single-walled carbon nanotubes.

Nanomaterial Labeling

Nanomaterials possess unique physical and chemical properties. As a result, they hold great promise for further enhancing the sensitivity of immunoassays. The types of nanomaterials are diverse and continue to be developed. In this section, we introduce several types of nanomaterials commonly used as optical signal tags in immunoassays, including noble metal nanoparticles, QDs, upconversion nanomaterials, and carbon-based nanomaterials, with a particular focus on plasmon resonance and luminescence tags. Nanomaterials that serve solely as matrix materials for loading tags and primarily function to amplify signals, are not covered in this section.

Localized Surface Plasmon Resonance Labeling

Localized surface plasmon resonance (LSPR) refers to the collective oscillation of conduction electrons in a metallic nanoparticle, typically composed of gold or silver, when excited by light at a specific wavelength. In brief, when electromagnetic waves irradiate noble metal nanoparticles, the free electrons on the nanoparticle's surface oscillate collectively. Resonance occurs when the oscillation frequency of these electrons matches the frequency of the incident electromagnetic field. This resonance causes the metal nanoparticle to absorb the incident light. In turn, the oscillating electrons emit electromagnetic waves, resulting in scattering light. Both the resonance absorption and resonance scattering occur at the same frequency, contributing to an extinction spectrum that can be measured using ultraviolet-visible (UV-Vis) absorbance spectroscopy. A nanoparticle solution typically exhibits a color complementary to the maximum absorption wavelength. The scattering spectrum can be measured using a fluorescence spectrometer or dark-field microscope. For example, spherical GNPs (70 nm in diameter) show maximum absorption of green light at 540 nm, making the solution appear reddish-purple. The maximum scattering wavelength coincides with the maximum absorption wavelength, so the gold particles appear green under dark-field microscopy. LSPR is influenced by factors such as particle size and shape, dielectric environment, and inter-particle spacing. Changes in these factors, induced by immunoreactions, can be utilized to detect the target. Light absorption is commonly employed in colorimetric immunoassays, while scattering is typically used in scattering-based immunoassays.

Colorimetry-Based Immunoassays Immunoreactions induce changes in the refractive index of the environment surrounding noble metal nanoparticles, which in turn lead to a shift in the LSPR absorption peak, denoted as $\Delta\lambda_{max}$. There are generally three mechanisms that cause $\Delta\lambda_{max}$: (1) Antigen binding to the GNP surface directly results in a shift in $\Delta\lambda_{max}$, with the magnitude of the shift increasing as antigen concentration rises. By monitoring the LSPR spectra of individual GNPs before and after antigen binding, a single-particle colorimetric immunoassay can be developed by correlating $\Delta\lambda_{max}$ with antigen concentration. For instance, Sim et al. reported that spherical GNPs (30 nm) exhibited a minimum detectable concentration of PSA of 1 pg/mL, with $\Delta\lambda_{max}$ observed to be 2.85 nm at this concentration.[64] To enhance sensitivity, they replaced the spherical GNPs with 50 nm × 18 nm gold nanorods, which enabled the detection of

PSA at a much lower concentration of 111 aM, with a $\Delta\lambda_{max}$ of 2.79 nm.[65] (2) Immunocomplexes can induce the aggregation or deaggregation of GNPs, leading to a noticeable change in the color of the solution. Acetylcholinesterase (AChE) catalyzes the conversion of acetylthiocholine into thiocholine, which, through electrostatic interactions, causes GNPs to aggregate. This aggregation results in a red shift in the absorption wavelength, from 520 nm to 700 nm, causing the solution color to shift from red to purple. In an ELISA, AChE was used as an enzyme label, and GNPs served as the chromogenic substrate, enabling the detection of enterovirus at a sensitivity level comparable to PCR.[66] CuO-labeled antibodies formed immunocomplexes with antigens, releasing Cu^{2+} ions in the process. The released Cu^{2+} ions induced the aggregation of gold nanoparticles that were modified with azide or alkynyl groups. The extent of this aggregation, and consequently the color change of the solution, correlated with the antigen concentration.[67] The interaction between phenylboronic acid and citric acid can cause aggregation of citric acid-modified GNPs, while hydrogen peroxide oxidation of boronic acid prevented this aggregation. In this case, glucose oxidase was employed as a label for the immunocomplex, generating hydrogen peroxide that triggers a color change in the solution from purple to red.[68] (3) The reaction induces morphological changes in gold nanoparticles when hydrogen peroxide reduces $HAuCl_4$ to gold atoms that then deposit onto 5 nm gold nanoparticles and form larger particles. This increase in size causes the color of the solution changing from colorless to red. At lower concentrations of hydrogen peroxide, the gold atom generation rate slows down, and the even synthesis of nanoparticle cannot keep up. More importantly, the nanoparticles aggregate, causing the color to change from red to blue. These phenomena can be applied to immunoassays. For example, glucose oxidase can be used as a label to generate high concentrations of hydrogen peroxide, promoting uniform gold nanoparticle synthesis. On the other hand, HRP consumes hydrogen peroxide and produces uneven or aggregated GNPs. In addition to deposition to alter nanoparticle size, GNPs also are etched to change their shape. This shape change, in turn, affects the solution color. For instance, TMB^{2+} generated by HRP catalysis,[69] as well as Cu^{2+} and other reagents,[70] cause a color change through etching of gold nanorods. Therefore, the change is useful in ELISA immunoassays.

Scattering-Based Immunoassays Scattering analysis encompasses elastic scattering and resonance scattering. Elastic scattering is not related to LSPR, but it is included here as a form of scattering. Dynamic light scattering (DLS) technology measures the temporal changes in scattering intensity of freely diffusing nanoparticles in solution. Particle size distribution is calculated through the intensity correlation function. GNPs and gold nanorods are used to label capture and detection antibodies, respectively. Upon formation of immunocomplexes, the resulting particle size distribution significantly differs from that of the unreacted particles. This variation correlates with antigen concentration. Liu et al. developed a dynamic light scattering-based homogenous immunoassay with a LOD of 0.1 ng/mL for PSA in buffer solution.[71]. However, GNPs are prone to adsorbing large molecules and aggregating, which can distort particle size measurements and limit the method's applicability in real sample analysis. Li et al. modified MnO_2 nanosheet-wrapped GNPs with detection antibodies.[72] These nanosheets formed immunocomplexes with corresponding capture antibodies immobilized on the substrate. Cysteine rapidly reduced Mn^{4+} to Mn^{2+}, causing the dissociation of the nanosheets and the subsequent release of GNPs into the solution. The concentration of GNPs, measured by DDLS, correlates with the antigen concentration. As the antigen concentration increases, the concentration of GNPs also increases. The dynamic range of this immunoassay spans from 1 to 100 pM, with a LOD of approximately 1 fM.

Noble metal-based resonance scattering immunoassays are performed at both the ensemble level and the single-particle level. The corresponding detection techniques for these assays are fluorescence spectroscopy and dark-field microscopy, respectively. In 2009, Huang et al. labeled detection antibodies with 60 nm silver nanoparticles.[73] The labeled antibodies formed a sandwich-structure immunocomplex with antigens and antibodies immobilized on a glass substrate. The resonance scattering spectrum of the silver nanoparticles on the glass surface was measured using a fluorescence spectrometer, with scattering intensity at 400 nm correlating with antigen concentration. The dynamic range for IgG was 10–1000 ng/mL, with a LOD of 1.46 ng/mL. In 2017, they employed both gold and silver nanoparticles as labels and developed a dual-component immunoassay using dark-field microscopy, LSPR-based counting immunoassays[74]. The dynamic range for AFP and CEA was 0.5–10 ng/mL. To further improve detection specificity and sensitivity, Xu et al. replaced the glass substrate with a gold surface, where gold nanoparticles were immobilized via immunoreactions.[75] This modification shifted the maximum absorption peak of LSPR from 530 nm to 676 nm and increased scattering intensity by 20.5 times. The resulting immunoassay demonstrated a dynamic range of 10^{-3} to 10^{3} ng/mL and a LOD of 1 pg/mL (6.67 fM) for IgG.

Luminescent Nanomaterial Labeling

Common luminescent nanomaterials, such as QDs, carbon dots, nanoclusters, and rare-earth nanomaterials, have garnered significant attention in various fields. In the context of DIs, this section primarily focuses on the recent advancements in the application of QDs and rare-earth nanomaterials. These materials have shown promising potential for enhancing the sensitivity and performance of immunoassays.

QD-Based Immunoassays QDs are luminescent nanocrystals synthesized from semiconductor materials, with sizes typically ranging from 1 to 10 nm. The emission wavelength of QDs is size-dependent, and they are commonly designated based on their peak emission wavelength. For instance, QD585 refers to a QD with a maximum emission wavelength of 585 nm. Notably, QDs of varying sizes can be excited by the same wavelength of light, which forms the basis for their application in multicomponent immunoassays. Additionally, QDs exhibit several advantages over organic fluorescent dyes, such as superior photostability and a higher quantum yield.

In the early stages of QD-based immunoassays, QDs were primarily used as fluorescent tags. It was expected that their superior fluorescence properties would enhance the performance of immunoassays. The process of QD-based immunoassay was similar to that of ELISA. It involved fixing antibodies onto a glass surface or plate, followed by antigen binding and QD-labeled antibody attachment. After washing, fluorescence intensity was detected using a scanner or spectrometer. However, the performance did not meet the expectations, with LODs comparable to those of fluorescence dye-based assays, typically at the ng/mL or nM level. Despite these limitations, the use of QDs as tags not only aimed to enhance sensitivity but also contributed to the diversification of immunoassay types. This led to the development of various multicomponent and homogeneous immunoassays. The unique advantages of QDs in multicomponent immunoassays lie in their ability to be excited by a single light source, significantly simplifying the instrumentation. Moreover, QDs can be employed as either donors or acceptors in resonance energy transfer (RET). When combined with quenchers or luminescent materials, QD-based immunoassays can be designed to turn on or off in response to target binding. For instance, when QDs are linked to graphene through an immunocomplex, the fluorescence intensity of the QDs is quenched by the graphene, and the decrease in fluorescence correlates with antigen concentration. In another example, plasmonic energy transfer occurs between QDs and GNPs when they are brought into close proximity by an immunocomplex, resulting in a decrease in the scattering intensity of the GNPs. Additionally, QDs can function as receptors, with lanthanide nanomaterials acting as donors. In QD-lanthanide nanomaterial-based immunoassays, LODs as low as sub-ng/mL levels have been achieved, with the ability to simultaneously detect up to six targets.[76] In DIs, QDs serve as highly effective markers for multicomponent homogeneous detection. This topic will be further explored in the section on homogeneous DIs.

Rare-Earth Upconversion Luminescent Nanomaterial-Based Immunoassays

Rare-earth upconversion luminescent nanomaterials (UCNPs) are typically synthesized from oxides, fluorides, and halide oxides doped with rare-earth ions such as Er^{3+}, Eu^{3+}, and Yb^{3+}. These nanoparticles have the unique ability to convert excitation photons into shorter wavelengths of light. UCNPs offer several advantages, including low toxicity, excellent photostability, narrow emission bands, long fluorescence lifetimes, and resistance to photobleaching. When excited by near-infrared light, they exhibit additional benefits, such as enhanced tissue penetration, minimal autofluorescence from biological tissues, and reduced tissue damage. Due to these properties, UCNPs have found applications in heterogeneous immunoassays, lateral flow immunoassay (LFIA) test strips, and single-molecule immunoassays, demonstrating comparable detection performance to other labeling techniques. In LFIA, the sensitivity of UCNPs is at least ten times higher than that of GNPs and fluorescent microsphere-based assays, as well as microplate ELISA.[77] However, the most promising application of UCNPs may be in homogeneous immunoassays, where their unique optical properties can offer significant advantages over traditional methods.

Luminescence resonance energy transfer (LRET) is a key strategy used in homogeneous immunoassays, where UCNPs serve as energy donors and QDs or GNPs act as acceptors. The detectable signal can either be emitted by the QDs, which receive the energy transfer, or by prequenched UCNPs. Figure 1.20A illustrates an immunoassay based on UCNP quenching, where $NaYF_4$:Yb, Er UCNPs and GNPs label two antibodies. Upon excitation by near-infrared light, UCNPs emit both green and red

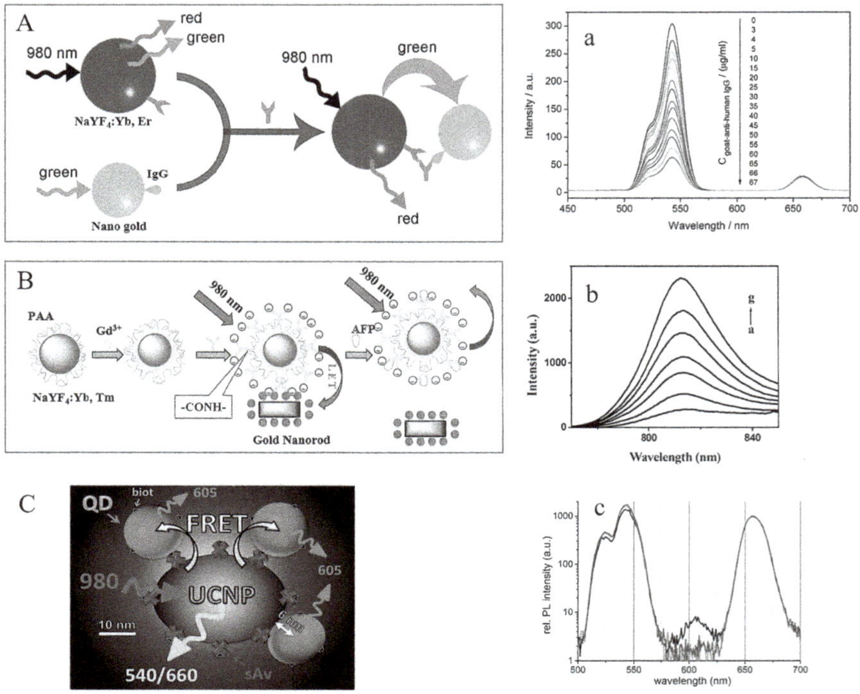

FIGURE 1.20 Luminescence resonance energy transfer-based homogeneous immunoassays using UCPNs as the donor and using GNPs (A), gold nanorods (B), and QDs(C) as the acceptors.

Source of Figure 1.20A: Reproduced from Wang, M.; Hong, W.; Mi, C, C. et al. Immunoassay of goat antihuman immunoglobulin g antibody based on luminescence resonance energy transfer between near-infrared responsive NaYF4:Yb, Er upconversion fluorescent nanoparticles and gold nanoparticles, *Analytical Chemistry*, 2009, 81, 8783–8789 (ref 78). Copyright 2009 American Chemical Society.

Source of Figure 1.20B: Reproduced from Chen, H.Q.; Guan, Y.Y.; Wang, S.Z. et al. Turn-on detection of a cancer marker based on near-infrared luminescence energy transfer from NaYF4: Yb,Tm/NaGdF4 core-shell upconverting nanoparticles to gold nanorods, *Langmuir*, 2014, 30, 13085–13091 (ref 79). Copyright 2014 American Chemical Society.

Source of Figure 1.20C: Reproduced from Mattsson, L.; Wegner, K.D.; Hildebrandt, N.; Soukka, T. *RSC Advances*, 2015, 5, 13270–13277 (ref 80), with permission of the Royal Society of Chemistry.

light. However, after the formation of the immunocomplex, the green emission is quenched due to RET. For efficient LRET, the green emission peak of the UCNPs must overlap with the absorption peak of the gold nanoparticles. The extent of quenching is linearly correlated with antigen concentration within a specific range. For example, the dynamic range for IgG was between 3–67 μg/mL (Figure 1.20A), with a LOD of 0.88 μg/mL[78]. Figure 1.20B presents a quenching-recovery immunoassay scheme, where gold rods are electrostatically bound to $NaYF_4$:Yb, Tm–$NaGdF_4$ core-shell UCNPs. In this setup, the UCNPs were initially quenched by the gold rods. Upon the addition of AFP, the gold rods dissociated from the UCNPs, restoring their fluorescence. The dynamic range for AFP was 0.18 to 11.44 ng/mL, with an LOD of 0.16 ng/mL.[79] Figure 1.20C shows a QD emission-based immunoassay, where QDs receive energy transfer from UCNPs and subsequently emit fluorescence.[80] Like FRET, LRET is distance-dependent, and its efficiency decreases when applied to large-molecule complexes, such as sandwich-structured immunocomplexes. UCNPs suffer from limitations such as lower luminescent efficiency, poor uniformity, and suboptimal dispersion. Their relatively large size (often several hundred nanometers) and nonspecific adsorption also hinder their performance. To enhance sensitivity and broaden their applicability, future research should focus on synthesizing high-luminescence, monodisperse UCNPs with improved uniformity and better dispersion.

Other Labeling

Other tags such as surface-enhanced Raman spectroscopy (SERS), electrochemiluminescence, bioluminescence, mass spectrometry, and electrochemistry are also commonly employed in immunoassays. However, due to their limited relevance to DIs, these are not covered in this book. More details can be found in the relevant literature.[81–84]

DEVELOPMENT TREND IN IMMUNOASSAYS

Precision medicine is often considered as the third medical revolution and a key development trend in modern healthcare. Its core consists of precision diagnosis and precision treatment, with precision diagnosis serving as the prerequisite and foundation for precision treatment. Precision diagnosis also provides critical technological support and an essential guarantee for the successful implementation of precision medicine. Accurate measurement of protein levels in blood is crucial for realizing precision diagnosis. As one of the most commonly used protein detection techniques, immunoassay plays a vital role in various fields, including tumor marker detection, prognosis monitoring, and drug analysis. However, the sensitivity levels of most existing immunoassays often do not meet the stringent requirements of precision medicine. Additionally, many multicomponent immunoassays are conducted using serial or parallel single-component assays. In this context, high-sensitivity, multicomponent immunoassays represent an inevitable trend in the evolution of precision medicine.

Looking back at the development of immunoassays, the primary objective has always been to achieve higher sensitivity and lower LOD. Reducing the LOD has been a longstanding goal for analytical chemists and remains a key focus across all analytical methods. Zherdev et al. summarized the LOD ranges of various immunoassays.[85] It is evident that there has been a gradual trend toward decreasing LODs over time. Among these technologies, only a few have achieved widespread application, such as ELISA, immunochromatography, and polarized fluorescence immunoassays. Radioimmunoassays, which saw rapid development in the second half of the 20th century, were widely used at the time but are now rarely employed. Agglutination immunoassays, due to their simplicity and speed, continue to be used for samples with high concentrations. While certain techniques offer very low LODs, they have not become widely adopted in clinical practice. Nonetheless, the pursuit of high-sensitivity detection remains a central objective for laboratory researchers. Developing highly sensitive immunoassays is crucial for identifying new biomarkers, enabling early disease screening, and supporting bedside diagnostics.

Approaches with high sensitivity can detect low-abundance molecules and proteins that are beyond the detection capabilities of current ones. Studying the correlations between changes in the concentrations of these low-abundance molecules and proteins and disease progression may help identify new biomarkers for severe diseases and construct precision biomarker panels, which, in turn, offers new opportunities for disease diagnosis. Furthermore, the existing methods are often limited in detecting changes that occur outside the normal range. Reliable and precise measurements are essential for constructing early screening tests for tumors, monitoring the dynamic changes in tumor biomarker levels, establishing correlations between biomarker levels and tumor progression, and assessing the likelihood of disease presence.

Changes in protein biomarker concentrations can be used to differentiate between healthy and diseased states and to track disease progression. High-sensitivity detection of protein level changes holds significant promise for identifying abnormal expression during the very early stages of severe diseases. While the traditional immunoassays are effective for diagnosing diseases with obvious symptoms, they fall short in meeting the demands of early screening for severe diseases, early diagnosis of neurological disorders, and early detection of cardiovascular diseases. These methods do not yet meet the requirements for precision medicine. The traditional immunoassays typically have lower sensitivity and higher LODs and thereby they detect blood antigen molecules only in the nM to pM range. In contrast, during the early stages of severe diseases, antigen concentrations in blood can be at the fM level or even lower, well beyond the detection capability of the traditional immunoassays, let alone precision measurement. For example, PSA levels in post-surgical pancreatic cancer patients range from 0.1 to 100 pg/mL; p24 antigen concentrations in early stage HIV patients range from 0.01 to 10 pg/mL; and the minimum concentration of Alzheimer's-related β-amyloid protein can be as low as 0.3 pg/mL. Such low concentrations are well outside the detectable range of the traditional immunoassays. Therefore, increasing the sensitivity of immunoassays to the fM level is essential for achieving precision medicine.

High-sensitivity detection also requires smaller sample volumes, making sampling more convenient and reducing matrix interference, which in turn enables more accurate measurements. For example, biomarkers for brain diseases are present at higher concentrations in cerebrospinal fluid but at lower levels in blood. High-sensitivity detection can directly measure these biomarkers in blood, eliminating the need for time-consuming and labor-intensive cerebrospinal fluid collection.

In summary, the advent of ultrahigh-sensitivity DIs provides a timely response to these challenges.

REFERENCES

1. Wild D. *The immunoassay handbook* (4th ed.), Oxford: Elsevier Ltd, 2013.
2. Gong FL. *Medical immunology* (2nd ed.), Beijing: Science Press, 2004.
3. Jiao K, Zhang SS. *Enzyme-linked immunosorbent assay technology and applications*, Beijing: Chemical Industry Press, 2004.
4. Berson SA, Yalow RA. Quantitative aspects of the reaction between insulin and insulin-binding antibody, *Journal of Clinical Investigation*, 1959, 38, 1996–2016.
5. Miles LEM, Hales CN. Labelled antibodies and immunological assay system, *Nature*, 1968, 219, 186–189.

6. Engvall E, Perlmann P. Enzyme-linked immunosorbent assay (ELISA) quantitative assay of immunoglobulin-G, *Immunochemistry*, 1971, 8, 871–874.
7. Vanweeme BK, Schuurs AHW. Immunoassay using antigen enzyme conjugates, *FEBS Letters*, 1971, 15, 232–236.
8. Belanger L, Sylvestre C, Dufour D. Enzyme linked immunoassay for alpha fetoprotein by competitive and sandwich, *Clinica Chimica Acta*, 1973, 48, 15–18.
9. Kohler G, Milstein C. Continuous cultures of fused cells secreting antibody of predefined specificity, *Nature*, 1975, 256, 495–497.
10. Schroeder HR, Vogelhut PO, Carrico RJ, et al. Competitive protein binding assay for biotin monitored by chemiluminescence, *Analytical Chemistry*, 1976, 48, 1933–1937.
11. Sano T, Smith Cl, Cantor Cr. Immuno-PCR: very sensitive antigen-detection by means of specific antibody-DNA conjugates, *Science*, 1992, 258, 120–122.
12. Rissin DM, Kan CW, Campbell TG. Single-molecule enzyme-linked immunosorbent assay detects serum proteins at subfemtomolar concentrations, *Nature Biotechnology*, 2010, 28, 595–599.
13. Lequin RM. Enzyme immunoassay (EIA)/enzyme-linked immunosorbent assay (ELISA), *Clinical Chemistry*, 2005, 51, 2415–2418.
14. Wu AHB. A selected history and future of immunoassay development and applications in clinical chemistry, *Clinica Chimica Acta*, 2006, 369, 119–124.
15. Ullman EF. Homogeneous immunoassays: historical perspective and future promise, *Journal of Chemical Education*, 1999, 76, 781–788.
16. Ueda H, Tsumoto K, Kubota K, et al. Open sandwich ELISA: a novel immunoassay based on the interchain interaction of antibody variable region, *Nature Biotechnology*, 1996, 14, 1714–1718.
17. Ren J, Wang LH, Han XY. Organic silicone sol–gel polymer as a noncovalent carrier of receptor proteins for label-free optical biosensor application, *ACS Applied Materials & Interfaces*, 2013, 5, 386–394.
18. Sam S, Touahir L, Andresa JS. Semiquantitative study of the EDC/NHS activation of acid terminal groups at modified porous silicon surfaces, *Langmuir* 2010, 26, 809–814.
19. Ouerghi O, Touhami A, Jaffrezic-Renault N. Impedimetric immunosensor using avidin–biotin for antibody immobilization, *Bioelectrochemistry*, 2002, 56, 131– 133.
20. Dong Y, Phillips KS, Cheng Q. Immunosensing of staphylococcus enterotoxin B (SEB) in milk with PDMS microfluidic systems using reinforced supported bilayer membranes (r-SBMs), *Lab on a Chip*, 2006, 6, 675–681.
21. Trilling AK, Beekwilder J, Zuilhof H. Antibody orientation on biosensor surfaces: a minireview, *Analyst*, 2013, 138, 1619–1627.
22. Liu M, Zhao H, Quan X, et al. Distance-independent quenching of quantum dots by nanoscale-graphene in self-assembled sandwich immunoassay, *Chemical Communications*, 2010, 46, 7909–7911.

23. Ullman EF, Kirakossian H, Switchenko AC, et al. Luminescent oxygen channeling assay (LOCI(TM)): sensitive, broadly applicable homogeneous immunoassay method, *Clinical Chemistry*, 1996, 42, 1518–1526.
24. Goldman ER, Medintz IL, Whitley JL. A hybrid quantum dot-antibody fragment fluorescence resonance energy transfer-based TNT sensor, *Journal of American Chemical Society*, 2005, 127, 6744–7675.
25. Kokko T, Kokko L, Lolvgren T, et al. Homogeneous noncompetitive immunoassay for 17-estradiol based on fluorescence resonance energy transfer, *Analytical Chemistry*, 2007, 79, 5935–5940.
26. Abe R, Ohashi H, Iijima I, et al. "Quenchbodies": quench-based antibody probes that show antigen-dependent fluorescence, *Journal of the American Chemical Society*, 2011, 133, 17386–17394.
27. Fukuyama M, Nakamura A, Nishiyama K, et al. Noncompetitive fluorescence polarization immunoassay for protein determination, *Analytical Chemistry*, 2020, 92, 14393–14397.
28. Wang JJ, Huang XY, Liu H, et al. Fluorescence and scattering light cross correlation spectroscopy and its applications in homogeneous immunoassay, *Analytical Chemistry*, 2017, 89, 5230–5237.
29. Zhang HY, Yang S, Beier RC, et al. Simple, high efficiency detection of microcystins and nodularin-R in water by fluorescence polarization immunoassay, *Analytica Chimica Acta*, 2017, 992, 119–127.
30. Wang J, Liu H, Huang X, et al. Homogeneous immunoassay for the cancer marker alpha-fetoprotein using single wavelength excitation fluorescence cross-correlation spectroscopy and CdSe/ZnS quantum dots and fluorescent dyes as labels, *Microchimica Acta*, 2016, 183, 749–755.
31. Lan T, Dong C, Huang X, et al. A sensitive, universal and homogeneous method for determination of biomarkers in biofluids by resonance light scattering correlation spectroscopy (RLSCS), *Talanta*, 2013, 116, 501–507.
32. Chen MJ, Wu YS, Lin GF, et al. Quantum-dot-based homogeneous time-resolved fluoroimmunoassay of alpha-fetoprotein, *Analytica Chimica Acta*, 2012, 741, 100–105.
33. (a) Dong J, Oka Y, Jeong HJ, et al. Detection and destruction of HER2-positive cancer cells by ultra quenchbody-siRNA complex, *Biotechnology and Bioengineering*, 2020, 117, 1259–1269. (b) Liu XJ, Huang CH, Dong XL, et al. Asynchrony of spectral blue-shifts of quantum dot based digital homogeneous immunoassay, *Chemical Communications*, 2018, 54, 13103–13106
34. Bristow RE, Smith A, Zhang Z, et al. Ovarian malignancy risk stratification of the adnexal mass using a multivariate index assay, *Gynecologic Oncology*, 2013, 128, 252–259.
35. Dunbar SA, Jacobson JW. Quantitative, multiplexed detection of Salmonella and other pathogens by Luminex® xMAP™ suspension array, *Salmonella*, 2007, 1–19.
36. Cohen JD, Li L, Wang Y, et al. Detection and localization of surgically resectable cancers with a multi-analyte blood test, *Science*, 2018, 359, 926–930.

37. Wang J, Li SJ. A brief report on the results of the national survey of nuclear medicine in 2018, *Chinese Journal of Nuclear Medicine and Molecular Imaging*, 2018, 38, 813–814.
38. Kasetsirikul S, Umer M, Soda N, et al. Detection of the SARS-CoV-2 humanized antibody with paper-based ELISA, *Analyst*, 2020, 145, 7680–7686.
39. Zhang Z, Zhu N, Zou Y, et al. A novel and sensitive chemiluminescence immunoassay based on AuNCs@ pepsin@ luminol for simultaneous detection of tetrabromobisphenol a bis (2-hydroxyethyl) ether and tetrabromobisphenol a mono (hydroxyethyl) ether, *Analytica Chimica Acta*, 2018, 1035, 168–174.
40. Zong C, Wu J, Wang C, et al. Chemiluminescence imaging immunoassay of multiple tumor markers for cancer screening, *Analytical chemistry*, 2012, 84, 2410–2415.
41. Zaitsu K, Ohkura Y. New fluorogenic substrates for horseradish peroxidase: rapid and sensitive assay for hydrogen peroxide and the peroxidase, *Analytical Biochemistry*, 1980, 109–113.
42. Sun J, Zhao JH, Bao XF, et al. Alkaline phosphatase assay based on the chromogenic interaction of diethanolamine with 4-aminophenol, *Analytical Chemistry*, 2018, 90, 6339–6345.
43. Zhao JH, Wang S, Lu SS, et al. Fluorometric and colorimetric dual-readout immunoassay based on an alkaline phosphatase-triggered reaction, *Analytical Chemistry*, 2019, 91, 7828–7834.
44. Hou FH, Zhang Q, Yang JP, et al. Development of a microplate reader compatible microfluidic chip for ELISA, *Biomedical Microdevices*, 2012, 14, 729–737.
45. Chen CX, Zhao JH, Lu YZ, et al. Fluorescence immunoassay based on the phosphate-triggered fluorescence turn-on detection of alkaline phosphatase, *Analytical Chemistry*, 2018, 90, 3505–3511.
46. Zhao D, Li J, Peng CY, et al. Fluorescent immunoassay based on the alkaline phosphatase-triggered in situ fluorogenic reaction of o-phenylenediamine and ascorbic acid, *Analytical Chemistry*, 2019, 91, 2978–2984.
47. Chen CX, Zhao D, Wang B, et al. Alkaline phosphatase-triggered in situ formation of silicon-containing nanoparticles for a fluorometric and colorimetric dual-channel immunoassay, *Analytical Chemistry*, 2020, 92, 4639–4646.
48. Bronstein I, Edwards B, Voyta JC. 1,2-dioxetanes: novel chemiluminescent enzyme substrates. Applications to immunoassays, *Journal of Bioluminescence and Chemiluminescence*, 1989, 4, 99–111.
49. BronsteIn I, Voyta JC, Thorpe GH, et al. Chemiluminescent assay of alkaline phosphatase applied in an ultrasensitive enzyme immunoassay of thyrotropin, *Clinical Chemistry*, 1989, 35, 1441–1446.
50. Snyder AP, Blyth DA, Parsons JA. Ion mobility spectrometry as an immunoassay detection technique, *Parsons Journal of Microbiological Methods*, 1996, 27, 81–88.
51. Rotman B. Measurement of activity of single molecules of β-D-galactosidase, *Proceedings of the National Academy of Sciences of the United States of America*, 1961, 47, 1981–1991.

52. Hu Q, Ma KF, Mei YQ. Metal-to-ligand charge-transfer: applications to visual detection of β-galactosidase activity and sandwich immunoassay, *Talanta*, 2017, 167, 253–259.
53. Imagawa M, Hashida S, Ohta Y, et al. Evaluation of β-D-galactosidase from *Escherichia coli* and horseradish peroxidase as labels by sandwich enzyme immunoassay technique, *Annals of Clinical Biochemistry: An International Journal of Biochemistry and Laboratory Medicine*, 1984, 21, 310–317.
54. Porstmann B, Porstmann T, Nugel E, et al. Which of the commonly used marker enzymes gives the best results in colorimetric and fluorimetric enzyme immunoassays: horseradish peroxidase, alkaline phosphatase or β-galactosidase? *Journal of Immunological Methods*, 1985, 79, 27–37.
55. Hosoda H, Takasaki W, Tsukamoto R, et al. Sensitivity of steroid enzyme immunoassays. Comparison of alkaline phosphatase, β-galactosidase and horseradish peroxidase as labels in a colorimetric assay system. *Chemical & Pharmaceutical Bulletin*, 1987, 35, 3336–3342.
56. Hosoda H, Tsukamoto R, Nambara T. Sensitivity of steroid enzyme immunoassays—comparison of four label enzymes in an assay system using a monoclonal anti-steroid antibody, *Chemical & Pharmaceutical Bulletin*, 1989, 37, 1834–1837.
57. Grandke J, Oberleitner L, Resch-Genger U, et al. Quality assurance in immunoassay performance—comparison of different enzyme immunoassays for the determination of caffeine in consumer products, *Analytical and Bioanalytical Chemistry*, 2013, 405, 1601–1611.
58. Kumari GL, Dhir, RN. Comparative studies with penicillinase, horseradish peroxidase, and alkaline phosphatase as enzyme labels in developing enzyme immunoassay of cortisol, *Journal of Immunoassay & Immunochemistry*, 2003, 24, 173–190.
59. Yu SC, Yu F, Liu L, et al. Which one of the two common reporter systems is more suitable for chemiluminescent enzyme immunoassay: alkaline phosphatase or horseradish peroxidase? *Luminescence*, 2016, 31, 888–892.
60. Gao L, Zhuang J, Nie L, et al. Intrinsic peroxidase-like activity of ferromagnetic nanoparticles, *Nature Nanotechnology*, 2007, 2, 577–583.
61. Wang YD, Xianyu YL. Nanobody and nanozyme-enabled immunoassays with enhanced specificity and sensitivity, *Small Methods*, 2022, 6, 2101576.
62. Huang T, Hu XL, Wang M, et al. Ionic liquid-assisted chemiluminescent immunoassay of prostate specific antigen using nanoceria as an alkaline phosphatase-like nanozyme label, *Chemical Communications*, 2021, 57, 3054–3057.
63. Geissler D, Stufler S, Lohmannsroben HG, et al. Six-color time-resolved forster resonance energy transfer for ultrasensitive multiplexed biosensing, *Journal of the American Chemical Society*, 2013, 135, 1102–1109.
64. Cao C, Sim SJ. Resonant Rayleigh light scattering response of individual Au nanoparticles to antigen-antibody interaction, *Lab on A Chip*, 2009, 9, 1836–1839.

65. Truong PL, Cao C, Park S, et al. A new method for non-labeling attomolar detection of diseases based on an individual gold nanorod immunosensor. *Lab on a Chip*, 2011, 11, 2591.
66. Liu DB, Wang ZT, Jin A, et al. Acetylcholinesterase-catalyzed hydrolysis allows ultrasensitive detection of pathogens with the naked eye, *Angewandte Chemie-International Edition*, 2014, 52, 14065–14069.
67. Qu WS, Liu YY, Liu DB, et al. Copper-mediated amplification allows readout of immunoassays by the naked eye, *Angewandte Chemie-International Edition*, 2011, 50, 3442–3445.
68. Yang YC, Tseng WL. 1,4-benzenediboronic-acid-induced aggregation of gold nanoparticles: application to hydrogen peroxide detection and biotin avidin-mediated immunoassay with naked-eye detection, *Analytical Chemistry*, 2016, 88, 5355–5362.
69. Ma XM, Lin Y, Guo LH, et al. A universal multicolor immunosensor for semiquantitative visual detection of biomarkers with the naked eyes, *Biosensors & Bioelectronics*, 2017, 87, 122–128.
70. Ma XM, Chen ZT, Kannan P, et al. Gold nanorods as colorful chromogenic substrates for semiquantitative detection of nucleic acids, proteins, and small molecules with the naked eye, *Analytical Chemistry*, 2016, 88, 3227–3234.
71. Liu X, Dai Q, Austin L, et al. A one-step homogeneous immunoassay for cancer biomarker detection using gold nanoparticle probes coupled with dynamic light scattering, *Journal of the American Chemical Society*, 2008, 130, 2780–2782.
72. Li C, Ma JH, Fan QX, Tao YQ, Li GX. Dynamic light scattering (DLS)-based immunoassay for ultra-sensitive detection of tumor marker protein, *Chemical Communications*, 2016, 52, 7850–7853.
73. Ling J, Li YF, Huang CZ. Visual sandwich immunoassay system on the basis of plasmon resonance scattering signals of silver nanoparticles, *Analytical Chemistry*, 2009, 81, 1707–1714.
74. Ma J, Zhan L, Li RS, et al. Color-encoded assays for the simultaneous quantification of dual cancer biomarkers, *Analytical Chemistry*, 2017, 89, 8484–8489.
75. Xiong Y, Fu T, Zhang DX, Zhang SP, Xu HX. Superradiative plasmonic nanoantenna biosensors enable sensitive immunoassay using the naked eye, *Nanoscale*, 2021, 13, 2429–2435.
76. Bhuckory S, Mattera L, Wegner KD, et al. Direct conjugation of antibodies to the ZnS shell of quantum dots for FRET immunoassays with low picomolar detection limits, *Chemical Communications*, 2016, 52, 14423–14425.
77. Hampl J, Hall M, Mufti NA, et al. Upconverting phosphor reporters in immunochromatographic assays, *Analytical Biochemistry*, 2001, 288, 176–187.
78. Wang M, Hong W, Mi CC, et al. Immunoassay of goat antihuman immunoglobulin G antibody based on luminescence resonance energy transfer between near-infrared responsive NaYF4:Yb, Erupconversion fluorescent nanoparticles and gold nanoparticles, *Analytical Chemistry*, 2009, 81, 8783–8789.

79. Chen HQ, Guan YY, Wang SZ, et al. Turn-on detection of a cancer marker based on near-infrared luminescence energy transfer from NaYF4:Yb,Tm/NaGdF4 Core–shell upconverting nanoparticles to gold nanorods, *Langmuir*, 2014, 30, 13085–13091.
80. Mattsson L, Wegner KD, Hildebrandt N, Soukka T. Upconverting nanoparticle to quantum dot FRET for homogeneous double-nano biosensors, *RSC Advances*, 2015, 5, 13270–13277.
81. Wang ZY, Zong SF, Wu L, et al. SERS-activated platforms for immunoassay: probes, encoding methods, and applications, *Chemical Reviews*, 2017, 119, 7910–7963.
82. Zhao LX, Sun L, Chu XG. Chemiluminescence immunoassay, TrAC-Trends in analytical, *Chemistry*, 2009, 28, 404–415.
83. Prathap MUA, Rodriguez CI, Sadak O, et al. Ultrasensitive electrochemical immunoassay for melanoma cells using mesoporous polyaniline, *Chemical Communications*, 2018, 54, 710–714.
84. Liu R, Wu P, Yang L, et al. Inductively coupled plasma mass spectrometry-based immunoassay: a review, *Mass Spectrometry Reviews*, 2014, 33, 373–393.
85. Zherdev AV, Dzantiev BB. Detection limits of immunoanalytical systems: limiting factors and methods of reduction, *Journal of Analytical Chemistry*, 2022, 77, 391–401.

CHAPTER 2

Principles of Digital Immunoassays*

OVERVIEW OF DIGITAL IMMUNOASSAYS

Conception

Digital immunoassay (DI) is an immunoassay technique that quantifies the number of immunocomplex molecules either directly or indirectly. When the detection and statistical objects are individual molecules, the DI is also referred to as single-molecule counting immunoassay. In contrast, conventional immunoassays quantify immunocomplexes by measuring their signal strength, where the signal intensity varies continuously with the changes in complex content. This is a continuous signal, also known as analog signal. DIs, on the other hand, record the number of molecules, which is a discrete, noncontinuous signal. In information technology, the discretization of signals is the foundation and prerequisite for digital signals. DIs are named after digital signals because they record discrete signals. The term "digital" reflects the discrete nature of the signals captured by DIs. Although DIs count molecules, the signal intensity of the immunocomplexes affects the accuracy of molecular recognition, thereby influencing the accuracy of the counting. The main advantage of DI is its

* The content in the sections titled "Quantification Method of Digital Immunoassay" and "Basic Theory of Digital Immunoassays" are mainly adapted from Liu, X.J.; Zhang, Q.Q.; Zong, C.H.; Gai, H.W. Digital immunoassay for proteins: theory, methodology, and clinical applications, *Analytical Chemistry*, 2025, 97, 17, 9077–9110 (ref 1a). Copyright 2025 American Chemical Society.

DOI: 10.1201/9781003729211-2

extremely high sensitivity. It is at least 1,000 times more sensitive than traditional immunoassays, reaching single-molecule detection levels. This makes it suitable for applications such as early cancer detection and the detection of rare antigens in blood, addressing the need for highly sensitive testing. It is considered a frontier technology in immunoassay in recent years and is regarded as the next-generation immune technology.

Figure 2.1 illustrates the similarities and differences between traditional immunoassays and DIs. Taking a sandwich-structured immunoassay as an example, both traditional and DIs require the formation of a sandwich immunocomplex, where the target antigen binds to both capture antibodies and detection antibodies. followed by labeling with a signal tag. Subsequently, in traditional immunoassays, the bulk of the sample is directly detected, and the signal intensity is positively correlated with the sample concentration within a certain range. The concentration of the unknown sample is then quantified by a pre-established standard curve. In DIs, the immunocomplexes are dispersed into numerous (up to several million) small spaces, with each microspace either containing the target molecule (positive space, denoted as 1) or not (negative space, denoted as 0). If the number of negative spaces is counted, Poisson distribution is used to perform absolute quantification of the antigen molecules. If counting immunocomplexes within positive microspaces, the number of antigen molecules in each positive space must be identified, as not every positive space contains only one molecule. Usually, a standard curve must be constructed in advance due to variability in the number of immunocomplexes per positive space. The quantification method for positive spaces is also a single-molecule counting method.

Application

The greatest advantage of DIs is their ultrahigh detection sensitivity, with detection limits (LODs) as low as 10^{-21} M (aM), equivalent to detecting dozens of molecules in a 100 mL sample. The significant improvement in sensitivity can change the way that severe diseases are managed, shifting from a passive response only after disease onset to proactive prevention before disease occurrence. Figure 2.2 illustrates the relationship between biomarker concentration, disease stages, sample sources, and the role of different immunoassay techniques under various conditions.

The purpose of measuring biomarkers in healthy individuals is to establish a healthy baseline. Establishing the concentration baseline of biomarkers in healthy individuals is a prerequisite for early disease diagnosis.

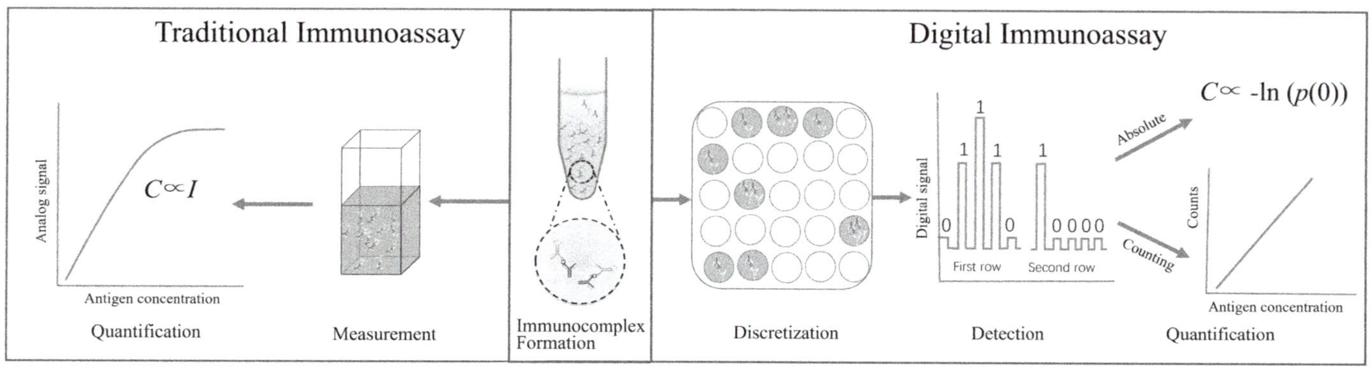

FIGURE 2.1 Comparison of procedures between DI and traditional immunoassay.

Source: Reproduced from Liu, X.J.; Zhang, Q.Q.; Zong, C.H.; Gai, H.W. Digital immunoassay for proteins: theory, methodology, and clinical applications, *Analytical Chemistry*, 2025, 97, 9077–9110 (ref 1a). Copyright 2025 American Chemical Society.

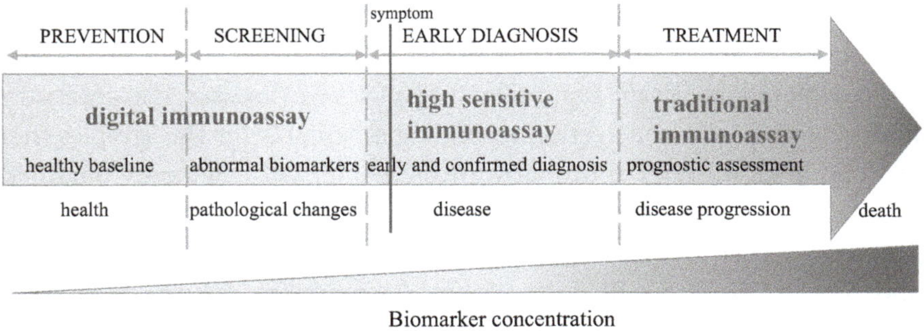

FIGURE 2.2 Summary of biomarker levels across disease stages and sample types, along with immunoassays capable of detecting these levels.

Only by establishing baselines and thresholds can changes in concentration be monitored for early detection. This is also the approach used to determine indicator abnormalities in clinical diagnostics. The average or median value of a particular marker in a healthy population is taken as the basis for determining the occurrence of related diseases, with values above the threshold considered positive. Ideally, the concentration range of all biomarkers in healthy individuals should be established. Some researchers have attempted to monitor the level fluctuations of 15 cytokines in the blood of healthy individuals over a 14-week period and found that Interleukin-15 (IL-15) and Tumor Necrosis Factor-α (TNF-α) showed little change in the measured population, while certain cytokines such as IL-5 and interferon-γ showed considerable individual differences.[1b] Some biomarkers in healthy individuals have very low concentrations, making them difficult to detect using traditional methods. Only DI can detect them. When determining the occurrence of diseases based on different biomarkers, it is not only necessary to rely on the average value of the population but also consider the changes in the individual.

The onset of symptoms is a key transition point in disease management. Most people consider the onset of symptoms as the point when they are "sick". In reality, many severe diseases have already entered the late stages by the time symptoms appear, making treatment much more difficult. Early screening for abnormal indicators before the onset of symptoms and diagnosis as early as possible can significantly reduce the mortality rate of severe diseases. For example, early detection and diagnosis of cancer before tumor invasion and metastasis greatly increase the possibility of successful surgery. However, it is not easy to effectively detect abnormal biomarkers before the appearance of symptoms. This is partly due to insufficient detection sensitivity and partly because the markers lack strong specificity.

To achieve large-scale, rapid screening of severe diseases, it is necessary to develop biomarker detection technologies suitable for the samples that are obtained through minimally invasive methods. Samples obtained through minimally invasive methods, such as blood, urine, sweat, and saliva, are easier to obtain than cerebrospinal fluid (CSF) or tumor tissue, and patients are more likely to accept the sampling, but the biomarker concentrations in these samples are much lower.

The ultrahigh sensitivity of DIs makes it suitable for constructing healthy baselines, detecting low-invasive samples, small volume samples, and complex samples, as well as early disease screening, early detection, and monitoring of therapeutic effects.

Low-Invasive Sample Detection: Blood, saliva, urine, and tear samples are easily secured through noninvasive or minimally invasive ways, but their biomarker content is low. CSF needs to be obtained through lumbar puncture, making it an invasive sample. The protein biomarkers for neurological diseases in CSF are high, but the blood-brain barrier prevents most protein biomarkers from entering the peripheral blood, resulting in very low levels of these biomarkers in the blood. Before the advent of DI, they could not be detected. For example, neurofilament light chain protein (NFL) is a biomarker for axonal injury and is associated with various neurological diseases.[2] Accurate measurement of its concentration is crucial for diagnosis, prognosis, and monitoring. The concentration of NFL in peripheral blood is only one-fortieth of that in CSF, and both enzyme-linked immunosorbent assay (ELISA) and electrochemiluminescence techniques lack the sensitivity to measure NFL accurately in blood. The sensitivity of DI is 126 times higher than that of ELISA and 25 times higher than that of electrochemical techniques, respectively. Such high sensitivity enables reliable quantification of NFL in serum under both physiological and pathological conditions,[3] thus laying the foundation for assessing the progression of neurological diseases through serum NFL levels.

Small Volume Sample Detection: In conventional measurement, at least 100 μL of a sample are required to ensure repeatability and accuracy. However, under certain conditions, only less than 25 μL of sample can be obtained, such as pediatric samples, small animal samples, tear fluid, and sweat. When monitoring the treatment effects of retinal vascular diseases, only 12 μL of aqueous humor can be collected at a time. Measuring the level of vascular endothelial growth factor (VEGF) in such a small volume is challenging with conventional immunoassays. The high sensitivity of DI allows for a 12-fold dilution of the sample while still ensuring sufficient measurement effectiveness.[4]

Measurement of Complex Samples: Matrix interference in complex samples can severely affect measurement accuracy. Serum and plasma samples typically need to be diluted at least four times to ensure recovery rate and measurement accuracy. If the dilution factor is too high, the concentration becomes too low, which affects the accuracy of the measurements. The high sensitivity of DI allows for significant dilution of complex samples, reducing matrix effects while still ensuring accurate measurement. For example, in testing for clostridium difficile toxins in stool, traditional methods like nucleic acid amplification have a high false-positive rate and lack clinical diagnostic specificity. DI can directly measure the toxins in a 20-fold diluted filtrate,[5] improving diagnostic accuracy.

Early Disease Detection: As shown in Figure 2.2, before symptoms of severe diseases appear, biomarker concentrations are extremely low. Currently, only ultrasensitive DI technology can detect abnormal changes in biomarker levels. Early disease detection requires not only high sensitivity but also high specificity and sensitivity of the biomarkers. At present, biomarkers for neurological diseases and viral diseases have higher sensitivity and specificity, making them more suitable for early detection. For example, in familial Alzheimer's disease, NFL levels in serum significantly increase 6.8 years before symptoms appear.[6] Similarly, in the first two weeks of HIV infection, the p24 antigen concentration ranges from 50×10^{-18} M to 15×10^{-15} M. If detected early, HIV can be diagnosed in the early stage.[7]

Monitoring Therapeutic Effects: After treatment, biomarker concentrations may decrease below the LOD of conventional techniques, and only high-sensitivity methods can monitor changes in biomarker concentrations and track therapeutic effects. After prostatectomy, there may still be trace amounts of prostate specific antigen (PSA) in the body, which could indicate prostate tumor metastasis. Conventional techniques cannot detect these low levels, but DI can. Furthermore, TNF-α antibody drugs reduce TNF-α concentrations in the body to the pg/mL level, below or close to the LODs of most immunoassays. Even if the target protein can be detected, the coefficient of variation (CV) is very high. It is challenging to accurately assess the treatment effect using such data. DI has an LOD in the fg/mL range, enabling the accurate assessment of target protein concentration changes during the treatment cycle.

Development

The initial pattern of DI was single-molecule counting immunoassay. Single-molecule counting immunoassay directly detects and counts individual molecules, while DI does not necessarily detect individual molecules. The earliest assay can be traced back to single-molecule capillary electrophoresis, which distinguished immunocomplexes from free antibodies based on electrophoretic mobility and performed single-molecule imaging detection during electrophoresis.[8] In 2006, Singulex reported the single-molecule counting immunoassay device, Erenna, marking the commercialization of DI.

The concept of digital enzyme immunoassay was introduced by Quanterix in 2010, where single enzyme-labeled immunocomplexes were partitioned into an array of microwells. The introduction of the hard partition into counting immunoassay enriched the field of DI. Microfluidic technology enabled the generation of millions of droplets per second, each less than 10 fL in volume. In 2013, the droplets could encapsulate single enzyme-labeled molecules and be used to count PSA, with LOD as low as 46 fM.[9] This was the first report of droplets being used in DI.

In 2016, Akama et al. proposed the time-based partition for enzyme-catalyzed product, where fluorescent dyes were deposited onto immunocomplexes through enzyme reactions, and signals were detected in real-time using a traditional flow cytometer.[10] This method does not require any hard partition microspaces.

In 2018, we proposed a soft partition-based homogeneous DI method using quantum dots (QDs) as tags, leveraging the spectral blue-shift asynchronous behavior of QDs to differentiate immunocomplexes from free QDs.[11, 12] In the same year, a homogeneous immunoassay based on particle diffusion for immunocomplex recognition was reported. The Brownian motion trajectories of magnetic particles in specific binding, nonspecific binding, and free states can be used to identify immunocomplexes.[13] In 2020, we introduced a total counting approach to overcome random sampling errors and achieved 100% partitioning through solvent evaporation-induced microsphere deposition.[14]

In 2021, solid-phase nanopores were applied in DI. Target molecules generate electrical signals as they pass through nanopore, with each electrical signal corresponding to a target molecule. This time partition-based DI transforms electrical nanopore signals into digital counts.[15] In the same year, we proposed using microsphere lenses to enhance the imaging capabilities of low numerical aperture objectives, enabling DIs. This method allows the detection of individual QDs, which serve as target tags, even with a 20× objective lens. This lays the foundation for improving statistical throughput and miniaturizing instruments.[16]

In 2022, Liu et al. used bubbles created by laser-heated plasma nanoparticles to separate virus particles.[17] This hard partition is different from microwell arrays and micro-droplets, as the bubble microspace only exists temporarily and its size is dependent on laser intensity, pulse time, and flow rate. That same year, we reported the three-component homogeneous DI[18] based on the two-component DI[19] developed in 2021, advancing the field of multicomponent homogeneous immunoassays.

Dr. Hongwei Gai has been involved in single-molecule detection since 2000 and is one of the earliest researchers in China to enter this field. At the end of 2006, Professor Gai became an independent principal investigator at the Bioengineering Center of Hunan University, where he focused on the optical properties of individual QDs and their applications. He discovered the asynchronous blue shift in QD spectra and developed methods to regulate this shift, making significant progress in single nanoparticle scattering imaging and ultrahigh-resolution imaging.[20] In late 2011, he moved to Jiangsu Normal University, where he established a team focusing high sensitive detection. Professor Gai developed a series of quantitative methods using aggregated QD oligomer, which laid the foundation for the development of a unique and novel homogeneous DI. He further advanced dual-component and three-component DIs.[21] This book is based on the author's 20 years of research in single-molecule detection and nearly 10 years of experience in single-molecule counting and quantitative analysis.

Commercialization

DIs are considered the next-generation immunoassay technology and have become a hot investment topic in the *in vitro* diagnostic (IVD) in recent years. Internationally, four companies have been involved in the development of single-molecule immunoassay products, in chronological order: Singulex and Quanterix from the United States, Mecwins from Spain, and Precision System Science (PSC) from Japan. Singulex ceased operations in the summer of 2019. Mecwins and PSC have only recently entered the field of ultrasensitive detection. Some companies in China also claim to be developing DI technologies, but no public reports have been released yet. The major events in the commercialization process are as follows:

SINGULEX

Founded in 2004.

In 2006, Singulex launched the single-molecule immunoassay instrument, Erenna, and secured $9.35 million in venture capital. The company moved its headquarters to California.

In 2007, the company raised $19 million in venture capital from OrbiMed.

In 2009, the company raised another $19 million in venture capital from Japan's JAFCO.

In 2012, Singulex filed for an IPO.

In 2015, Merck Group acquired Singulex's life sciences research business, including the Erenna platform technology.

In 2016, the company sold 20% of its shares to Grifols, raising $50 million. It was reported for adding unnecessary testing costs to medical and military health insurance claims.

In 2018, Singulex paid a $1.25 million fine for violating the U.S. False Claims Act.

In July 2019, the company closed and laid off 71 employees. Over its history, Singulex raised a total of approximately $230 million over nine rounds of funding.

QUANTERIX

Founded in 2007 by Professor David Walt of Tufts University.

In 2008, the company raised $15 million in venture capital, including ARCH Venture Partners.

In 2010, Quanterix published a paper in Nature Biotechnology,[22] introducing digital enzyme immunoassay technology. The company developed the Single Molecule Array (SiMoA) device.

In 2012, the company raised $2 million through bond financing and $18.5 million in venture capital from bioMerieux.

In 2014, Quanterix launched the first generation fully automated single-molecule immunoassay instrument, the HD-1.

In 2016, the company raised $46 million in investments from Trinitas Capital and others.

In 2017, Quanterix went public on NASDAQ, raising $64 million.

In 2018, the company launched the semi-automated SR-X system.

In 2019, Quanterix launched the HD-X, an upgraded version of the HD-1.

From 2019 to 2021, the company raised $434 million in post-IPO equity investments, with a total of $533 million raised over 11 rounds of funding.

MECWINS

Founded in 2008 by Professors Javier Tamayo and Montserrat Calleja at the Madrid Institute of Micro and Nanoelectronics.

In 2018, Grifols and CRB jointly invested €4 million for the development of ultrasensitive diagnostic technologies.

In June 2021, Mecwins partnered with Quidel to jointly develop point-of-care (POC) devices with detection limits lower than the pg/mL level.

In 2021, the company developed a plasma-based digital biomolecule ultrasensitive detection platform (AVAC).

Looking at these existing commercial instruments, the assays are all heterogeneous immunoassay techniques. Either immunocomplexes are generated on magnetic beads and then separated magnetically, or they are immobilized on glass substrates and then washed off. This makes automation of sample measurement more challenging, with fully automated instruments being large and nearly impossible to miniaturize for bedside use. Moreover, the hard partition mode of microarrays causes many target molecules to be undetected, limiting further improvements in detection sensitivity.

Operation Unit

DIs are divided into four main unit operations based on the procedure: immunocomplex formation, compartmentalization (partition), recognition and amplification, and counting (Figure 2.3A).

Immunocomplex Formation: The target antigen is captured through an immunoreaction, and the efficiency of capture determines the number of

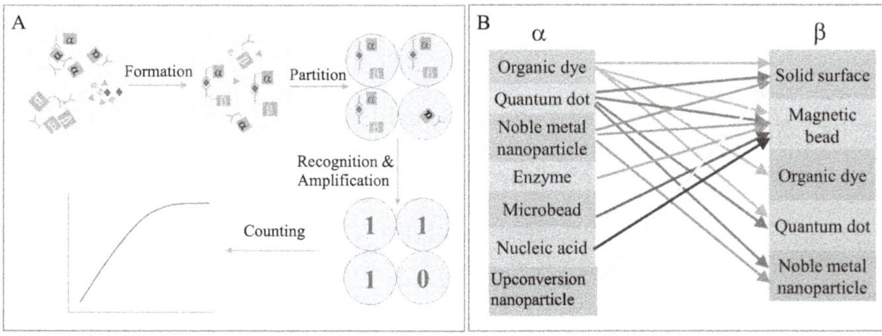

FIGURE 2.3 General process of DIs (A) and reported combination of α tags and β tags (B).

Source of Figure 2.3B: Reproduced from Liu, X.J.; Zhang, Q.Q.; Zong, C.H.; Gai, H.W. Digital immunoassay for proteins: theory, methodology, and clinical applications, *Analytical Chemistry*, 2025, 97, 9077–9110 (ref 1a). Copyright 2025 American Chemical Society).

immunocomplexes to be formed, which in turn affects the LODs. Immunocomplexes formed in DIs are typically of a sandwich structure, i.e., capture antibody - antigen - detection antibody. Both capture and detection antibodies can be coupled with the required functional tags, as shown in Figure 2.3B, where α is used for signal generation and β is used for separation. For example, when signal tags are needed, a detection tag α (such as a dye, quantum dot, or enzyme) can be coupled with the detection antibody; when immunocomplexes need to be separated, separation functional tags β (such as magnetic beads, interfaces) can be coupled to the capture antibody. The line between α and β in Figure 2.3B represents reported coupling combinations. It is observed that DIs using magnetic beads as separation carriers are the most common, with nearly all detection tags having been combined with magnetic beads to form immunocomplexes. The second most common β tag is surface.

Immunocomplex Partition: Partition refers to the process of dispersing immunocomplex molecules into distinguishable, countable, and statistically detectable dispersed states in space or time. This is also known as compartment or discretization and is a key technology in DIs. Partition can be classified into two categories: hard discretization and soft discretization. Hard discretization involves isolating target molecules into countable microspaces, including microwell array discretization, droplet discretization, and bubble discretization. In hard discretization, microspaces are isolated by two phase interfaces. Soft discretization involves creating nonoverlapping dispersed states without physical microspaces, with molecules being in a continuous phase, and is divided into temporal and spatial discretization. Discretization efficiency refers to the ratio of the number of immunocomplex molecules available for detection to the total number of immunocomplexes. For hard discretization, it is the ratio of immunocomplexes entering microspaces, while for soft discretization, it is the ratio of immunocomplexes deposited on a glass slide or passing through a detector. Discretization efficiency significantly affects quantitative results. Hard discretization not only disperses but also concentrates and separates the immunocomplexes. For samples with fM concentrations, if individual molecules are discretized into 10 μm droplets (0.5 pL in volume, with a concentration of 3 pM), the concentration in the microspace is raised to pM levels, making detection easier. Matrix molecules in the sample are also dispersed into microspaces, reducing the total number of matrix molecules in each microspace, increasing the ratio of sample molecules to matrix molecules, thus reducing background noise and improving detection efficiency. Microwell hard discretization cannot fully encapsulate 100% of the immunocomplex molecules. Soft discretization methods are relatively simple but lack control, and most spatial soft discretization methods also face the problem that not all immunocomplexes are deposited and discretized.

Recognition and Signal Amplification of Immunocomplexes: This involves distinguishing immunocomplexes using signals such as light, electricity, or magnetism, often combined with signal amplification techniques. The ideal recognition method should be able to accurately identify immunocomplexes from interference and free probes, and distinguish between specific and nonspecific binding. Signal amplification technologies used in DIs include enzyme-catalyzed amplification and nucleic acid amplification.

Counting Immunocomplexes: Counting-based quantification is the characteristic of DIs, avoiding errors from macrosignal strength fluctuations. The objects being counted can be positive spaces, negative spaces, or the immunocomplexes themselves or their tags. The objects being counted vary slightly depending on the experimental conditions. These counting objects are collectively referred to as discrete units. In hard discretization, the discrete unit is a single microspace; in soft discretization, the discrete unit is a single target molecule. If the discretization efficiency reaches 100%, and individual immunocomplexes cannot be distinguished within a positive discrete space, negative spaces should be counted using Poisson distribution for absolute quantification. If individual immunocomplexes can be distinguished in positive discrete spaces, the number of immunocomplexes can be counted. When discretization efficiency is less than 100%, the number of immunocomplexes in positive discrete spaces should be quantified using a standard curve.

Figure 2.4 illustrates the design of the main units of DI. First, decide if separation is necessary after the immunocomplexes form, and choose between homogeneous or heterogeneous analysis. In homogeneous analysis, you'll need to select the detection method, such as flow cytometry or surface deposition detection. Next, choose the types of tags for both the detection and capture antibodies, which are typically different ($\alpha \neq \beta$). For heterogeneous immunoassays, start by determining the separation method, such as magnetic bead separation or interface elution. Then, choose the appropriate discretization method, like microwell or droplet, and select the corresponding signal recognition and detection techniques based on the chosen method.

Classification

We classify representative literature on DIs according to two dimensions: discretization and separation, dividing them into four categories shown in the four quadrants of Figure 2.5.

The horizontal axis represents hard discretization and soft discretization. DIs are based on the counting of individual immunocomplexes, where the discretization of immunocomplexes into countable forms is one of the key steps. According to the nature of the discretization technology, DIs are divided into hard discretization DIs and soft discretization DIs. Hard discretization includes microwell array discretization, droplet discretization, and bubble discretization. Soft discretization is divided into temporal and spatial discretization.

The vertical axis classifies whether target molecules are separated from the matrix before detection, into homogeneous and heterogeneous immunoassays. In homogeneous immunoassays, because immunocomplexes and free antibodies are not separated, constructing recognition strategies between them is the key technology. Existing homogeneous DI recognition methods mainly include light intensity difference, spectral difference, diffusion difference, and binding difference. Heterogeneous immunoassays are further classified based on the separation method: magnetic bead separation, interface separation, microsphere separation, and electrophoresis separation; based on the tag on the detection antibody, they can be enzyme, QD, metal nanoparticles, microspheres, nucleic acids, dyes, upconversion luminescence materials, SERS, etc.

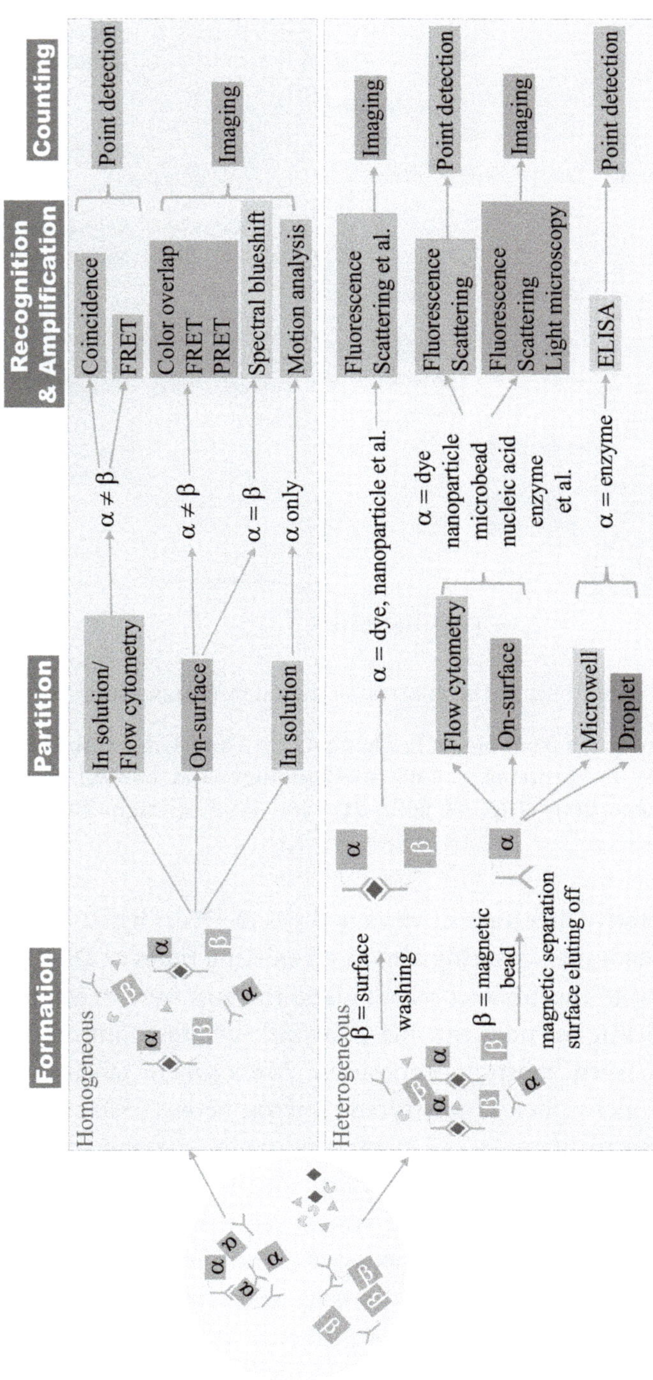

FIGURE 2.4 Major operational units of a DI.

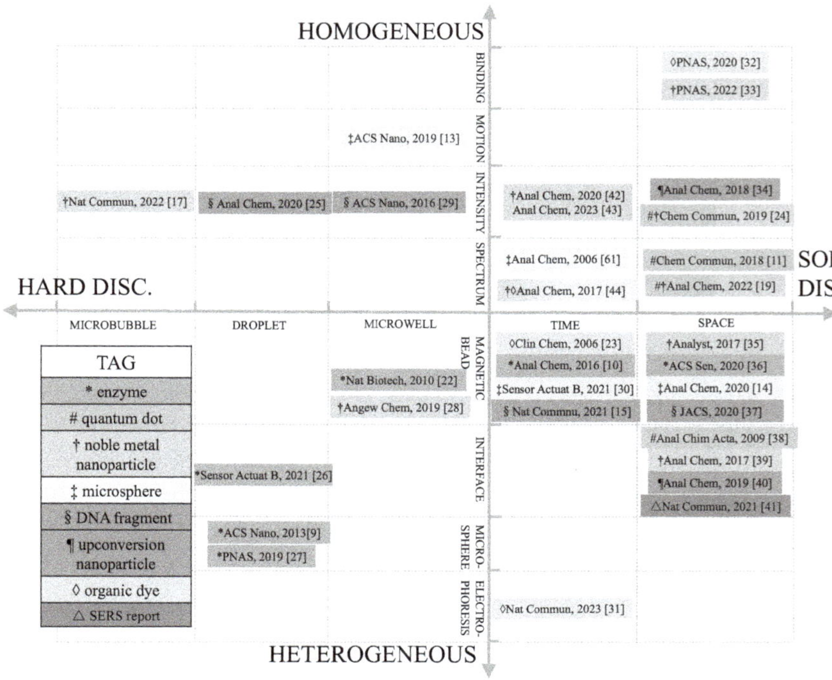

FIGURE 2.5 Milestones in the progress of digital immunoassays.

Source: Reproduced from Liu, X.J.; Zhang, Q.Q.; Zong, C.H.; Gai, H.W. Digital immunoassay for proteins: theory, methodology, and clinical applications, *Analytical Chemistry*, 2025, 97, 9077–9110 (ref 1a). Copyright 2025 American Chemical Society.

Representative literature of various types of DIs is listed in the corresponding quadrants, covering almost all existing types of DIs. The background color of the boxes corresponds to the tags used in the literature. Enzyme labeling includes enzyme-catalyzed substrate luminescence and enzyme-catalyzed substrate deposition. Microsphere labeling includes fluorescent microspheres, polystyrene microspheres (PSs), and magnetic bead labeling in literature.[13] Quanterix's digital enzyme immunoassay, which utilizes microwell discretization, magnetic bead separation, and enzyme signal amplification, is a typical representative of nonhomogeneous hard discretization immunoassays in the third quadrant.[22] Sigulex's single-molecule immunoassay technology, which uses magnetic bead separation, flow discretization, and dye probes, is a representative of soft discretization nonhomogeneous analysis in the fourth quadrant.[23] Mecwins' technology uses interface separation, gold nanoparticles probes, and spatial discretization, which belongs to soft discretization nonhomogeneous analysis. Our research group, using QDs as detection probes and spatial discretization, implements homogeneous DIs, mostly falling in the first quadrant.[11,19,24] The challenges with hard discretization include limitations in the number of microwells, loading efficiency, and difficulties in fabrication. The problem with heterogeneous assays is the numerous steps and complex sample processing, making them hard to miniaturize. Therefore, the ideal DI should combine soft discretization with a homogeneous mode. We believe that the development trend of DI should move from the third quadrant in Figure 2.5 to the first quadrant.

QUANTIFICATION METHODS OF DIGITAL IMMUNOASSAYS

In DIs, there are two counting-based quantitative methods. One method counts the negative spaces and then determines absolute quantities of antigen using Poisson distribution (absolute quantification). The other method counts the immunocomplexes (positive spaces) and then quantifies antigen using a pre-established calibration curve.

Absolute Quantification

Absolute quantification doesn't need a pre-established calibration curve. Instead, it determines the number of target molecules using Poisson distribution, based on the proportion of negative discrete microspaces. While this approach seems straightforward, it requires perfect partitioning of all targets—a 100% discretization efficiency. However, achieving this level of efficiency is challenging with most discretization techniques. For example, in microwell discretization, even with magnetic separation and fluid sweeping, the efficiency usually hovers around 50%. Similarly, in soft discretization, the efficiency of depositing sample molecules onto a substrate, equaling to the discretization efficiency in hard discretization, is often low, with achieving 100% efficiency still a challenge. Droplet discretization is notable for its potential to achieve near-100% efficiency, which is why the Poisson distribution is effectively used for absolute quantification in digital droplet PCR.

Poisson Distribution

In DIs, discretization is essentially to distribute m sample molecules into numerous microspaces. The average number of target molecules per microspace (λ) is linked to the molar concentration (C) of sample and the number of microspace (n), as depicted in Equation 2.1:

$$\lambda = \frac{m}{n} = C \times V_d \times N_A. \tag{2.1}$$

Here, V_d represents the volume of a microspace, N_A denotes Avogadro's constant, and λ is an unknown value. Once λ is determined, the sample concentration is calculated using Equation 2.1. Assuming all sample molecules are evenly distributed into microspaces, meaning a 100% discretization efficiency, the loading of target molecules into a specific microspace happens randomly. Therefore, the value of λ is random. When m is sufficiently large, the discretization process follows statistical principles. In statistics, if an experiment is repeated m times with a probability p for a particular outcome in each trial, the probability of the outcome occurring k times follows a binomial distribution. The distribution of target molecules into microspaces is similar to these independent trials, where m represents the total number of target molecules, and the event is the partitioning of molecules into a specific microspace. The probability of a molecule being in a specific microspace (p) is 1/n. The probability distribution (P(k)), which represents the likelihood of k molecules in specific microspaces, follows a binomial distribution as expressed in Equation 2.2:

$$P(k) = \binom{m}{k} p^k q^{m-k}, 0 \le k \le m. \tag{2.2}$$

Here, $\binom{m}{k} = \frac{m!}{(m-k)!k!}$, $p = \frac{1}{n}$, $q = 1 - p$. The expected value (μ) is equal to λ. It can be calculated according to Equations 2.3a and 2.3b, respectively.

$$\mu = mp = \lambda \tag{2.3a}$$

$$\sigma^2 = mpq \tag{2.3b}$$

When m is significantly large and p is sufficiently small (i.e. n is large), the binomial distribution converges to Poisson distribution. In general, when m is ≥ 20 and p is ≤ 0.05, the binomial distribution is closely approximated by Poisson distribution. Under such conditions, the number of target molecules in a specific microspace is a random variable, denoted as

X_m. The statistical distribution of X_m follows Poisson distribution, as described by Equation 2.4:

$$P(X_m = k) = e^{-\lambda} \frac{\lambda^k}{k!}. \tag{2.4}$$

Then, $\mu = \sigma^2 = \lambda$. When a microspace does not contain any target molecules (a negative microspace), k = 0. The probability, P(0), which represents the proportion of negative microspaces, is calculated by Equation 2.5a:

$$P(0) = e^{-\lambda}. \tag{2.5a}$$

After applying natural logarithm to *P*(0), Equation 2.5a is transformed into Equation 2.5b:

$$\ln(P(0)) = -\lambda. \tag{2.5b}$$

Equation 2.6 is derived by combining Equations 2.1 and 2.5b. The target concentration is then calculated using Equation 2.6:

$$C = \frac{\lambda}{N_A \times V_d} = -\frac{\ln[p(0)]}{N_A \times V_d}. \tag{2.6}$$

Given a known number of immunocomplexes, the number of discrete spaces, n, affects the distribution of molecules in each space. To intuitively understand the Poisson distribution probability, Figure 2.5 shows the distribution of 50,000 immunocomplexes in each space with the total number of discrete spaces change. Assume that all 50,000 immunocomplexes are distributed across discrete spaces (40 pL/space). It's important to note that a different total number of discrete spaces results in a different sample volume (the sample volume is the total number of discrete spaces multiplied by the volume of each space), which also means a change in sample concentration. When the number of discrete spaces is 20,000, the expected number of molecules per space is 2.5 (50,000/20,000 = 2.5). According to the Poisson distribution, the probability P(0) = 0.0821 is calculated using Equation 2.2, and λ is 2.4998 using Equation 2.5b, which is very close to the expected value. When the number of discrete spaces is 50,000, the expected number of molecules per space is 1. The probability P(0) = 0.368

is calculated, and λ is 0.9996, which is also very close to the expected value. As the number of discrete spaces increases, the probability of empty spaces increases. Theoretically, there must be at least one empty discrete space to ensure P(0) ≠ 0; otherwise, it will be impossible to calculate the sample concentration. If the number of empty spaces is too small, even slight measurement fluctuations will significantly affect the quantification results. The total number of discrete spaces does not necessarily need to exceed the number of sample molecules, nor does it need to guarantee that there is only one target molecule per space. As shown in Figure 2.6, it is also not guaranteed that each space will contain exactly one target molecule.

For unknown samples, the concentration is calculated using equation 2.1 and 2.5b, where P(0) is the ratio of the number of negative discrete spaces to the total number of discrete spaces. The larger this ratio, the lower the sample concentration. The theoretical upper limit of absolute quantification is 1/n, and the lower limit is (n−1)/n. The larger the total number of spaces, the wider the dynamic range. However, when the sample

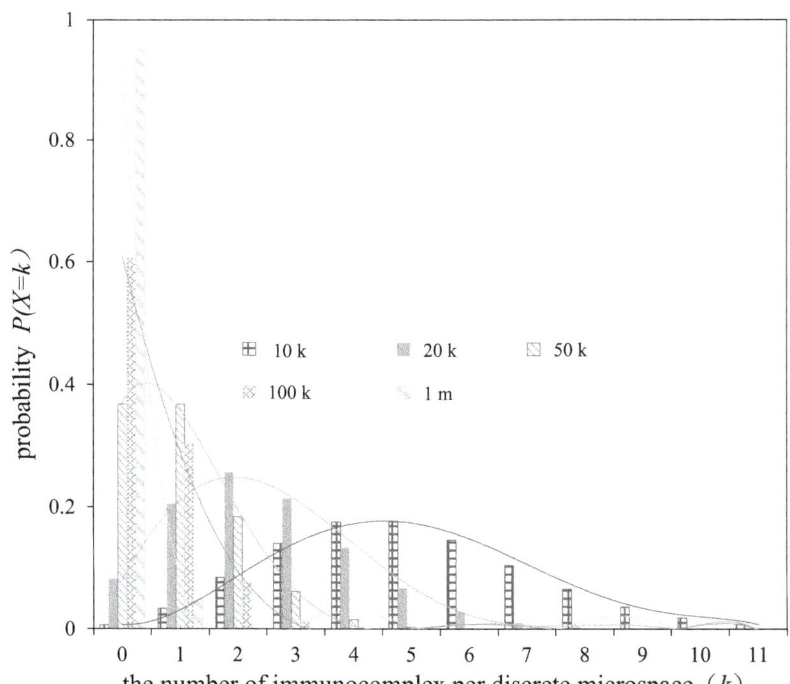

FIGURE 2.6 Poisson distributions of 50,000 immunocomplexes partitioned into varying numbers of spaces.

concentration approaches the two extremes of the detection range, even small changes in negative spaces can significantly affect the quantitative results. Therefore, the practical dynamic range is usually considered to be 60% of the theoretical dynamic range, typically spanning 2–3 orders of magnitude. To expand the dynamic range, both simulated and numerical signals are combined. When the target concentration is low, the molecular density is low, and individual molecules can be distinguished, so counting quantification is used. When the target concentration is high, molecules overlap, and individual molecules cannot be distinguished or counted. In this case, quantification is done using methods similar to analog signal-based quantification, based on the intensity of signals from a unit area, a single bead, or a single signal event. This approach can expand the dynamic range to six to eight orders of magnitude.

Quantitative Accuracy of Poisson Distribution Estimate

The discretization of sample molecules is random. $P(X = k)$ serves as a statistical estimator, with $P(0)$ being a key parameter of interest. The range within which $P(0)$ is estimated is defined as the confidence interval. For binomial distributions, this confidence interval often approximates a normal distribution. The confidence interval calculated using Equation 2.7 is called the Standard Interval (SI), while the interval calculated using Equation 2.8 is referred to as the Wilson interval (WI):

$$SI = P \pm \alpha \times \sqrt{P \times \frac{1-P}{n}}, \tag{2.7}$$

$$WI = \frac{P + \frac{\alpha^2}{2n} \pm \alpha \sqrt{\frac{P(1-P) + \frac{\alpha^2}{4n}}{n}}}{\left(1 + \frac{\alpha^2}{n}\right)}. \tag{2.8}$$

Here, for a 95% confidence interval, α equals 1.96; P stands for the probability of negative microspaces.

The precision of Poisson distribution estimates can be examined from three perspectives: the reliability of the confidence interval, the uncertainty within the confidence interval, and the uncertainty of λ. The reliability of a confidence interval is assessed by its coverage probability, which is the likelihood that the true value falls within the interval. The confidence level represents the expected coverage probability, so a confidence interval is considered more reliable if its actual coverage probability closely matches the confidence level. The coverage probability of the SI can vary. Brown et al. studied how the coverage probability of SI relates to sample sizes.[45] Figure 2.7A shows if P = 0.005, even for large sample sizes (n = 2000), some intervals' coverage probability is significantly lower than the confidence level. Figure 2.7B and 2.7C exhibits how the coverage probabilities of SI and WI changes with P when the sample size is fixed at 50. The SI calculated by equation 2.7 shows significant bias when P(0) was close to 0 or 1, whereas the WI using equation 2.8 performs better in these cases. Therefore, in such situations, the WI is a more appropriate choice for representing the confidence interval.

Figure 2.8 depicts how the lengths of WI and SI change as n varies, with P(0) set at 0.999 (Figure 2.8A), 0.99 (Figure 2.8B), and 0.9, respectively. Generally, as n increases, the lengths of both WI and SI decrease. This trend indicates that uncertainty decreases with a larger n, as uncertainty is measured by the length of the confidence interval divided by P(0). Table

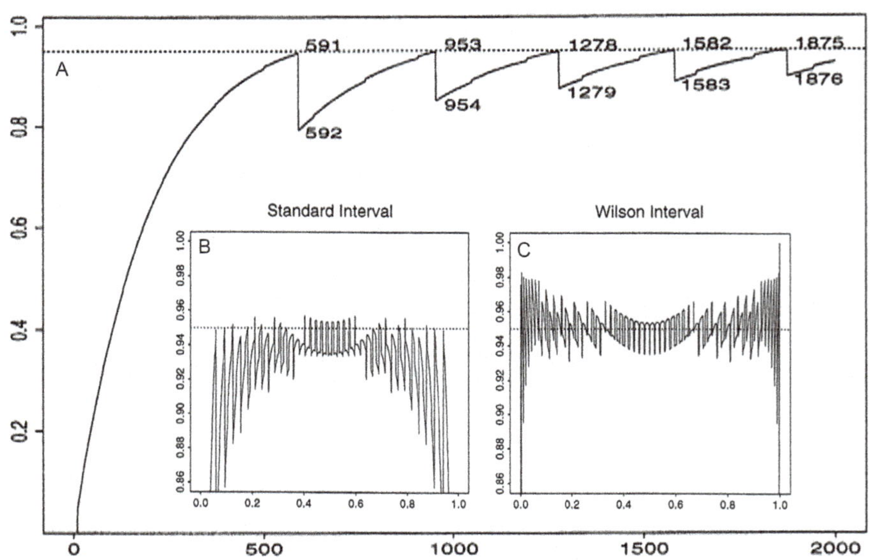

FIGURE 2.7 Oscillation in coverage intervals: (A) SI with P = 0.005, (B) SI with n = 50 and (C) WI with n = 50.

Source: Reprinted with the permission of the Institute of Mathematical Statistics (ref 45: Brown LD, Cai TT, DasGupta A. Interval estimation for a binomial proportion. *Statistical Science*, 2001, 16, 101–133).

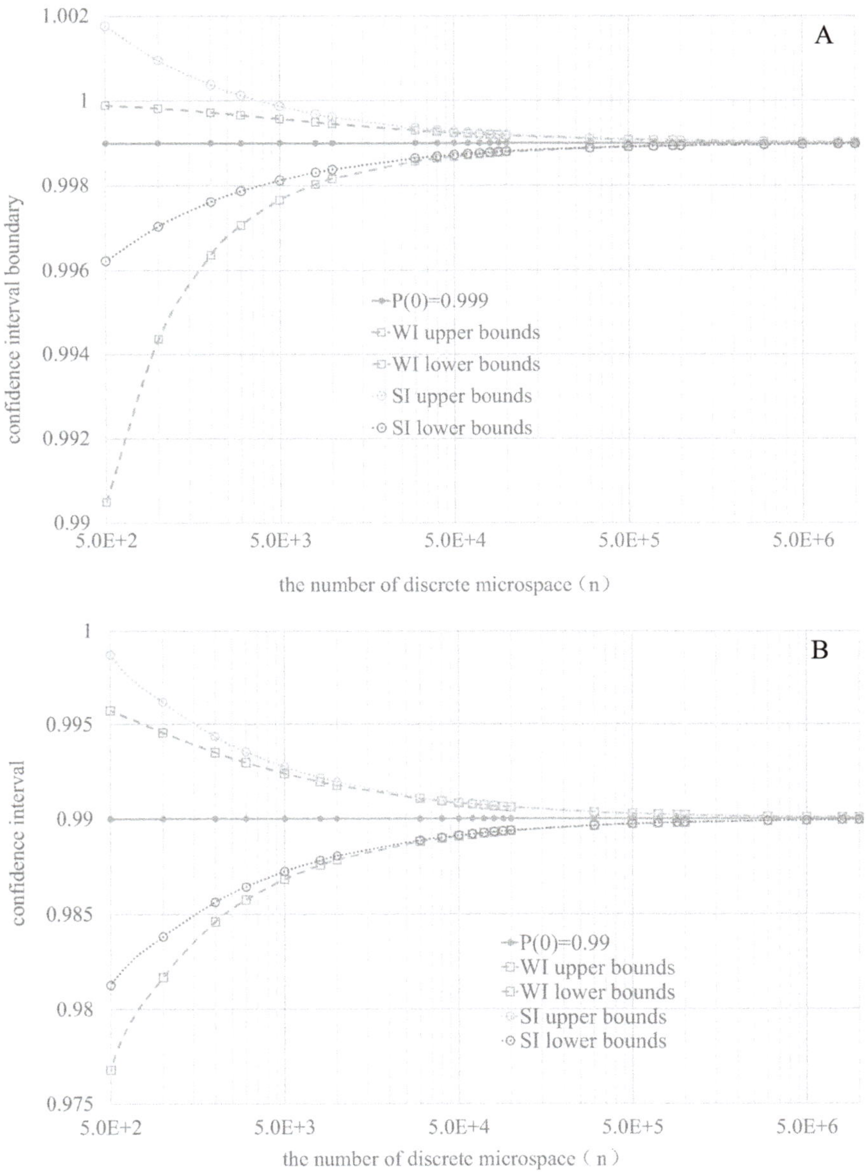

FIGURE 2.8 Comparison of sizes between WI (A) and SI (B).

2.1 provides comparison of uncertainties between WI and SI at different P(0) values. When P(0) is close to 1, a noticeable difference in uncertainty between the two confidence intervals is observed for smaller n. Typically, the difference between the lower bounds of WI and SI is greater than the difference between their upper bounds, meaning WI generally has slightly

TABLE 2.1 Comparison of Uncertainties between WI and SI

n		P(0) Uncertainty 0.999	0.99	0.9
500	WI	0.0094	0.019	0.0586
	SI	0.0055	0.017	0.0584
2,000	WI	0.00337	0.009	0.0292
	SI	0.00277	0.0088	0.0292
5,000	WI	0.0019	0.00041	0.0185
	SI	0.0018	0.00056	0.0185
50,000	WI	0.00056	0.00176	0.00584
	SI	0.00056	0.00176	0.00584
500,000	WI	0.000176	0.00557	0.00185
	SI	0.000175	0.00557	0.00185

higher uncertainty compared to SI. However, these differences diminish as n increases or as P decreases. When P(0) is 0.5, there is little difference between the lengths of the two confidence intervals. Even with a small n (e.g., 500), WI and SI almost overlap. In summary: When n exceeds 5000, SI is preferred over WI. For n below 2000 and P(0) > 0.9 or P(0) < 0.1, WI is more suitable. For n between 2000 and 5000, use SI if P(0)<0.9; otherwise, WI is preferable.

Further, the SI confidence interval of P(0) is substituted into equation 2.5a to calculate the uncertainty of λ. As shown in Figure 2.9, the greater the number of discrete spaces, the smaller the uncertainty. When the total number of discrete spaces is fixed, $\lambda = 1.6$, and the uncertainty is minimized. At this point, P(0) = 0.2, which means that under the defined discrete space conditions, the measurement uncertainty of λ is minimized when the proportion of negative microspaces is 20%. If the measurement uncertainty is required to be less than 10%, n = 5,000, then P(0) must satisfy 0.019 < P(0) < 0.687. In this case, the number of negative spaces must satisfy 95 < the number of negative microspaces < 3435. When n = 300,000, P(0) must satisfy P(0) < 0.995, meaning the number of positive microspaces must be greater than 1500. When n = 1,000,000, P(0) < 0.999, meaning the number of positive microspaces must be greater than 1,000.

Calibration Curve-Based Quantification

Absolute quantification doesn't require a standard curve, making it straightforward and easy to perform. However, it relies on all target

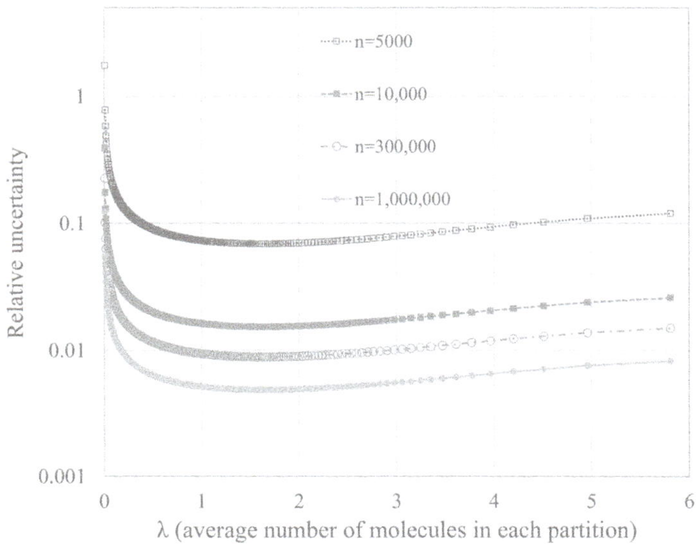

FIGURE 2.9 Relative uncertainty of the average number of molecules partitioned into a single microspace (λ).

Source: Redrawn with permission from Quan, PL; Sauzade, M; Brouzes, E; dPCR: a technology review, *Sensors*, 2018, 18 (4), 1271 (ref 46).

molecules being fully involved in droplet dispersion, with a 100% discretization efficiency. Achieving this level of efficiency is challenging, and it can only be accomplished under conditions of droplet discretization and evaporation to dryness through soft discretization. When discretization efficiency drops below 100%, quantification is done by counting the number of positive discrete microspaces, and a calibration curve must be established. This method is commonly used in single-molecule counting immunoassays and is divided into two modes: indirect counting and direct counting.

Indirect Counting

When the quantity of molecules in a positive discrete microspace is unknown, indirect counting is used. This method requires knowing the proportion of negative discrete microspaces. Using this proportion, the number of molecules in the positive microspaces is inferred from a Poisson distribution, which allows for the determination of the target concentration. However, this calculated concentration only reflects the molecules that have been partitioned into microspaces and doesn't account for any sample loss during discretization and pre-processing. Therefore, a standard curve must be established beforehand. An example of this approach is quantifying an enzyme-linked assay using microwell discretization. Magnetic beads—the empty bead and the target beads containing galactosidase-labeled immunocomplexes—are placed in microwells, with one bead per well. The enzyme in the immunocomplex catalyzes a substrate to produce luminescence, indicating the presence of targets. Although the exact number of targets per bead can't be precisely measured, the Poisson distribution allows for quantification through indirect counting. The specific steps for the calculation process are as follows:

1. Count the total number of positive microwells for a series of standard samples with various concentrations.
2. Subtract the number of the false-positive microwells from the total positive count at each concentration.
3. Calculate the proportion of positive microwells, P(active).
4. Determine the probability of negative microwells, $P(0) = 1 - P(\text{active})$.
5. Compute the number of enzyme molecules λ per microwell, which is also the number of enzyme molecules per magnetic bead, using Equation 2.31b.
6. Plot the standard curve with the concentration of the standard samples on the x-axis and λ on the y-axis.

Table 2.2 presents the measurement data and data processing steps from ref. 22. The second column shows the measured number of luminous microwells. The third column indicates the ratio of luminous microwells to total magnetic beads. The fifth column contains the λ value calculated using equation 2.5b. The sixth column shows the background-subtracted value. The first and sixth columns are used as the x and y coordinates, respectively, to plot the standard curve. A similar quantitative method can be applied using only magnetic beads, without microwells.[37]

Direct Counting

When the quantity of target molecules within a discrete microspace is accurately measured, direct counting is used for quantification. This method is commonly employed in soft discretization-based detection

TABLE 2.2 Process of Indirect Counting-Based Quantification: Step-by-Step Overview

1	**2**	**3**	**4**	**5**	**6**
Enzyme Concentration (aM)	Average Number of Positive Beads	Positive Ratio P(active)	Negative Probability Distribution P(0)	Average Number of Enzymes per Bead (λ)	Corrected Number of Enzymes per Bead
Calculated Value	Measured Value	The Values in the 2nd Column Divided by the Total Number of Beads	P(0) =1−P(active)	ln(P(0)) = −·λ···	The Value in the 5th Column Minus the Background
0	1	0.0016%	0.999984	0.000016	0
0.35	3	0.0086%	0.999914	0.000086	0.00007
0.7	5	0.0099%	0.999901	0.000099	0.000083
3.5	22	0.0413%	0.999587	0.00041309	0.000397
7	38	0.0713%	0.999287	0.00071325	0.000697
35	237	0.4461%	0.995539	0.00447098	0.004455
70	385	0.8183%	0.991817	0.00821666	0.008201
350	1,787	3.3802%	0.966198	0.0343865	0.034371
700	4,036	7.5865%	0.924135	0.07889711	0.078881
3,500	15,634	30.6479%	0.693521	0.36597376	0.365958
7,000	24,836	44.5296%	0.554704	0.58932064	0.589305

Source: The data is sourced from ref. 22.

assays and is particularly effective with organic dyes and QDs as label tags. These tags have unique properties that allow for precise counting of the tags within a microspace. For organic dyes, photobleaching caused a stepwise decrease in fluorescence intensity, with the number of steps corresponding to the number of dye molecules present in the aggregate. Figure 2.10A shows the photobleaching process of Cy5 molecules bound to DNA. The stepwise intensity decrease indicates that one molecule is photobleached at a time. In panel (i), there are 9 steps, indicating that a total of 9 Cy5 molecules are sequentially bleached.[47] By recording the bleaching process, the number of molecules in each discrete unit can be quantified. Unlike organic dyes, the fluorescence intensity of a single QD exhibits strong blinking, and it is highly resistant to photobleaching, making it difficult to determine the number of QDs through bleaching steps. We discovered the asynchronous blue shift of QD spectra, where the first-order spectrum of discrete units continuously splits during the blue-shift process. The number of splits can be used to quantify the number of QDs.[48] Figure 2.10B illustrates the identification of a 4-mer QD complex. The first-order spectra in column C undergoes blue shift under continuous illumination. Due to the asynchronous blue shift of the QDs, the first-order spectrum of each QD splits from the aggregate spectrum. The number of splitting corresponds to the number of QDs. Once the number of molecules in each positive microdomain is confirmed, the number of positive molecules at different concentrations can be directly counted to construct a standard curve.

BASIC THEORY OF DIGITAL IMMUNOASSAYS

Traditionally, the dynamic range of immunoassays was thought to span about nine-fold above and below the equilibrium dissociation constant (K_D), ranging from $K_D/9$ to $9K_D$. For a typical antigen-antibody interaction with a K_D of 1 nM, this translates to a dynamic range of approximately 111 pM to 9 nM. However, many studies have shown that immunoassays can detect targets at concentrations well below K_D, for example, DIs are able to detect molecules at aM level. This apparent discrepancy between theoretical expectations and empirical results raises a critical question: Is the measurement reliable when antigen concentrations fall below $K_D/9$? To address this, we need to consider three key theoretical issues when dealing with ultralow antigen concentrations, such as those at the aM level: (1) Can immunocomplexes form, and can their number exceed the LOD

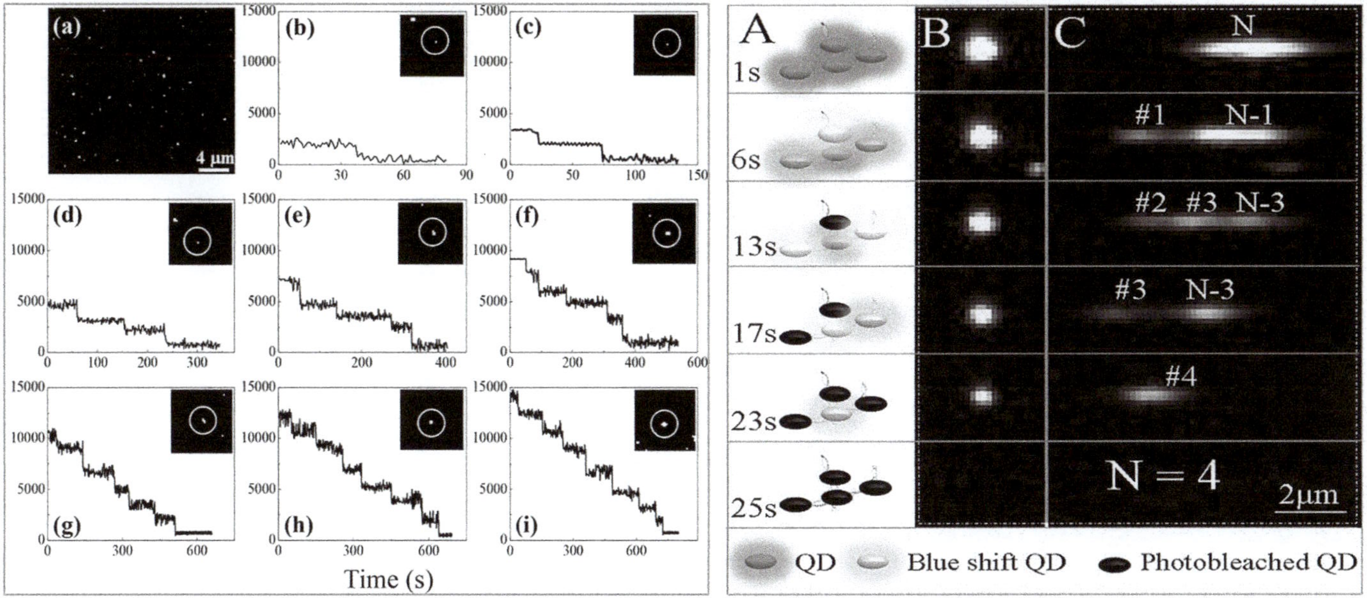

FIGURE 2.10 Quantification based on photobleaching steps of (A) organic dyes and (B) asynchronous blue shift of QDs.

Source of Figure 2.10A: Reproduced from Shi, X.B.; He, Y.; Gao, W.L.; Liu, X.Y.; Ye, Z.J.; Liu, H.; Xiao, L.H. Quantifying the degree of aggregation from fluorescent dye-conjugated DNA probe by single molecule photobleaching technology for the ultrasensitive detection of adenosine. *Analytical Chemistry*, 2018, 90, 3661–3665 (ref 47). Copyright 2018 American Chemical SocietySource of Figure 2.10B: Reproduced from Shi, X.B.; Dong, S.L.; Li,M.M.; Liu, X.J.; Zhang, Q.Q.; Zhao, W.F.; Zong, C.H.; Zhang, Y.W.; Gai, H.W. *Chemical Communications*, 2015, 51, 2353–2356 (ref 48), with permission of The Royal Society of Chemistry.

at such low concentrations? (2) Does the duration of the immunoreaction meet the requirements for practical measurement? (3) How does interference from sample matrices affect the LOD? These questions relate to the thermodynamics, kinetics, and measurement errors involved. In the following discussion, we will explore these three issues in detail.

Thermodynamics

Let's start by discussing the immunoreaction that produces the antibody-antigen immunocomplex, as described in Equation 2.9. We'll then move on to the formation of the sandwich-structured compound:

$$Ab_1 + Ag \underset{k_{off}}{\overset{k_{on}}{\rightleftharpoons}} Ab_1Ag. \tag{2.9}$$

Here, Ab_1, Ag, and Ab_1Ag stand for the capture antibody, the antigen, and the immunocomplex, respectively. k_{on} and k_{off} represent the association rate constant and the dissociation rate constant, respectively.

The equilibrium dissociation constant for this immunoreaction is derived from the law of mass action, as indicated in Equation 2.10:

$$K_D = \frac{1}{K_{eq}} = \frac{[Ab_1][Ag]}{[Ab_1Ag]}. \tag{2.10}$$

Here, K_D is the equilibrium dissociation constant with unit of mol/L, and K_{eq} is the equilibrium constant (L/mol). $[Ab_1]$, $[Ag]$, and $[Ab_1Ag]$ are the concentrations of the capture antibody, the antigen, and the immunocomplex at equilibrium, respectively.

The total concentrations of Ab_1 and Ag, which are equal to their respective initial concentrations denoted as $[Ab_1]_0$ and $[Ag]_0$, are given by Equation 2.11 and 2.12, respectively.

$$[Ab_1]_0 = [Ab_1] + [Ab_1Ag] \tag{2.11}$$

$$[Ag]_0 = [Ag] + [Ab_1Ag] \tag{2.12}$$

There are two methods to calculate the concentration of immunocomplexes: one is the simplified data process, and the other is the direct data process.

Simplified Data Process

The simplified treatment is more commonly used in traditional immunoassay theory. It directly substitutes equation 2.11 into equation 2.10 without considering the impact of the generated complex on the antigen concentration, and solves the equation as follows (Equation 2.13):

$$[Ab_1Ag] = \frac{K_{eq}[Ab_1]_0[Ag]}{(1 + K_{eq}[Ag])}. \tag{2.13}$$

As shown in Figure 2.11, the curve of $[Ab_1Ag]$ versus log[Ag] forms an S-shape. Generally, an affinity with a K_D value smaller than 10 nM is considered high, while an affinity with a K_D value greater than 1 μM is considered low. When [Ag] is very small, $1 + K_{eq}[Ag]$ is approximately equal to 1 in equation (2.13). Under these conditions, $[Ab_1Ag]$ and [Ag] are linearly related. When [Ag] is very large, $1 + K_{eq}[Ag]$ is approximately equal to $K_{eq}[Ag]$. At this point, $[Ab_1Ag]$ becomes independent of [Ag], and the binding curve reaches saturation. This simplified data processing method qualitatively describes the antigen-antibody binding process and is suitable for detection at the ensemble level. However, it is not appropriate for cases involving low antigen concentrations. For instance, when the antibody concentration is 10^{-8} M, K_{eq} is 10^9 M^{-1}, and the antigen concentration is 10^{-19} M, the denominator is approximately 1 and the numerator is 10^{-18} M. These values imply that the concentration of the antigen-antibody complex exceeds the antigen concentration, which is obviously impossible.

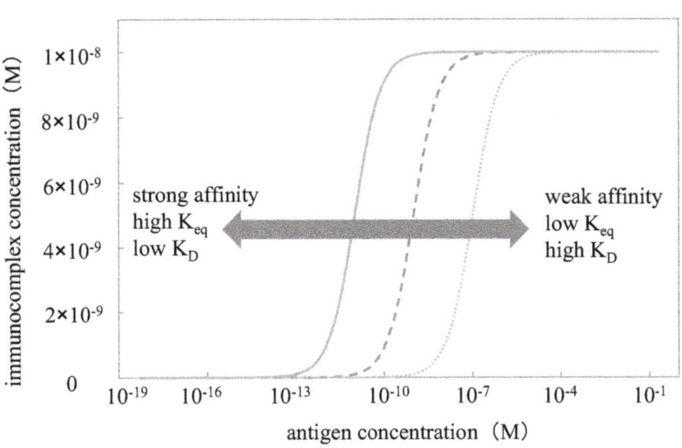

FIGURE 2.11 Immunobinding curve.

Therefore, when the antigen concentration is extremely low, the direct data processing method must be used.

Direct Data Process

In direct data process, the impact of the generated complex on the antigen concentration is taken into account. Substituting Equations 2.11 and 2.12 into Equation 2.10 results in Equation 2.14:

$$K_D = \frac{\left([Ab_1]_0 - [Ab_1Ag]\right)\left([Ag]_0 - [Ab_1Ag]\right)}{[Ab_1Ag]}. \tag{2.14}$$

Rearranging Equation 2.14 yields Equation 2.15:

$$[Ab_1Ag]^2 - \left(K_D + [Ag]_0 + [Ab_1]_0\right)[Ab_1Ag] + [Ab_1]_0[Ag]_0 = 0. \tag{2.15}$$

The connection between $[Ab_1Ag]$ and $[Ag]_0$ can be established by solving the quadratic Equation 2.15. Table 2.3 presents the calculated ratio of immunocomplex to $[Ag]_0$ (here, 1 fM) under different K_D values and initial concentrations of capture antibodies. Figure 2.12 illustrates the relationship between $[Ab_1Ag]$ and $[Ag]_0$ when antigen concentrations range from 10^{-21} to 10^{-4} M. To make the description clear, we use letter plus number (XN, X = A, C, D; N = 1, 2, 3) to represent various immunoreaction conditions. A, C, and D represent K_D of 10^{-6}, 10^{-9}, and 10^{-11} mol/L, respectively. 1, 2, and 3 correspond to $[Ab_1]0$ of 10^{-6}, 10^{-9}, and 10^{-10} mol/L, respectively. When the K_D remains constant, the observation that the saturated concentration of Ab_1Ag decreases in the order of X1, X2, and X3 suggest that the value of the saturated concentration of Ab_1Ag increases as the $[Ab_1]_0$ increases. When the $[Ab_1]_0$ remains constant, the saturated concentration of Ab_1Ag tends to be similar across the different conditions of X1. This pattern persists across the conditions of X2 or X3 as well. These

TABLE 2.3 Calculated Concentration Ratio of Ab_1Ag to Ag

K_D (M)$[Ab_1]$ (M)	10^{-6}	10^{-9}	10^{-10}
10^{-6}	50% (A1)	0.1% (A2)	0.01% (A3)
10^{-9}	100% (C1)	50% (C2)	9% (C3)
10^{-11}	100% (D1)	100% (D2)	91% (D3)

Note: The total concentration of Ag is 1 fM.

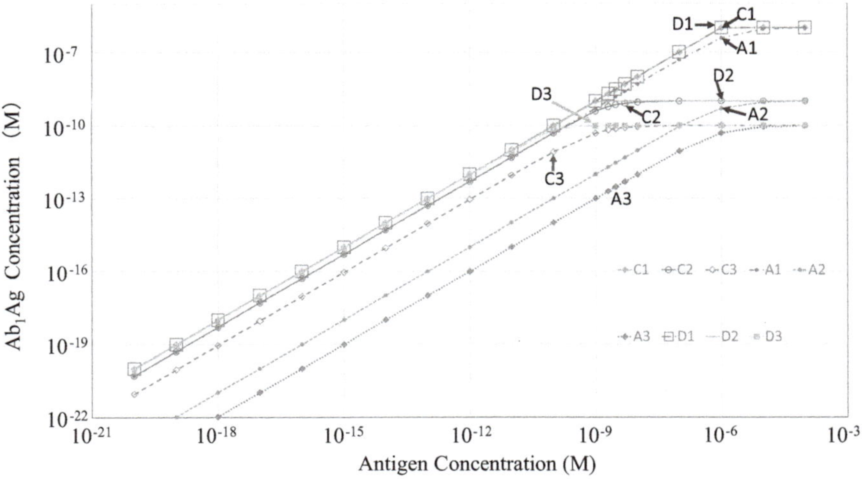

FIGURE 2.12 Relationship between the concentrations of total Ag and Ab_1Ag.

Source: Reproduced from Liu, X.J.; Zhang, Q.Q.; Zong, C.H.; Gai, H.W. Digital immunoassay for proteins: theory, methodology, and clinical applications, *Analytical Chemistry*, 2025, 97, 9077–9110 (ref 1a). Copyright 2025 American Chemical Society.

observations indicate that the saturated concentrations of Ab_1Ag are unaffected by the changes of K_D; instead, it primarily depends on $[Ab_1]_0$. However, the concentration of Ag needed for saturation decreases in the order of D3, C3, and A3, indicating that the concentration of Ag required to saturate Ab1Ag increases with the K_D values. This is because a smaller K_D means a stronger binding between the Ab_1 and Ag, and therefore it is easier to form the Ab_1Ag complex. Let's consider two scenarios: In Scenario 1, when $[Ag]_0$ and $[Ab_1]_0$ are the same under the conditions of A1 and C1, and the K_D under the A1 condition is 1,000 times bigger than that of the C1 condition, the value of $[Ab_1Ag]$ formed under the C1 condition will be twice that formed under the A1 condition. In Scenario 2, when K_D and $[Ag]_0$ are identical in the conditions of A1 and A2, and $[Ab_1]_0$ under the A1 condition is 1000 times bigger than that under the A2 condition, the quantity of $[Ab_1Ag]$ formed in the A1 condition will be 500 times greater than that in the A2 condition. This suggests that increasing the concentration of Ab_1 is more effective in increasing the amount of Ab_1Ag than screening Ab_1 with lower K_D. Backing to ultrasensitive DIs, we aim to know the minimum values of K_D, $[Ag]_0$, and $[Ab_1]_0$ are needed to form detectable levels of immunocomplex. Let's examine two scenarios in a 100 μL aliquot containing 1 aM Ag. The A3 condition characterized by a high K_D and low $[Ag]_0$ represents the least favorable condition for immunoreaction. Here, six immunocomplex molecules form. If noise is not considered, such a quantity of immunocomplex molecules can be detected using single molecule detection techniques. The C2 condition stands for typical immunoreaction conditions, 30 immunocomplexes form. Therefore, returning to the initial question 1, the answer is positive for the Ab_1Ag immunocomplex: the number of formed immunocomplexes exceeds the LOD.

Next, let's calculate the number of sandwich-structured immunocomplexes. We can use two different methods for this. One simple method is to consider Ab_1Ag as the antigen and treat the detection antibody (Ab_2) as Ab_1, as outlined in Equation 2.16:

$$Ab_1Ag + Ab_2^* \underset{k_{off2}}{\overset{k_{on2}}{\rightleftharpoons}} Ab_1AgAb_2^*. \tag{2.16}$$

Here, Ab_2^* represents the signal tagged Ab_2, with k_{on2} and k_{off2} denoting the association and dissociation rate constants, respectively. Equation 2.17 is derived in the same manner as Equation 2.15:

$$\left[Ab_1AgAb_2^*\right]^2 - \left(K_{D2} + \left[Ab_1Ag\right]_0 + \left[Ab_2^*\right]_0\right)\left[Ab_1AgAb_2^*\right] + \left[Ab_2^*\right]_0\left[Ab_1Ag\right]_0 = 0. \tag{2.17}$$

Here, K_{D2} is the equilibrium dissociation constant between Ab_1Ag and Ab_2^* with unit of mol/L. $\left[Ab_2^*\right]$ and $\left[A_{b1}AgAb_2^*\right]$ represent the concentrations of Ab_2^* and $Ab_1AgAb_2^*$ at equilibrium, respectively. $\left[Ab_2^*\right]_0$ stands for the initial concentrations of Ab_2^*. The results in Table 2.3 remain valid, indicating that the number of sandwich-structured immunocomplexes is sufficient for detection even at extremely low antigen concentrations. However, it is important to note that excess Ab_2^* can increase background signals and impact sensing performance. In most DIs, the excess Ab_2^* is removed. This separation of Ab_2^* from the immunocomplex shifts the immunoreaction equilibrium toward dissociation, thereby reducing the quantity of immunocomplexes.

A comprehensive approach is to consider the probability of forming the sandwich-structured immunocomplex in reactions beyond those described in Equations 2.9 and 2.16, as shown in Figure 2.13.

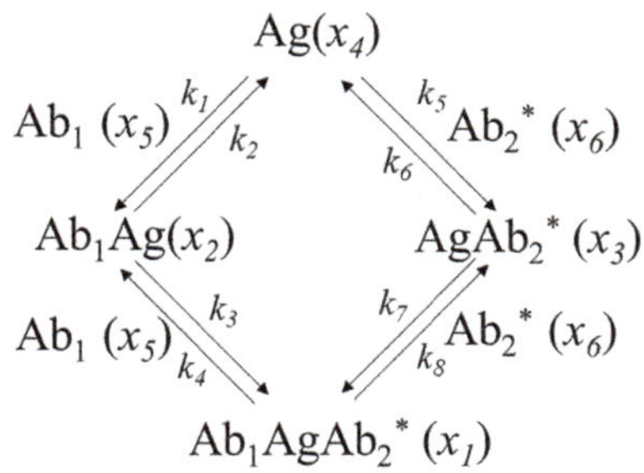

FIGURE 2.13 Possible reactions involved in a sandwich-structured immunoassay.

Source: Reproduced from Liu, X.J.; Zhang, Q.Q.; Zong, C.H.; Gai, H.W. Digital immunoassay for proteins: theory, methodology, and clinical applications, *Analytical Chemistry*, 2025, 97, 9077–9110 (ref 1a). Copyright 2025 American Chemical Society.

The quantity of sandwich-structured immunocomplexes is determined through numerical calculations. For simplicity, let [Ab_1AgAb_2*] = x1, [Ab_1Ag] = x2, [$AgAb_2$*] = x3, [Ag] = x4, [Ab_1] = x5, and [Ab_2*] = x6. Once all reactions reach equilibrium, the equations in Equation group 2.18 are obtained:

$$k_3 \cdot x_2 \cdot x_6 + k_7 \cdot x_3 \cdot x_5 = k_4 \cdot x_1 + k_8 \cdot x_1, \quad (2.18a)$$

$$k_1 \cdot x_4 \cdot x_5 + k_4 \cdot x_1 = k_2 \cdot x_2 + k_3 \cdot x_2 \cdot x_6, \quad (2.18b)$$

$$k_{15} \cdot x_4 \cdot x_6 + k_8 \cdot x_1 = k_6 \cdot x_3 + k_7 \cdot x_3 \cdot x_5, \quad (2.18c)$$

$$p = x_1 + x_2 + x_3 + x_4, \quad (2.18d)$$

$$q_1 = x_1 + x_2 + x_5, \quad (2.18e)$$

$$q_2 = x_1 + x_3 + x_6. \quad (2.18f)$$

Here, p represents the total concentration of antigen, q1 and q2 stand for the initial concentration of capture antibody and detection antibody, respectively. k_1 to k_8 refers to the reactions rate constants. The quantity of sandwich-structured immunocomplexes is estimated by numerically solving Equation group 2.18, given the rate constants and total concentrations. We assume that all reactions have the same dissociation constants, and that the concentrations of the capture and detection antibodies are equal. When k_a and k_d range from 10^4 to 10^6 M^{-1} s^{-1} and 10^{-3} to 10^{-6} s^{-1}, respectively, the quantity of sandwich-structured immunocomplexes is calculated to be approximately 100%, 50%, less than 10%, and 0.1% of $[Ag]_0$ under the C1, D1, D2, and D3 conditions, the A1 and C2 conditions, the C3 condition, and the A2 condition, respectively. This calculation shows that, even when considering all possible reactions, around 15 immunocomplexes form in a 100 μL aliquot containing 1 aM antigen under typical immunoassay conditions (C2). Therefore, the answer to question 1 remains positive for sandwich-structured immunocomplexes, as the number of formed immunocomplexes exceeds LOD under typical immunological conditions.

Kinetics

In the previous section, we examined whether sufficient immunocomplexes could form for detection when the antigen is at extremely low concentrations, without considering the time required to reach equilibrium. This time determines the duration of the measurement. The length of the measurement period is crucial for evaluating efficiency. Two main factors influence the time needed to achieve immunoreaction equilibrium: the speed of the immunoreaction itself and the rate of molecular diffusion.

Immunokinetics

The study of immunokinetics focuses on how quickly immunocomplexes form. We start by deriving the equation for the formation rate of antibody-antigen immunocomplexes and then extend this to include sandwich-structured immunocomplexes. Equation 2.19 is derived from Equations 2.9–2.11:

$$\frac{\partial[\mathrm{Ab_1Ag}]}{\partial t} = \mathrm{k_{on}}\left([\mathrm{Ab_1}]_0 - [\mathrm{Ab_1Ag}]\right)\left([\mathrm{Ag}]_0 - [\mathrm{Ab_1Ag}]\right) - \mathrm{k_{off}}[\mathrm{Ab_1Ag}]. \tag{2.19}$$

Equation 2.19 solved through numerical calculation is depicted in Figure 2.14 (the data was from the ref. 49) at a given reaction rate constants. When the values of K_D remain constant, a higher kon results in a bigger formation rate of immunocomplex, indicating that k_{on} primarily influences the formation rate of immunocomplex. To understand that, how much the antigen is consumed after a 1000-second reaction is calculated. Approximately 67%–89%, 20%, and 2.3% of the antigen converts into immunocomplexes for the k_{on} of 10^6 M^{-1} s^{-1}, 10^5, and 10^4 M^{-1} s^{-1}, respectively. This implies that even with the k_{on} of 10^4 M^{-1} s^{-1} and an

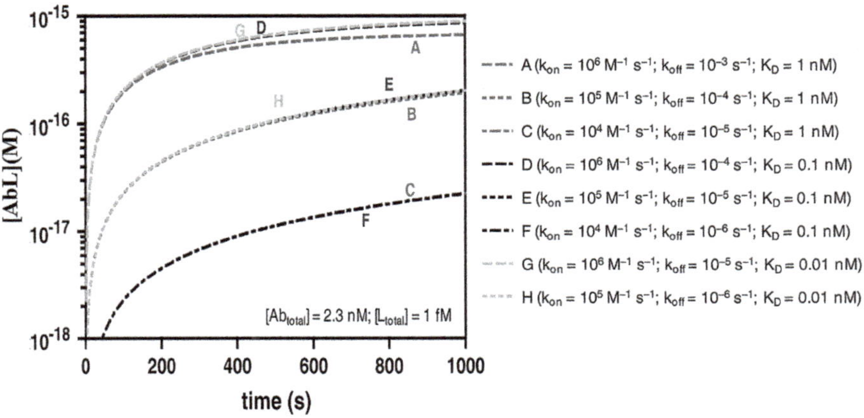

FIGURE 2.14 Change in immunocomplex concentration over reaction time, assuming 2.3 nM antibody and 1 fM antigen.

Source: Reprinted from J. Immun. Meth., Vol. 378, Chang L.; Rissin D.M.; Fournier D.R.; Piech T.; Patel P.P.; Wilson D.H.; Duffy DC. Single Molecule Enzyme-Linked Immunosorbent Assays: Theoretical Considerations, pp. 102–115 (ref 49). Copyright 2012, with permission from Elsevier.

antigen concentration of 1 fM, approximately 0.023 fM immunocomplexes form after a 1000-second reaction. 3 Undoubtedly, this concentration exceeds the LOD of DIs.

Considering all the involved reactions in Figure 2.13, a group of differential Equation 2.20a for sandwich-structured immunocomplex is derived from the group of Equation 2.18a:

$$\frac{\partial x_1}{\partial t} = k_3 \cdot x_2 \cdot x_6 + k_7 \cdot x_3 \cdot x_5 - k_4 \cdot x_1 - k_8 \cdot x_1, \tag{2.20a}$$

$$\frac{\partial x_2}{\partial t} = k_1 \cdot x_4 \cdot x_5 + k_4 \cdot x_1 - k_2 \cdot x_2 - k_3 \cdot x_2 \cdot x_6, \tag{2.20b}$$

$$\frac{\partial x_3}{\partial t} = k_{15} \cdot x_4 \cdot x_6 + k_8 \cdot x_1 - k_6 \cdot x_3 - k_7 \cdot x_3 \cdot x_5. \tag{2.20c}$$

Equations 2.19 and 2.20a is solved numerically or it is simplified into Equation 2.21 when $[Ag] << [Ab_1]$:[50]

$$\frac{\partial [Ab_1Ag]}{\partial t} = k_{on}[Ab_1]_0([Ag]_0 - [Ab_1Ag]) - k_{off}[Ab_1Ag]. \tag{2.21}$$

Integrating Equation 2.21 yields Equation 2.21a that is rearranged into Equation 2.21b:

$$\int_0^t [Ab_1Ag]\,dt = \frac{[Ag]_0[Ab_1]_0 k_{on}}{[Ab_1]_0 k_{on} + k_{off}} e^{-([Ab_1]_0 k_{on} + k_{off})t}\Big|_0^t, \tag{2.21a}$$

$$[Ab_1Ag](t) = \frac{[Ag]_0[Ab_1]_0 k_{on}}{[Ab_1]_0[Ag]_0 + k_{off}}\left(1 - e^{-([Ab_1]_0 k_{on} + k_{off})t}\right). \tag{2.21b}$$

The time required to achieve 99% equilibrium ($t_{99\%}$) is defined in Equation 2.22a and calculated according to its rearranged version, Equation 2.2b:

$$\frac{[Ab_1Ag](t_{99\%})}{[Ab_1Ag]_{eq}} = 0.99 = 1 - e^{-([Ab_1]_0 k_{on} + k_{off})t_{99\%}}, \tag{2.22a}$$

$$t_{99\%} = \frac{4.605}{[Ab_1]_0 k_{on} + k_{off}}. \tag{2.22b}$$

Equation 2.22a indicates that the time required to achieve equilibrium decreases with the initial concentration of antibody increase and is unaffected by the concentration of antigen. However, an increase in the initial concentration of antibody also enlarges background signal. Under the condition A of Figure 2.14, $t_{99\%}$ is calculated to be 3744 seconds.

Under typical conditions of immunoreaction (C2), where both the initial concentrations of the antibodies and K_D are in the nM range, the immunocomplexes formed within a 20-minute reaction is sufficient for DI detection. The $t_{99\%}$ is calculated to be approximately 1 hour, which meets the requirements of traditional diagnostic testing. This implies that the answer to question 2 is indeed yes. However, antigen must diffuse to the vicinity of antibody before immunoreaction occurs. It should be mentioned that the previous conclusion does not take diffusion kinetics into account. The following discussion will address how molecular diffusion impacts the overall detection time.

Diffusion Kinetics

The antibody is either attached to microspheres or a flat surface in DIs. Let's consider the scenario where the antigen diffuses to the microsphere as shown in Figure 2.15A.[51] The diffusion coefficient of microspheres with a diameter of 1 μm is around 0.49 μm²/s, while that of proteins ranges from 10 to 100 μm²/s. Due to the significant difference in diffusion coefficients between microspheres and proteins, microspheres can be considered stationary. Assuming that the microspheres do not aggregate in solution, they are regarded as a periodic medium that centers around the microspheres within a radius of R. In a solution with a volume V_s containing N microspheres, R is calculated by Equation 2.2:

$$R = \sqrt[3]{\frac{3V_S}{4\pi N}}. \tag{2.23}$$

In this spherical coordinate system, the average density of antigen that diffuses and binds to the microsphere surface is referred to as $<n_b>$ and is calculated using Equation 2.24:[51]

$$r_0 < r < R \frac{\partial[Ag]}{\partial t} = D_m \nabla^2 [Ag]. \tag{2.24}$$

Here r is the coordinate distance to the center of the microsphere.

When $r = r_0$, Equation 2.25a is deduced from Equation 2.15 and further reorganized into Equation 2.25b:

$$N_A D_m \frac{\partial[Ag]}{\partial t} = k_{on}^{3D} [Ag](n_{ab} - \langle n_b \rangle) - k_{off}^{3D} \langle n_b \rangle, \tag{2.25a}$$

$$\frac{\partial\langle n_b \rangle}{\partial t} = k_{on}^{3D} [Ag](n_{ab} - \langle n_b \rangle) - k_{off}^{3D} \langle n_b \rangle. \tag{2.25b}$$

Here D_m is the antigen diffusion coefficient, N_A is Avogadro's constant, nab is the surface density of antibody, and are the 3D association and dissociation rate constants, respectively.

When r = R, Equation 2.2 is derived:

$$\frac{\partial[Ag]}{\partial t} = 0. \tag{2.26}$$

When t = 0, Equation 2.27 is obtained:

$$[Ag] = [Ag]_0, \langle n_b \rangle = 0. \tag{2.27}$$

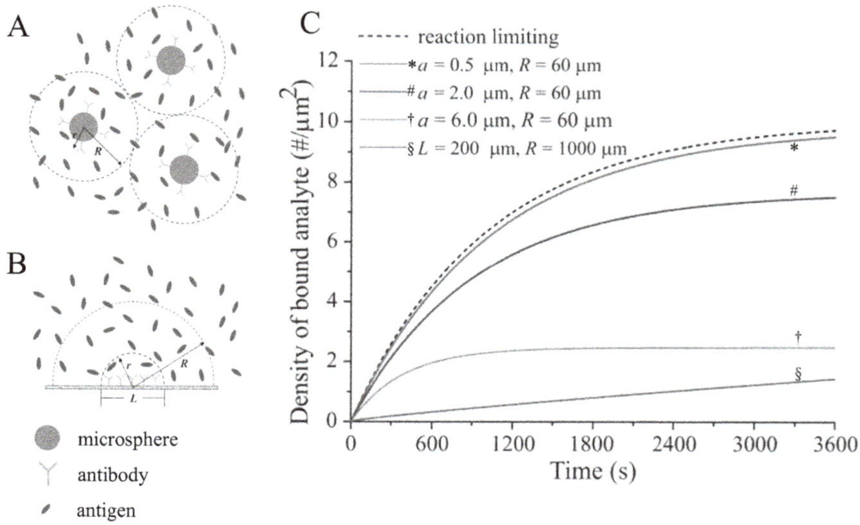

FIGURE 2.15 Schemes of antigens binding to (A) microbeads and (B) solid surfaces (C) and the number of antigen molecules bound to microspheres with varying diameters and surfaces over time.

Source: Reprinted from *Biosensors and Bioelectronics*, Vol. 154, Wu D.; Voldman J. An integrated model for bead-based immunoassays, pp. 112070 (ref 5). Copyright 2020, with permission from Elsevier.

When r equals R, the influence of other microspheres is taken into account. When the impact of diffusion kinetics on the binding rate of antigen is ignored, the binding rate of antigen is solely controlled by the reaction rate, as depicted in Equation 2.28, representing an extreme case of binding efficiency:

$$\langle n_b \rangle = \frac{[Ag]_0 n_{ab}}{k_D^{3D}} \left(1 - e^{-k_{off}^{3D} t}\right). \tag{2.28}$$

Then, let's consider the scenario where the antigen diffuses to a flat surface. If binding duration is not considerably longer than diffusion duration, the sensing surface can be considered as a large microsphere with a characteristic length radius of L/2 as depicted in Figure 2.15B. $\langle n_b \rangle$is calculated still using Equation 2.28. In this case, r_0 is equal to L/2, and the Equation 2.29 is derived:

$$R = \sqrt[3]{\frac{3V_s}{2\pi}}. \tag{2.29}$$

Figure 2.15C illustrates the numerical simulation results of Equation 2.24 and compares the time needed for antigen to bind to a flat surface with that to microsphere with different sizes. The binding process for small microsphere (R = 0.5 μm) closely follows the reaction rate-controlled binding process calculated using Equation 2.28 (dashed line). This suggests that, for small microspheres, the binding time is mainly determined by the reaction rate with minimal influence from diffusion. In contrast, the binding curves for large microspheres (R=2 μm and 6 μm) are no longer solely dependent on the reaction rate, and diffusion kinetics begin to impact the binding process. The 6 μm microsphere with a higher antibody loading achieves equilibrium more rapidly but exhibits a lower binding density of antigen. On the other hand, a flat surface (L = 200 μm) shows the lowest binding efficiency and requires the longest time to reach equilibrium. When fM-leveled antigen molecules interact with antibodies immobilized on a flat surface, it would take approximately 3 months to achieve equilibrium through diffusion alone, significantly reducing detection effectiveness. This is the major reason for replacing flat surface with microspheres as the interface for immunoreactions. Taking both molecule diffusion and immunokinetics into account, the response to question 2 is somewhat positive, depending on the size of microsphere.

Measurement Accuracy

In the previous two sections, the conclusion has been drawn that even when the concentration of antigen is extremely low, a sufficient number of antigens can bind to antibodies and form immunocomplex molecules within a specific immunoreaction time, thereby enabling DI detection. In this section, we will investigate whether the quantity of bound antigen molecules is enough to surpass noise and generate an effective signal. Suppose 50 target molecules are captured and detected, but there are 1,000 background signals originating from nonspecific adsorption. The lower limit of the background noise is Poisson noise, $\sqrt{1000} \approx 32$; however, due to experimental conditions, the noise level may actually be 70~100 molecules. The signals of target molecules are overwhelmed by the noise and cannot be used for quantitative analysis.

The total signal captured by a detector (I_{tot}) is the sum of the signal derived from target molecules (referred to as target signal), background signal, and noise. The background signal sources predominantly comprise the baseline and nonspecific binding. The noise mainly consists of baseline noise, nonspecific binding noise, and target signal noise. All the relationships can be expressed in Equation 2.30:

$$I_{tot} = I_s + I_{bl} + I_{NSB} + \sqrt{\sigma_s^2 + \sigma_{bl}^2 + \sigma_{NSB}^2}. \quad (2.30)$$

Here, I_s, I_{bl}, and I_{NSB} stand for the intensities of target signal, baseline signal, and nonspecific binding signal (NSB), respectively. σ_s, σ_b, and σ_{NSB} represent their corresponding noises. Baseline noise consists of various instrumental noises, such as thermal noise and f noise. Baseline noise exhibits a positive correlation with I_{bl}, so do target signal noise and I_s.

Nonspecific Binding Signal

Nontarget molecules coexisting in sample bind to detection antibody and/or capture antibody with a certain binding constant. This produces signals that aren't related to the target. If their abundances are high, nontarget molecules generate a background signal that can't be omitted, even though they have weaker affinities toward antibodies. When K_{D1}>> [Ag] >> [Ab_1], the ratio of immunocomplexes to the total antibodies (f) is defined as in Equation 2.31a:

$$f = \frac{[Ab_1Ag]}{[Ab_1]_0} = \frac{[Ag]}{K_{D1} + [Ag]}. \quad (2.31a)$$

Logarithmically transforming f yields Equation 2.31b:

$$\log(f) \approx \log[Ag] - \log(K_{D1}). \quad (2.31b)$$

If nontarget molecules (T_2) with an association constant (K_{D2}) toward antibodies are considered, f is expressed as in Equation 2.31c:

$$f_{total} = \frac{[Ag]}{K_{D1}} + \frac{[T_2]}{K_{D2}}. \quad (2.31c)$$

When KD2>>[T2]>>[Ab1], taking the logarithm of Equation 2.31c results in Equation 2.31d:

$$\log(f_{total}) \approx \log\left([Ag] + \frac{K_{D1}}{K_{D2}}[T_2]\right) - \log(K_{D1}). \quad (2.31d)$$

Equation 2.31d indicates that T_2 affects signal. When $[T_2]$ is notably high, K_{D1} must be sufficiently low (high affinity) to suppress the undesired binding signal caused by T_2. Let's explore how low K_{D1} needs to be to ensure that the signal error induced by nontarget binding remains below 10% in a specific scenario where the concentrations of antigen and nontarget molecules (such as globulins) are 1 nM and 5 mM, respectively. Assuming K_{D2} is 500 mM, K_{D1} should be less than 9 nM to maintain the desired accuracy. Improving the affinity between antibody and target can enhance detection specificity. The alternative approach is to recognize the target through multiple antibodies. This is precisely why two distinct antibodies are employed in DIs. Each selectively binds to different parts of the target molecule, thereby forming a sandwich-structured immunocomplex.

A common strategy to eliminate nonspecific protein adsorption is to reduce nonspecific adsorption sites on material surfaces by means of physical adsorption or chemical bonding, thereby decreasing the total amount of adsorbed proteins. Blocking reagents include bovine serum albumin (BSA),[52] polyethylene glycol (PEG),[53] and lipid bilayers,[54] among others. In addition, some layer-by-layer assembly methods and novel material modification approaches have also made progress in reducing nonspecific adsorption.[55] In particular, modifications using dimethyldichlorosilane (DDS) and Tween 20 have demonstrated excellent blocking effects, achieving a 30-fold reduction in nonspecific protein adsorption compared to PEG modification.[56] However, residual nonspecific adsorption remains non-negligible, producing over 30 nonspecific adsorption signals within an imaging area of 2500 μm^2, thereby affecting the LOD.

In addition to surface modification and blocking adsorption sites to reduce nonspecific adsorption, developing recognition methods that can distinguish specific binding from nonspecific adsorption, and subsequently subtracting the contribution of nonspecific adsorption during data processing, is also an effective strategy to mitigate its impact. In 2020, Chatterjee et al. identified signals originating from nonspecific adsorption by exploiting the distinct temporal changes in signal intensity between specific and nonspecific binding.[32] The binding constant between the capture antibody immobilized on the surface and the antigen was greater than that between the Cy5-labeled detection antibody and the antigen ($K_d \geq 10$ nM, $k_{off} \geq 1\ min^{-1}$). The detection antibody repeatedly bound and dissociated from the antigen, resulting in a blinking phenomenon where fluorescence intensity alternated between bright and dark states—a process termed fingerprint dynamic monitoring. In contrast, detection antibodies nonspecifically adsorbed onto the surface do not dissociate, and their fluorescence intensity remained constant until photobleaching occurred, with no blinking observed. By analyzing the temporal changes in fluorescence intensity, it was possible to distinguish which fluorescence signals originated from antigen detection and which resulted from nonspecific adsorption. During quantitative analysis, nonspecific adsorption signals were subtracted directly, significantly improving both detection sensitivity and dynamic range. Using this method, the LOD for plasminogen activator inhibitor-1 (PAI-1), interleukin-6 (IL-6), vascular endothelial growth factor A (VEGF-A), and interleukin-34 (IL-34) were 0.68 fM, 1.2 fM, 3.5 fM, and 6.5 fM, respectively—55 to 383 times lower than those achieved with conventional ELISA, with the dynamic range improved by 3.5 orders of magnitude. The main drawback of this method was its low detection efficiency, or low sampling efficiency. The antigen capture efficiency ranged between 0.5% and 1.5%, and the imaging area of each field of view (100 μm × 100 μm) accounted for only 0.1% of the total sample chamber area. Each measurement typically acquired data from 9 fields of view, resulting in monitoring approximately 0.01% of the total sample per measurement.

Noji et al. employed the differences in diffusion behavior to distinguish nonspecific adsorption.[57] Capture antibodies were immobilized at the bottom of microwells, and detection antibodies were attached to magnetic beads. When biomarkers were present in the solution, the magnetic beads bound to the bottom of the microwells through immunoreactions. The magnetic beads within the microwells exhibited three types of motion: free diffusion, confined diffusion, and fixed binding. A microscope recorded the trajectories of bead movements, which reflected the different modes of motion. These motion modes in turn revealed whether the beads were specifically bound or nonspecifically adsorbed. Specifically bound beads exhibited confined Brownian motion, whereas nonspecifically adsorbed beads displayed extremely limited mobility.

Molecular Shot Noise

The noise stemming from the random fluctuations in the quantity of target molecules within the sampling volume is termed molecular shot noise (MSN), also known as Poisson noise.[58,59]

For a homogeneous binary solution containing target and solvent, the number of target molecules is denoted as n_A and the total number of molecules (target and solvent) as n_T. The probability (p) of capturing analyte molecules in a sample is calculated using Equation 2.23:

$$p = {}^{n_A}\!/_{n_T}. \tag{2.32}$$

The standard deviation of capturing analyte molecules (σ_A) in a sample is expressed in Equation 2.33:

$$\sigma_A^2 = p(1-p). \tag{2.33}$$

The distribution of p follows a binomial distribution as shown in Equation 2.34a:

$$\frac{\sigma_A}{n_A} = \sqrt{\left(\frac{1-p}{n_T p}\right)}. \tag{2.34a}$$

When the number of solvent molecules significantly outweighs that of target molecules, Equation 2.25a is simplified to Equation 2.34b:

$$\frac{\sigma_A}{n_A} = n_A^{-\frac{1}{2}}. \tag{2.34b}$$

This implies that MSN equals the square root of the number of target molecules in Equation 2.34c:

$$\sigma_{MSN} = \sqrt{n_A}. \tag{2.34c}$$

To achieve σ_{MSN} smaller than 10%, a minimum of 100 analyte molecules has to be detected in a sample. Let's calculate the value of σ_{MSN} for 5 aM analyte. The average number of analyte molecules per μL is approximately 3, and the probabilities of capturing 0, 1, and 2 analyte molecules in a 1 μL sample would be 5%, 15%, and 22%, respectively. When σ_{MSN} exceeds 10% in each sampling, it is impossible to accurately quantify analyte, although identify analyte remains feasible. To address Poisson noise, two strategies are executed: (1) Sufficient analyte should be analyzed to reduce sampling error, and (2) All analyte molecules in a sample have to be detected to entirely eliminate Poisson noise.

Our research group proposed a strategy to quantify all target molecules within the entire sample volume, shifting from sampling-based statistical quantification to complete statistical quantification. This approach eliminates errors caused by random sampling and enables absolute quantification without the need for a standard curve.[14] Figure 2.16A illustrates the difference between random sampling counts and complete statistical counts. Random counting involves measuring a subset of the whole, where Poisson noise is unavoidable; in contrast, nonrandom counting involves counting the entire population. To achieve complete counting, the entire sample was reacted with microspheres conjugated with detection antibodies and magnetic beads conjugated with capture antibodies, forming immunocomplexes. Magnetic separation was then used to remove free microspheres from the solution, enriching the magnetic beads and magnetic bead–microsphere complexes. Chemical elution broke apart the

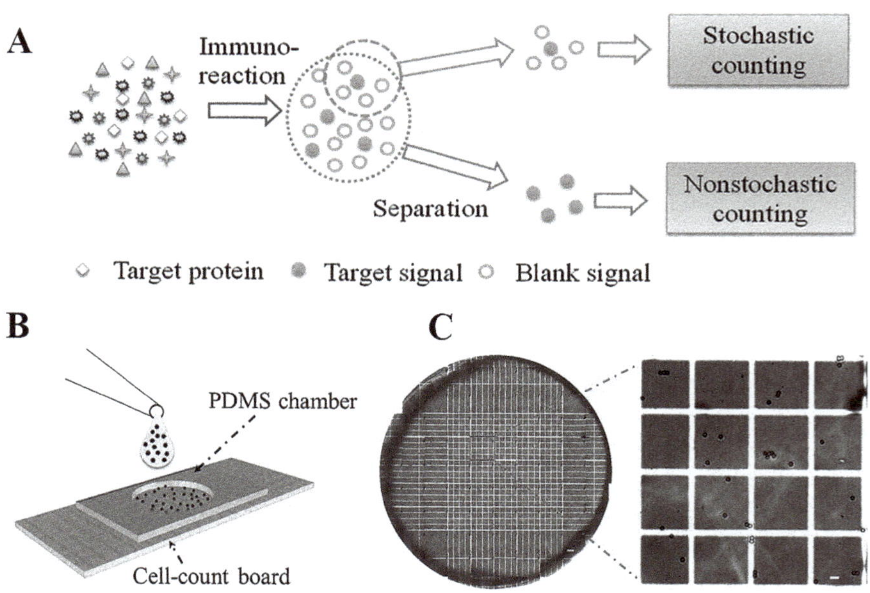

FIGURE 2.16 Single-molecule immunoassay using complete statistical quantification: (A) comparison between sampling-based statistical quantification and complete statistical quantification, (B) microsphere deposition by evaporation to dryness, and (C) combined images through grid-by-grid imaging.

Source: Reproduced from Zhang, Q.Q.; Zhang, X.B.; Li, J.J.; Gai, H.W. Nonstochastic protein counting analysis for precision biomarker detection: suppressing Poisson noise at ultralow concentration, *Analytical Chemistry*, 2020, 92, 654–658 (ref 14). Copyright 2020 American Chemical Society.

complexes, releasing free microspheres. A second magnetic separation removed the magnetic beads, retaining the microspheres. The number of microspheres corresponded directly to the number of antigen molecules. By counting all microspheres under a bright-field microscope, quantitative calculation was performed. To ensure comprehensive counting and minimize the microspheres missing, we deposited the microspheres onto a grid slide through multiple cycles of evaporation. The slide was imaged grid-by-grid and the images were stitched together for counting, as shown in Figures 2.16B and 2.16C.

Error Analysis

The relative error in a detector system (E_{tot}) is formulated as shown in Equation 2.35a:

$$E_{tot}(\%) = \left[E_{MSN} + \frac{I_{system}}{I_{tot}} \right] \times 100. \tag{2.35a}$$

E_{MSN} and I_{system} are calculated according to Equation 2.35b and 2.35c, respectively:

$$E_{MSN} = \frac{\sigma_A}{n_A} = \frac{1}{\sqrt{n_A}}, \tag{2.35b}$$

$$I_{system} = I_{bl} + I_{NSB} + \sqrt{\sigma_s^2 + \sigma_{bl}^2 + \sigma_{NSB}^2}. \tag{2.35c}$$

I_s is calculated using Equation 2.35d:

$$I_s = \xi[\text{Ab1Ag}]. \tag{2.35d}$$

Here, ξ represents the amplification factor. In DIs, ξ is regarded as either the antigen concentration or the count of antigen molecules.

In DIs, the noise stems mainly from background noise, nonspecific binding noise, and MSN. The impact of each type of noise on quantification accuracy changes with its concentration, and its proportion relative to the total noise. In ensemble detection with fM-level LOD, background noise makes up a significant portion of the total noise. However, in single-molecule detection, background noise is no longer the main factor due to significant advancements in modern detectors and imaging systems, which have greatly improved quality and reduced noise. Particularly when paired with fluorescent probes exhibiting high quantum yield, universal fluorescent microscopy can meet the prerequisites for single-molecule detection that the signal strength from single molecule ($_{Ism}$) surpasses six times the standard deviation of the noise (σ_{bl}) above the background intensity. If the nonspecific noise is small enough to be omitted, MSN becomes the principal contributor to the total noise, and therefore significantly influences sample quantification at low concentrations. Let's calculate the relative errors using Equation 2.35a and thereby elucidate how MSN decides the total error in a scenario where $K_D = 10^{-9}$ mol/L, $I_{bl} = 10$, $\sigma_{bl} = \sigma_s = 0.01$, and $\xi = I_{bl} + 6\sigma_{bl} = 10.06$.[60] Figure 2.17 shows the calculation process of errors. First, the concentration of immunocomplexes and the number of molecules is calculated based on the antigen concentration using Equation 2.19. Then, the total signal intensity, MSN, and overall system deviation are calculated using the corresponding formulas. Using the data from ref. 60, the calculated results show measurement errors are significantly large at low molecule numbers. For example, when detecting merely 10 antigen molecules, the relative error stemmed from MSN is calculated to be 32%, while the relative error from background noise is 10%. Consequently, a total relative error will be 42%. To guarantee precise quantification with a total relative error below 10%, the antigen count must surpass 131.

It's essential to note that the previous calculations did not account for NSB. However, NSB is an inevitable phenomenon in all immunoassays, though its extent may vary. When NSB is considered, the error calculation process is generally similar to the one described on the previous one, except that NSB terms are introduced into I_{tot} and background noise. Assuming the target molecule concentration is 1 fM, and the number of NSB molecules ranges from 0 to 14,000. The intensity of NSB is calculated

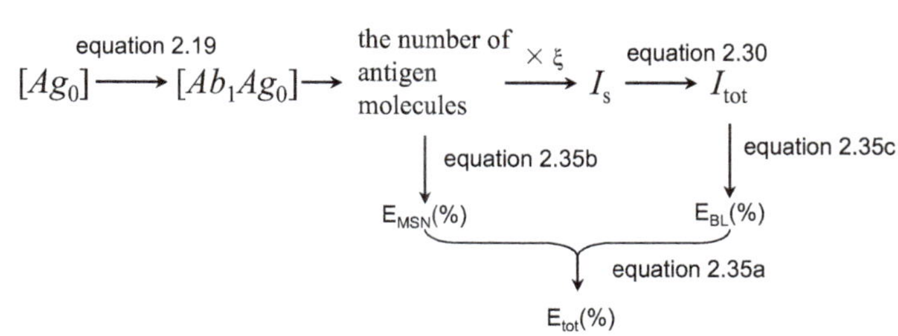

FIGURE 2.17 Error calculation process.

using $\xi = 10$, with other parameters identical to those in the earlier calculation process. To maintain a measurement error of less than 10%, the number of NSB molecules must be fewer than 2,850, which corresponds to approximately 94 aM in a 50 μL volume. When changing the target molecule concentration and recalculating the effect of NSB, it can be observed that in order to meet the requirement of less than 10% measurement error, the amount of NSB adsorption generally needs to be less than one-tenth of the target molecules.

REFERENCES

1. (a) Liu XJ, Zhang QQ, Zong CH, Gai HW. Digital immunoassay for proteins: theory, methodology, and clinical applications, *Analytical Chemistry*, 2025, 97, 9077–9110. (b) Barro C, Zetterberg H. The blood biomarkers puzzle-A review of protein biomarkers in neurodegenerative diseases, *Journal of Neuroscience Methods*, 2021, 361, 109281.
2. Kuhle J, Barro C, Andreasson U et al. Comparison of three analytical platforms for quantification of the neurofilament light chain in blood samples: ELISA, electrochemiluminescence immunoassay and Simoa, *Clinical Chemistry and Laboratory Medicine*, 2016, 54, 1655–1661.
3. Gopfert JC, Reiser A, Yanez VAC et al. Development and evaluation of an ultrasensitive free VEGF-A immunoassay for analysis of human aqueous humor, *Bioanalysis*, 2019, 11, 875–886.
4. Song LA, Zhao MW, Duff DC et al. Development and validation of digital enzyme-linked immunosorbent assays for ultrasensitive detection and quantification of clostridium difficile toxins in stool, *Journal of Clinical Microbiology*, 2015, 53, 3204–3212.
5. Wu DL, Dinh TL, Bausk BP, Walt DR. Long-term measurements of human inflammatory cytokines reveal complex baseline variations between individuals, *American Journal of Pathology*, 2017, 187, 2620–2626.
6. Preische O, Schultz SA, Apel A et al. Serum neurofilament dynamics predicts neurodegeneration and clinical progression in presymptomatic Alzheimer's disease, *Nature Medicine*, 2019, 25, 277–283.
7. Wilson DH, Rissin DM, Kan CW, Fournier DR, et al. The Simoa HD-1 analyzer: a novel fully automated digital immunoassay analyzer with single-molecule sensitivity and multiplexing. *JALA*, 2016, 21, 533–547.
8. Gai HW, Bai JL, Lin BC. Single molecules capillary electrophoresis, *Chinese Journal of Analytical Chemistry*, 2002, 30, 869–874.
9. Shim JU, Ranainghe RT, Smith CA, et al. Ultrarapid generation of femtoliter microfluidic droplets for single-molecule-counting immunoassays, *ACS Nano*, 2013, 7, 5955–5964.
10. Akama K, Shirai I, Suzuki S. Droplet-free digital enzyme-linked immunosorbent assay based on a tyramide signal amplification system, *Analytical Chemistry*, 2016, 88, 7123–7129.

11. Liu XJ, Huang CH, Dong XL, Liang AY, Zhang YS, Zhang QQ, Wang Q, Gai HW. Asynchrony of spectral blue-shifts of quantum dot based digital homogeneous immunoassay, *Chemical Communications*, 2018, 54, 13103–13106.
12. Liu XJ, Huang CH, Zong CH, Liang AY, Wu ZJ, Zhang YS, Zhang QQ, Zhao WF, Gai HW. A single-molecule homogeneous immunoassay by counting spatially "overlapping" two-color quantum dots with wide-field fluorescence microscopy, *ACS Sensors*, 2019, 3, 2644–2650.
13. Akama K, Iwanaga N, Yamawaki K, Okuda M, Jain K, Ueno H, Soga N, Minagawa Y, Noji H. Wash- and amplification-free digital immunoassay based on single-particle motion analysis, *ACS Nano* 2019, 13, 13116–13126.
14. Zhang QQ, Zhang XB, Li JJ, Gai HW. Nonstochastic protein counting analysis for precision biomarker detection: suppressing poisson noise at ultralow concentration, *Analytical Chemistry*, 2020, 92, 654–658.
15. He L, Tessier DR, Briggs K, Tsangaris M, Charron M, McConnell EM, Lomovtsev D, Tabard-Cossa V. Digital immunoassay for biomarker concentration quantification using solid-state nanopores, *Nature Communications*, 2021, 12, 5348.
16. Zhang QQ, Li JJ, Pan XY, Liu XJ, Gai HW. Low-numerical aperture microscope objective boosted by liquid-immersed dielectric microspheres for quantum dot-based digital immunoassays, *Analytical Chemistry*, 2021, 93, 12848–12853.
17. Liu Y, Ye H, Huynh H, Xie C, Kang P, Kahn JS, Qin Z. Digital plasmonic nanobubble detection for rapid and ultrasensitive virus diagnostics, *Nature Communications*, 2022, 13, 1687.
18. Liu XJ, Sun YY, Pan XY, Wu ZJ, Gai HW. Digital duplex homogeneous immunoassay by counting immuno-complex labeled with quantum dots, *Analytical Chemistry*, 2021, 93, 3089–3095.
19. Liu XJ, Lin XY, Pan XY, Gai HW. Multiplexed homogeneous immunoassay based on counting single immunocomplexes together with dark-field and fluorescence microscopy, *Analytical Chemistry*, 2022, 94, 5830–5837.
20. Liu XJ, Tu Y, Gai HW. Imaging of single molecules by wide-field optical microscopy, *Progress in Chemistry*, 2013, 25, 370–379.
21. Su YT, Gai HW. Single molecules counting immunoassay, *Laser & Optoelectronics Progress*, 2022, 59, 0617011.
22. Rissin DM, Kan CW, Campbell TG, Howes SC, Fournier DR, Song L, Piech T, Patel PP, Chang L, Rivnak AJ, Ferrell EP, Randall JD, Provuncher GK, Walt DR, Duffy DC. Single-molecule enzyme-linked immunosorbent assay detects serum proteins at subfemtomolar concentrations, *Nature Biotechnology*, 2010, 28, 595–600.
23. Wu AHB, Fukushima N, Puskas R, Todd J, Goix P. Development and preliminary clinical validation of a high sensitivity assay for cardiac troponin using a capillary flow (single molecule) fluorescence detector, *Clinical Chemistry*, 2006, 52, 2157–2159.

24. Liu XJ, Zhang YS, Liang AY, Ding HW, Gai HW. Plasmonic resonance energy transfer from a Au nanosphere to quantum dots at a single particle level and its homogenous immunoassay, *Chemical Communications*, 2019, 55, 11442–11445.
25. Byrnes SA, Huynh T, Chang TC, Anderson CE, McDermott JJ, Oncina CI, Weigl BH, Nichols KP. Wash-free, digital immunoassay in polydisperse droplets, *Analytical Chemistry*, 2020, 92, 3535–3543.
26. Qian SQ, Wu H, Huang B, Liu Q, Chen Y, Zheng B. Bead-free digital immunoassays on polydopamine patterned perfluorinated surfaces, *Sensors & Actuators: B Chemical*, 2021, 345, 130341.
27. Yelleswarapu V, Buser JR, Haber M, Baron J, Inapuri E, Issadore D. Mobile platform for rapid sub–picogram-per-milliliter, multiplexed, digital droplet detection of proteins, *Proceedings of the National Academy of Sciences of The United States of America*, 2019, 116, 4489–4495.
28. Chen H, Li Z, Zhang LZ, Saway P, Shi JB, Wang P. Quantitation of femtomolar protein biomarkers using a simple microbubbling digital assay via bright-field smartphone imaging, *Angewandte Chemie International Edition*, 2019, 58, 13922–13928.
29. Kim D, Garner OB, Ozcan A, Carlo DD. Homogeneous entropy-driven amplified detection of biomolecular interactions, *ACS Nano* 2016, 10, 7467–7475.
30. Xu ST, Wu JC, Chen C, Zhang JY, Wang Y, Xu H. A micro-chamber free digital bio detection method via the "sphere-labeled-sphere" strategy, *Sensors & Actuators: B Chemical* 337 2021. 129794.
31. Krainer G, Saar KL, Arter WE et al. Direct digital sensing of protein biomarkers in solution, *Nature Communications*, 2023, 14, 653.
32. Chatterjee T, Knappik A, Sandford E, et al., Direct kinetic fingerprinting and digital counting of single protein molecules, *Proceedings of the National Academy of Sciences of The United States of America*, 2020, 117, 22815–22822.
33. Zeng Q, Zhou XY, Yang YT, Sun Y, Wang JA, Zhai CH, Li JH, Yu H. Dynamic single-molecule sensing by actively tuning binding kinetics for ultrasensitive biomarker detection, *Proceedings of the National Academy of Sciences of The United States of America*, 2022, 119, e2120379119.
34. Li X, Wei L, Pan LL, Yi ZY, Wang X, Ye ZJ, Xiao LH, Li H-W, Wang JF. Homogeneous immunosorbent assay based on single-particle enumeration using upconversion nanoparticles for the sensitive detection of cancer biomarkers, *Analytical Chemistry*, 2018, 90, 4807–4814.
35. Wu X, Li T, Tao GY, Lin RY, Pei XJ, Liu F, Li N. A universal and enzyme-free immunoassay platform for biomarker detection based on gold nanoparticle enumeration with dark-field microscope, *Analyst*, 2017, 142, 4201–4205.
36. Maley AM, Garden PM, Walt DR. Simplified digital enzyme-linked immunosorbent assay using tyramide signal amplification and fibrin hydrogels, *ACS Sensors*, 2020, 5, 3037–3042.

37. Wu C, Garden PM, Walt DR. Ultrasensitive detection of attomolar protein concentrations by dropcast single molecule assays, *Journal of the American Chemical Society*, 2020, 142, 12314–12323.
38. Jiang DF, Liu CX, Wang L, Jiang W. Fluorescence single-molecule counting assays for protein quantification using epi-fluorescence microscopy with quantum dots labeling, *Analytica Chimica Acta*, 2010, 662, 170–176.
39. Ma J, Zhan L, Li RS, et al. Color-encoded assays for the simultaneous quantification of dual cancer biomarkers, *Analytical Chemistry*, 2017, 89, 8484–8489.
40. Mickert MJ, Farka Z, Kostiv U, Hlavacek A, Horak D, Skladal P, Gorris HH. Measurement of sub-femtomolar concentrations of prostate-specific antigen through single-molecule counting with an upconversion-linked immunosorbent assay, *Analytical Chemistry*, 2019, 91, 9435–9441.
41. Li JR, Wuethrich A, Sina AAI, Cheng H-H, Wang YL, Behren A, Mainwaring PN, Trau M. A digital single-molecule nanopillar SERS platform for predicting and monitoring immune toxicities in immunotherapy, *Nature Communications*, 2021, 12, 1087.
42. Huang ZL, Li ZY, Jiang M, Liu R, Lv Y. Homogeneous multiplex immunoassay for one-step pancreatic cancer biomarker evaluation. *Analytical Chemistry*, 2020, 92, 16105–16112.
43. Zhang JH, Liu MJ, Zhou F, Yan HL, Zhou YG. Homogeneous electrochemical immunoassay using an aggregation–collision strategy for alpha-fetoprotein detection. *Analytical Chemistry*, 2023, 95, 3045–3053.
44. Wang JJ, Huang XY, Liu H, Dong CQ, Ren JC. Fluorescence and scattering light cross correlation spectroscopy and its applications in homogeneous immunoassay. *Analytical Chemistry*, 2017, 89, 5230–5237.
45. Brown LD, Cai TT, DasGupta A. Interval estimation for a binomial proportion, *Statistical Science*, 2001, 16, 101–133.
46. Quan PL, Sauzade M, Brouzes E. dPCR: A technology review, *Sensors*, 2018, 18, 1271.
47. Shi XB, He Y, Gao WL, Liu XY, Ye ZJ, Liu H, Xiao LH. Quantifying the degree of aggregation from fluorescent dye-conjugated DNA probe by single molecule photobleaching technology for the ultrasensitive detection of adenosine, *Analytical Chemistry*, 2018, 90, 3661–3665.
48. Shi XB, Dong SL, Li MM, Liu XJ, Zhang QQ, Zhao WF, Zong CH, Zhang YW, Gai HW. Counting quantum dot aggregates for the detection of biotinylated proteins. *Chemical Communications*, 2015, 51, 2353–2356.
49. Chang L, Rissin DM, Fournier DR, Piech T, Patel PP, Wilson DH, Duffy DC. Single molecule enzyme-linked immunosorbent assays: theoretical considerations, *Journal of Immunological Methods*, 2012, 378, 102–115.
50. Wegner GJ, Wark AH, Lee HJ, Codner E, Saeki T, Fang SP, Corn RM. Real-time surface plasmon resonance imaging measurements for the multiplexed determination of protein adsorption/desorption kinetics and surface enzymatic reactions on peptide microarrays, *Analytical chemistry*, 2004, 76, 5677–5684.

51. Wu D, Voldman J. An integrated model for bead-based immunoassays, *Biosensors and Bioelectronics*, 2020, 154, 112070.
52. Jeyachandran YL, Mielczarski JA, Mielczarski E. Efficiency of blocking of non-specific interaction of different proteins by BSA adsorbed on hydrophobic and hydrophilic surfaces, *Journal of Colloid and Interface Science*, 2010, 341, 136–142.
53. Heyes CD, Groll J, Moller M, Nienhaus GU. Synthesis, patterning and applications of star-shaped poly(ethylene glycol) biofunctionalized surfaces, *Molecular Biosystems*, 2007, 3, 419–430.
54. Granéli A, Yeykal C, Prasad TK, Greene EC. Organized arrays of individual DNA molecules tethered to supported lipid bilayers, *Langmuir*, 2006, 22, 292–299.
55. Gnanasampanthan T, Beyer CD, Yu WF, et al. Effect of multilayer termination on nonspecific protein adsorption and antifouling activity of alginate-based layer-by-layer coatings, *Langmuir*, 2021, 37, 5950–5963.
56. Hua BY, Han KY, Zhou RB, et al. An improved surface passivation method for single-molecule studies, *Nature Methods*, 2014, 11, 1233–1236.
57. Akama K, Iwanaga N, Yamawaki K, Okuda M, Ain K, Ueno H, Soga N, Minagawa Y, Noji H. Wash- and amplification-free digital immunoassay based on single-particle motion analysis, *ACS Nano*, 2019, 13, 13116–13126.
58. Hungerford JM, Christian GD. Statistical sampling errors as intrinsic limits on detection in dilute solutions, *Analytical Chemisty*, 1986, 58, 2567–2568.
59. Chen DY, Dovichi NJ. Single-molecule detection in capillary electrophoresis: molecular shot noise as a fundamental limit to chemical analysis, *Analytical Chemistry*, 1996, 68, 690–769.
60. Woolley CF, Hayes MA, Mahanti P, Gilman SD, Taylor T. Theoretical limitations of quantification for noncompetitive sandwich immunoassays, *Analytical and Bioanalytical Chemistry*, 2015, 407, 8605–8615.
61. Agrawal A, Zhang CY, Byassee T, Tripp RA, Nie SM. Counting single native biomolecules and intact viruses with color-coded nanoparticles, *Analytical Chemistry*, 2006, 78, 1061–1070.

CHAPTER 3

Discretization in Digital Immunoassays

In digital signal analysis, the key step is discretization—converting continuous analog signals into digital form. This involves sampling the analog signals at a high enough frequency so that they are effectively discretized, followed by quantization, encoding, and transmission. This process transforms the analog signals into digital ones. Similarly, discretization is also the core technique in digital immunoassays (DIs). Traditional immunoassays quantify based on the signal intensity generated by the bulk of immunocomplexes, where the signal variation is continuous and analog. In DIs, however, by artificially increasing the distance between target molecules, they are transformed from a clustered, indistinguishable state into a dispersed and countable one. This process, known as discretization, is what sets DIs apart from other types of immunoassays. Depending on how the distance between molecules is increased and maintained, discretization in DIs is classified into two types: soft discretization and hard discretization. Soft discretization involves increasing the distance between target molecules through limited dilution, allowing them to exist in a dispersed, non-overlapping state. This approach essentially creates a virtual form of micro-compartmentalization. In contrast, hard discretization refers to the physical isolation of target molecules into micro-compartments, where each molecule is forced into the surrounding medium by actual physical boundaries. In other words, the discrete state and spacing of the molecules are maintained by these physical barriers.

DOI: 10.1201/9781003729211-3

Compared to hard discretization, soft discretization is simpler and easier to implement. However, because the virtual micro-compartments it relies on are not fixed in size and may overlap, detection sensitivity can be compromised. Hard discretization, on the other hand, offers significantly higher sensitivity. By confining immunocomplexes within fixed-volume physical micro-compartments, it effectively concentrates the target molecules while minimizing background interference. This confined environment makes it easier to detect targets than in bulk solution. As a result, hard discretization-based DIs can achieve exceptionally high sensitivity, with limits of detection (LODs) reaching the attomolar (aM) range.

SOFT DISCRETIZATION

Soft discretization involves gradually diluting the solution to increase the distance between immunocomplexes, ensuring they remain dispersed and non-overlapping during detection. This approach is often referred to as natural discretization. There are two primary modes to achieve soft discretization: temporal discretization, where individual molecules pass through the detection zone at different times (also known as flow-based discretization), and spatial discretization, where molecules are evenly distributed across a two-dimensional surface without overlapping. Detection techniques vary depending on the discretization mode used and are generally categorized into time-resolved point detection and spatially resolved imaging. Regardless of the detection techniques, soft discretization is well-suited for analyzing ultralow concentration samples within a confined space. However, as sample concentration increases, the efficiency of soft discretization declines. Target molecules become unevenly distributed, leading to variability and overlap in the virtual micro-compartments, which ultimately affects detection accuracy.

Time-Resolved Point Detection

Time-resolved point detection, also known as single-molecule flow-based detection, is commonly used for temporal discretization-based, single-molecule counting. In this approach, target molecules are driven by pressure or electrophoretic forces to pass through a detection point one by one. Each molecule is detected and recorded in sequence, producing a signal curve similar to a chromatogram or electropherogram. Quantification is based on counting the number of detected signals. A typical example of a time-resolved point detection is the Erenna Immunoassay System developed by Singulex.[1] Samples or standards are first incubated with capture antibody-coated magnetic beads in a 96-well plate. After incubation, unbound samples or standards are removed via magnetic separation, and the beads are thoroughly washed. Fluorescently labeled detection antibodies are then added and allowed to bind to the captured targets. Following another round of magnetic separation and washing, a dissociation buffer is introduced to disrupt the antibody-antigen interactions, releasing the detection antibodies from the beads. The resulting solution is transferred to a 384-well filter plate and centrifuged to remove the beads and buffer, leaving only the detection antibodies in the filter plate. This plate is then placed into the Erenna detector. Liquid from each well is drawn into a 100 μm capillary flow cell and passes through a detection window. As the labeled antibodies flow through the laser focal point, they emit fluorescence signals, which are captured by an objective lens and recorded by a photon counter. These signals are detected as discrete pulses, with each pulse corresponding to the detection of a single photon. By analyzing pulse count, peak intensity, and total photon number, the system delivers quantification with a dynamic range spanning over 4.5 orders of magnitude.

The main challenge in detecting a single dye molecule lies in overcoming background signals—specifically, light scattering from solvent molecules and fluorescence from impurities. To illustrate it, we consider a scenario involving 1 pL of water, which contains roughly 3×10^{13} water molecules in a $10 \times 10 \times 10$ μm^3 volume. The Raman scattering cross-section of a water molecule at 488 nm excitation is approximately 10^{-28} cm^2, while a typical fluorophore molecule has an absorption cross-section around 10^{-16} cm^2. This means the total Raman scattering from the solvent of water can easily exceed the absorption cross-section of a single dye molecule. Now imagine there's a single impurity molecule in that same volume—equivalent to a concentration of about 1.7×10^{-12} M. If it's strongly fluorescent, it can completely mask the signal from the target dye. Even if it fluoresces weakly, it still adds background noise, lowering the signal-to-noise ratio and making detection harder. Reducing the detection volume is one effective way to minimize background fluorescence. When the volume is decreased to around 0.2 fL ($1\ \mu m^3 = 1$ fL), detecting single molecules becomes significantly easier. However, this also lowers detection efficiency. Typically, the laser excitation spot is circular, with an e^{-2} diameter ranging from 10 to 50 μm, while the diameter of the sample stream (or sheath flow) is around 5 to 20 μm. The detection volume can be approximated as a cylinder, with its radius defined by the laser spot and its height by the sample stream diameter. Additional techniques to reduce background include laser photolysis, which breaks down fluorescent impurities. Flow rate is also critical—if the flow is too fast, multiple molecules may enter the detection region at once, compromising single-molecule resolution and reducing the accuracy of quantification.

Single fluorescent dye molecules are extremely small and emit weak signals, making them prone to signal overlap or being missed altogether. To enhance the fluorescence signal of individual target molecules, nanoparticles with high quantum yields and narrow emission spectra—such as quantum dots (QDs)—are used to label detection antibodies. In a dual-color QD labeling strategy, capture and detection antibodies were tagged with two distinct colors of QDs, respectively, forming corresponding capture and detection probes.[2] These probes are bound to target molecules to form sandwich-like immunocomplexes. As the capture probes, detection probes, and immunocomplexes flowed through a capillary detection zone, they were individually detected and counted. Capture and detection probes each generated a pulse signal in their respective detection channels, while immunocomplexes produced simultaneous signal bursts in both channels. This coincident signal in the dual channels allowed for clear identification and accurate quantification of the immunocomplexes.

To reduce the difficulty of single-molecule or single-particle fluorescence detection, tyramide signal amplification (TSA) has been applied in flow-based DIs.[3] Tyramide is a para-substituted phenolic compound that, under the catalysis of horseradish peroxidase (HRP), reacts with hydrogen peroxide to generate tyramide radicals. These highly reactive radicals covalently bind to aromatic residues of proteins. The fluorescent molecules are deposited onto the protein surface through tyramide pre-labeled with fluorescent dyes or biotin above the reaction. The process of HRP-catalyzed, TSA-based DIs is as the following: magnetic beads modified with capture antibodies were incubated with hepatitis B surface antigen (HBsAg), biotin-labeled detection antibodies, and HRP-conjugated streptavidin, in turn, to form HRP-labeled beads (target beads). Biotin-labeled tyramide substrate was then catalyzed by HRP to generate biotin-labeled tyramide radicals, which bound to proteins on the bead surface and heavily deposited around the HRP and immunocomplexes. Fluorescent dye-labeled streptavidin bound to the biotin carried by tyramide, completing the labeling of target beads with fluorescent dyes. They were then detected via flow cytometry or fluorescence microscopy. On one hand, HRP catalyzed the deposition of a large number of tyramide molecules onto the protein surface, effectively increasing the number of binding sites for fluorescent dyes. On the other hand, because tyramide radicals had an extremely short lifespan, the deposition occurred almost exclusively around the HRP enzyme, with minimal off-target labeling. This highly localized labeling significantly enhanced the local fluorescence signal. Together, these effects allowed for direct differentiation between target beads and blank beads without the need for droplet encapsulation or microarray-based discretization, significantly lowering the technical barriers to DIs. In the detection of HBsAg, this method achieved a linear dynamic range spanning four orders of magnitude and an LOD of 139 aM—over 20 times more sensitive than conventional ELISA—showcasing its strong potential for hepatitis B screening and prevention.

In addition to fluorescence-based flow detection, time-resolved point detection also includes electrical impedance pulse detection. This detection involves two key steps: (1) Target proteins are first converted into detectable micron-sized particles or particle aggregates through immunoreaction. (2) These particles are then detected and counted via electrical impedance pulses. Jiang et al. adopted a two-step strategy.[4] In the first step, antibody-modified microspheres reacted with target proteins off-chip, forming microsphere aggregates through immunoreaction. In the second step, the resulting solution was introduced into a microfluidic chip for electrical impedance pulse detection. As the microspheres passed through a narrow sensing channel on the chip, they obstructed the flow path, leading to changes in circuit impedance or reduced ion flow. Single microspheres, dimers, and trimers each produced distinct impedance pulse signals of varying intensities. Software was used to identify and count the number of dimers, which served as the basis for quantifying the target proteins. Using this approach, human ferritin was successfully detected in fetal bovine serum across a concentration range of 0.1 to 208 ng/mL.

Davis et al. integrated both steps—immunoreaction and electrical impedance pulse detection—into a single microfluidic chip.[5] Capture antibodies were immobilized on the bottom of the immunoreaction channel, while detection antibodies were attached to micron-sized particles. All immunoassay steps—including the capture of target protein, the binding of detection antibodies, and washing—were carried out within the immunoreaction region. The specifically bound microspheres were then eluted using an alkaline buffer and flowed sequentially through a narrow sensing slit, where they generated measurable impedance pulse signals.

The key advantage of this integrated microfluidic DI platform is its low sample consumption—less than 5 μL per test—and its seamless combination of immunoreaction and signal quantification. However, the platform's sensitivity for protein detection remained relatively limited. For instance, the LOD for IL-6 was approximately 50 pM. This reduced sensitivity was primarily attributed to the substantial steric hindrance posed by the micron-sized particles, which lowers binding efficiency. While reducing the particle size to the nanoscale can potentially mitigate steric effects and improve binding, nanoparticles are too small relative to the sensor slit, producing only ultra small changes in impedance. As a result, detecting their electrical signals becomes significantly more difficult.

Nanopores or nanochannels serve as effective sensing units by amplifying electrical impedance signals from nanoparticles, and they have been widely used for the detection and analysis of nucleic acids, individual particles, and single viruses. A 50 nm thick silicon nitride film was deposited on a silicon wafer, and 300 nm diameter through-holes were fabricated in the film.[6] To reduce the device's capacitance, suppress ionic noise, and enhance the signal-to-noise ratio of impedance pulses, a 5 μm thick polyimide protective layer was added. Both sides of the nanopores were sealed with PDMS to allow for fluid handling. Samples were introduced into the negative PDMS well, while the buffer solution was added to the positive PDMS well. Under physiological conditions, viruses such as influenza A, influenza B, adenovirus, coronavirus, and respiratory syncytial virus typically exhibit negatively charged surfaces. Driven by an applied electric field, these viruses passed through the nanopores, disrupting the ionic current and generating transient pulse signals. These ion flow pulses, produced by individual virus particles, were classified using machine learning, achieving 99% accuracy in identifying the five virus types. One of the main advantages of nanopore sensing for single-virus detection is that it does not require any labeling. However, when a single protein molecule passes through the nanopore, the resulting change in ionic current is too small to be detected. This limitation currently restricts the application of nanopore sensing in single-molecule immunoassays.

To enhance the ion flow signal of a single protein molecule passing through a nanopore, Vincent et al. developed a novel approach using star-shaped double-stranded DNA (dsDNA) nanostructures as probes (P1 and P2).[7] They designed a linking DNA strand, half of which was complementary to P1 and the other half to P2. This linking DNA connected P1 and P2, forming a dumbbell-shaped DNA nanostructure (DB). As DB, P1, and

P2 passed through the nanopore, they generated distinct electrochemical signals. While P1 and P2 produced single-peak signals, DB generated a double-peak signal. The difference in these signal patterns enabled precise counting of the DBs. By attaching antibodies to the linking DNA, the detection of target proteins was effectively transformed into the counting of linking DNA strands. The number of linking DNA molecules could be accurately quantified by counting the number of DBs with the nanopore sensor. For example, when detecting recombinant thyroid-stimulating hormone (rTSH), this nanopore-based detection platform achieved an LOD of approximately 20 pM.

In conventional single-molecule, flow-based detection, immunorecognition and target capture typically occur on solid-phase surfaces. However, these surfaces offer limited binding areas for target molecules, and nonspecific adsorption can easily lead to false positives. Additionally, certain analytes—such as exosomes or biomolecular aggregates—are not well-suited for solid-phase adsorption, as the process can damage the sample and compromise its integrity. To overcome these limitations, Tuomas et al. developed a single-molecule microfluidic sensing platform based entirely on liquid-phase reactions.[8] This platform combined electrophoretic separation with flow-based, single-molecule detection. Fluorescently labeled nucleic acid aptamers reacted with analytes directly in solution. The reaction mixture was then introduced into a microfluidic chip and passed through a free-flow electrophoresis region. Under an applied electric field, unbound aptamers—being smaller—migrated faster, while aptamers bound to analytes migrated more slowly due to increased size. This difference in migration speed led to distinct lateral shifts, allowing the two populations to be electrophoretically separated. At a specific lateral position, a laser confocal microscope combined with a single-photon counter was used to detect individual bound aptamer molecules. Using this platform, a wide range of analytes—including streptavidin, immunoglobulin E (IgE), α-synuclein fibrils, CD63+ exosomes, and protein aggregates such as FUS condensates—were successfully quantified at the single-molecule level. These analytes were tagged with fluorescent dyes like Atto488 or Alexa488 through their corresponding aptamers, demonstrating the versatility of this approach. Compared to traditional solid-phase single-molecule detection, this fully liquid-phase microfluidic sensing system eliminates nonspecific surface binding, reduces analysis time, enhances immunocapture efficiency and detection sensitivity, and broadens the scope of analytes that can be effectively detected.

Spatially Resolved Imaging Detection

In spatially resolved imaging detection, external force fields or specific binding reactions are used to deposit or immobilize target proteins or signal reporters onto a substrate surface. These target molecules are naturally dispersed such that the distance between adjacent molecules exceeds the resolution limit of optical imaging. High-sensitivity cameras can capture the spatial positions of individual molecules, and specialized software is then used to count them. The concentration of target molecules is ultimately determined using a standard calibration curve.

To achieve spatially resolved imaging, the target molecules must be deposited, immobilized, and spatially dispersed on the substrate. There are three primary methods for depositing molecules or particles onto a substrate: affinity-based deposition, electrostatic attraction, and solvent evaporation.

Affinity-Based Deposition

Biomolecules specifically interact with one another through hydrogen bonds, electrostatic interactions, hydrophobic forces, and coordination bonds. These specific binding interactions are collectively referred to as biological affinity. The relationship between molecules with mutual affinity is often compared to a "lock and key" mechanism. Common examples of high-specificity affinity pairs include antigen-antibody pairs, hormone-receptor pairs, and complementary nucleic acid strands.

As early as 1998, Seeger et al. utilized antigen-antibody interactions to detect spatially dispersed single C-actin protein molecules.[9] Capture antibodies were immobilized on a substrate surface, and the target protein was sequentially reacted with the capture antibody followed by a fluorescently labeled detection antibody, forming a capture antibody–target protein–fluorescent antibody immunocomplex. Due to the specificity of antigen-antibody binding, the fluorescent dye molecules were evenly distributed across the substrate. Single fluorescent molecules were then detected with high sensitivity using semiconductor laser confocal fluorescence spectroscopy. For example, in the detection of C-actin protein, a monoclonal chicken anti-C-actin IgG was covalently attached to the surface to capture the target protein, followed by reaction with a Cy5-labeled polyclonal IgY. In the sample group, single-molecule photon pulse signals were successfully observed, while no such signals were detected in the negative control group. This result initially demonstrated the potential of the method for diagnosing myocardial infarction. However, fluorescent dyes often emit weak signals, requiring advanced high-resolution microscopy techniques such as confocal or total internal reflection fluorescence (TIRF) microscopy for single-molecule detection. In addition, fluorescent dyes are prone to photobleaching under high-intensity excitation light, which limits their effectiveness as labeling agents for target molecules.

Fluorescent nanoparticles, such as QDs and upconversion nanoparticles, offer advantages like strong light absorption, narrow emission spectra, and excellent resistance to photobleaching, making them promising alternatives to traditional organic dyes. Our research group employed QDs as labeling tags for target proteins and utilized antigen-antibody affinity to achieve single-molecule spatial imaging of proteins.[10] To prepare the substrate, a glass slide was silanized to introduce amino groups. This amino-functionalized surface was then treated with glutaraldehyde to form an aldehyde-functionalized substrate. Capture antibodies of prostate specific antigen (PSA) were then covalently immobilized on the surface via reaction with the aldehyde groups. The modified substrate was sequentially incubated with PSA and QD-labeled detection antibodies, resulting in the formation of QD-labeled immunocomplexes on the substrate. Unbound or nonspecifically bound antibodies were thoroughly washed away. Thanks to the specificity of antigen-antibody interactions, the QD labels were selectively deposited on the glass surface. Wide-field fluorescence microscopy was then used to detect and count PSA molecules at the single-molecule level. Due to spatial steric effects between nanoparticles, QDs bound to the substrate created a localized exclusion zone around each particle. This steric hindrance effectively prevented nanoparticle aggregation and overlapping fluorescence signals, thereby improving the accuracy of spatial imaging detection. However, this same steric effect can also reduce the immunocapture efficiency per unit area, leading to decreased detection sensitivity.

Gorris et al. used upconversion nanoparticles (β-$NaYF_4$: 18 mol% Yb^{3+}, 2 mol% Er^{3+}) as labeling tags for proteins and applied a similar strategy to spatially disperse the particles for imaging PSA.[11] Capture antibodies were immobilized on a substrate surface, and the target antigen along with upconversion nanoparticle-labeled detection antibodies were sequentially introduced. Through antigen-antibody affinity interactions, immunocomplexes labeled with upconversion particles were formed on the substrate. High-resolution fluorescence images of the upconversion nanoparticles were acquired using inverted wide-field fluorescence microscopy. The images displayed bright green emission spots that were relatively uniformly distributed, indicating effective spatial dispersion of the nanoparticles facilitated by the antigen-antibody interactions. Cross-sectional analysis of the fluorescence intensity from individual upconversion particles revealed a high signal-to-noise ratio. By counting the number of fluorescence spots corresponding to the upconversion nanoparticles, quantitative detection of PSA was achieved. The platform demonstrated a detection range of 10 pg/mL to 1.0 ng/mL, with an LOD of 1.2 pg/mL.

Electrostatic Attraction-Based Deposition

Electrostatic attraction between oppositely charged particles is a common phenomenon in nature, with its strength depending on the amount of charge and the distance between particles. By adjusting the functional groups and the pH of the solution, the surfaces of the substrate and labeled particles can be engineered to carry opposite charges, enabling the spatial dispersion of target particles via electrostatic attraction. For example, amino-modified DNA sequence 1 was immobilized on magnetic beads to create capture beads, while thiol-modified DNA sequence 2 was attached to the gold nanorods to form detection probes. The target DNA hybridized with both DNA sequences, forming dsDNA complexes that linked the magnetic beads to the gold nanorods. After magnetic separation, the detection probes were eluted and concentrated. The quantity of detection probes was directly proportional to the concentration of the target DNA. To prepare the detection probes, the substrate was treated with 3-aminopropyltriethoxysilane, resulting in a positively charged surface. Because the detection probes carried DNA, they were negatively charged. This charge difference enabled uniform deposition of the detection probes onto the substrate via electrostatic attraction. Due to the strong scattering properties of the gold nanorods, the detection probes could be visualized under dark-field microscopy. By counting the gold nanorods in the dark-field images, the concentration of the target DNA was quantitatively determined.[12] By replacing DNA sequences 1 and 2 with capture and detection antibodies, He et al. achieved single-molecule-counting-based DI for PSA, with an LOD as low as 8 fg/mL.[13]

Charge attraction-based deposition is simple, easy to use, and widely applied for the spatial distribution of nanoscale signal-tagged particles. In our work, we utilized commercially available positively charged glass slides modified with poly-L-lysine from Thermo Scientific (Cat. No. 4951plus4) to carry out charge attraction-based deposition of QDs or gold nanoparticles. Red QDs (emission wavelength: 655 nm) were conjugated to capture antibodies, while green QDs (emission wavelength: 565 nm) were conjugated to detection antibodies. These conjugates were then incubated with antigens in serum to form sandwich immunocomplexes. The resulting mixture was directly applied to the positively charged glass slides. Because carboxylated QDs carried negative charges, they were rapidly and uniformly deposited onto the positively charged substrate via electrostatic attraction. The slides were then imaged using fluorescence microscopy. QDs that were not involved in immunocomplex appeared as distinct red or green fluorescence spots. In contrast, the red and green QDs inside an immunocomplex both still emitted their respective colors, but due to fluorescence overlap, the immunocomplex appeared as yellow spots.[14] The number of yellow spots showed a linear correlation with the logarithm of antigen concentration, allowing for quantification of target protein levels based on spot count. Using this deposition method, we successfully achieved high-sensitivity detection of both single and multiple analytes, including QD–antigen–QD and QD–antigen–gold nanoparticle immunocomplexes.[15–17]

Regardless of whether electrostatic or affinity-based methods are used to deposit nanoparticles, the sedimentation efficiency plays a crucial role in determining the sensitivity and accuracy of detection. In theory, higher sedimentation efficiency leads to better detection performance and a lower LOD. However, due to the resolution limits of optical microscopy (around 200 nm), nanoparticle deposition cannot be too dense. If the nanoparticles are packed too closely, their distance may fall below the 200 nm resolution threshold, causing their images to overlap and making them indistinguishable from one another. Additionally, the imaging area of cameras (e.g., CMOS: 13.2 mm × 13.2 mm) is relatively small, and at high magnification, the amount of digital signal captured in a single frame is limited. To gather statistically significant data, multi-frame imaging is typically used to capture more particles. To prevent spots to be imaged overlap, the spots are randomly selected, ensuring they are spaced far apart. However, this approach can sometimes result in missed target signals, which can compromise the accuracy of measurements.

Solvent Evaporation-Based Deposition

Both affinity-based and electrostatic sedimentation are governed by a dynamic equilibrium between the particles and substrate, where particles continuously bind and dissociate. Theoretically, it's impossible to sediment all particles—only partial sedimentation can be achieved. To achieve 100% sedimentation efficiency, we used solvent evaporation to force the particles onto the surface of a substrate.[18] Detection antibodies were linked 3.0 μm polystyrene microspheres (PSs) to form detection beads, while capture antibodies were modified onto 2.8 μm magnetic beads to form capture beads. These beads bound to the corresponding antigen to form immunocomplexes. The immunocomplexes, along with any excess capture beads, were collected under a magnetic field, and the immunocomplexes were broken using an acidic buffer, releasing the PS beads. The number of PS beads in the resulting solution was directly proportional to the concentration of the target protein. A 3 mm diameter hole was punched in a 1 mm-thick (polydimethylsiloxane) PDMS sheet using a hole puncher. This hole was aligned with the grid of a commercial cell counting slide and reversibly sealed, creating a counting chip with positioning functionality (Figure 3.1A). A certain volume of the solution was added into the microwell. Upon heating, the solvent (water) evaporated completely, causing the PS beads to sediment onto the bottom of the microwell. The forced deposition from solvent evaporation ensured maximum sedimentation efficiency, reaching 100%. After several rounds of adding the solution and evaporating the water, all released PS beads were fixed at the bottom of the counting device. Using the positioning grid, bright-field images of the PS beads were captured in specific areas using an inverted microscope. The total number of PS beads across all images was counted, which corresponded to the number of antigen molecules (Figure 3.1B). Using carcinoembryonic antigen (CEA) as a model, this method achieved a dynamic range of 5–500 aM and an LOD of 4.9 aM. By addressing issues like overlap and omissions in multi-frame imaging with the positioning grid, using microscale beads to overcome high-density signal limitations, and enhancing particle sedimentation efficiency through solvent evaporation, our method significantly boosted the statistical signal quantity in DIs, improved sensitivity, and reduced the LOD. However, solvent evaporation-based sedimentation also has its limitations—it's only applicable to microscale particle sedimentation. This is because solvent evaporation can lead to particle aggregation. Even in aggregates, optical microscopy can still resolve and count the microscale particles inside. However, when it comes to nanoparticles, conventional optical microscopy struggles to distinguish the individual particles within aggregates.

As mentioned earlier, solvent evaporation-induced particle aggregation limits the effectiveness of nanoparticle deposition. Allier et al. developed a

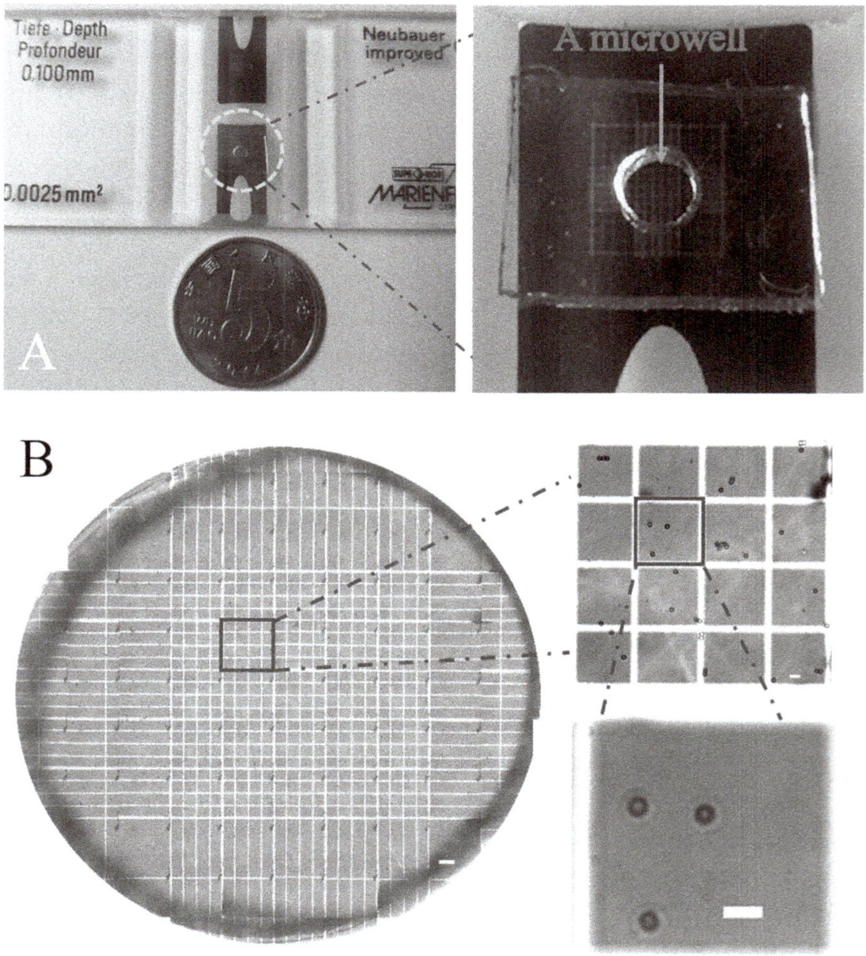

FIGURE 3.1 Solvent-induced deposition of polystyrene microspheres: (A) schematic of the counting chip and (B) bright-field image of the deposited microspheres.

Note: The scale bars represent 8 μm.

Source: Reproduced from Zhang, Q.Q.; Zhang, X.B.; Li, J.J.; Gai, H.W. Nonstochastic protein counting analysis for precision biomarker detection: suppressing Poisson noise at ultralow concentration, *Analytical Chemistry*, 2020, 92, 654–658 (ref 18). Copyright 2020 American Chemical Society.

method using polymer solution evaporation to form a continuous wetting film for enhancing optical image resolution.[19] In this approach, carboxylated PSs with a diameter of 200 nm were suspended in a solution of polyethylene glycol (PEG, molecular weight 600) and Tween 20, with the concentration of Tween 20 set at its critical micelle concentration. Microliter-scale droplets of this nanoparticle suspension were deposited onto a hydrophilic coverslip and allowed to dry at room temperature. As the water evaporated, a monolayer PEG film formed, and a small fraction of the nanoparticles were embedded in the film, acting as microlenses. Inspired by this, we applied a similar strategy by evaporating nanoparticles just a few nanometers in size in the medium of dextran to improve particle distribution uniformity and reduce aggregation during sedimentation (Figure 3.2A and B).[20] This deposition method was named Synergistic Sedimentation of Brownian Motion Suppression and Dehydration Transfer (SynSed). Consider a scenario where the concentrations and volumes of the QD solutions are 1.0 pM and 1 μL for SynSed, and 50 pM and 2 μL for electrostatic attraction sedimentation. The sedimentation efficiency of QDs using SynSed increased to 91.4%, compared to approximately 10% for electrostatic attraction-based deposition (Figure 3.2C-F). Additionally, the variation in QD distribution was limited to just 8.8%. By incorporating SynSed into DIs, we significantly improved the LOD, lowering it to 3.9 aM for CEA. This also expanded the dynamic range to cover three orders of magnitude in concentration, from 0.01 to 10 fM.

HARD DISCRETIZATION

Hard discretization refers to physically isolating individual proteins or signals into separate, confined microspaces. These microspaces typically have volumes in the fL range, which significantly increases molecular concentration and reduces background noise. The physical barriers also enable the use of signal amplification strategies, like enzymatic reactions and nucleic acid amplification, without crosstalk between adjacent digital signals, thus improving detection sensitivity. Depending on the type of physical barrier, hard discretization is classified into two main types: microarray well discretization and microdroplet discretization.

Microarray Well Discretization

Microarray well discretization involves dispersing immunocomplexes formed by target proteins into fL-scale microwells. Enzymes carried by the immunocomplexes catalyze the conversion of substrates within the microwells, generating fluorescent signals. Based on the materials used to construct the wells, microarray wells are classified into two types: optical fiber bundle-based microwell arrays and polymer-based microwell arrays.

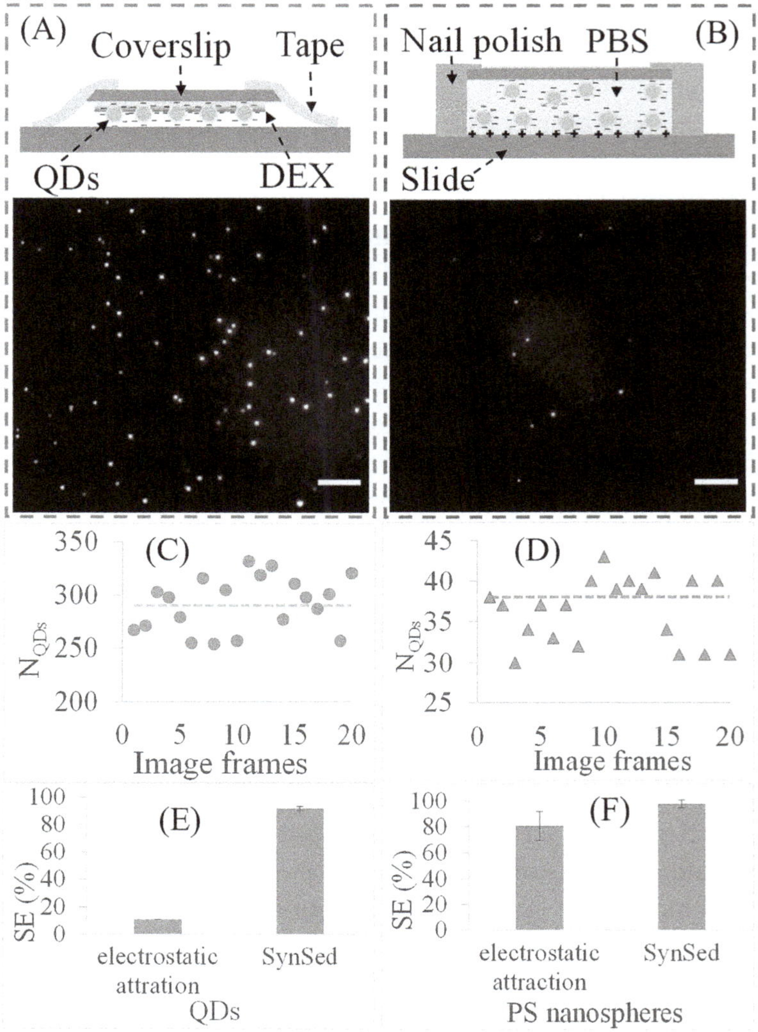

FIGURE 3.2 Deposition process and effect of SynSed and electrostatic attraction sedimentation: (A) using SynSed and (B) using electrostatic attraction sedimentation. (C) QD deposition count within 20-frame images using SynSed, (D) QD deposition count within 20-frame image using electrostatic attraction sedimentation, (E) QD sedimentation efficiency (SE) comparison between SynSed and electrostatic attraction sedimentation, and (F) PS SE comparison between SynSed and electrostatic attraction sedimentation.

Note: The scale bars represent 10 μm.

Source: Reproduced from Zhang, Q. Q.; Chai, W. W.; Pan, X. Y.; Gai, H. W. Amplification-free digital immunoassay down to the attomolar level by synergistic sedimentation of Brownian motion suppression and dehydration transfer, *Analytical Chemistry*, 2024, 96, 3054–3061 (ref 20). Copyright 2024 American Chemical Society.

The fabrication technologies for these two types differ significantly, and the properties like the size, number, and density of the microwells are all influenced by the fabrication methods. Achieving single-molecule or single-particle discretization using microwell arrays requires overcoming challenges related to fabrication techniques, discretization efficiency, and microwell sealing.

Techniques used for Microwell Array Fabrication

The chemical etching process for optical fiber bundle microwell arrays is described as follows:[21] The cross-section of a 5 cm-long optical fiber bundle is polished using diamond polishing films with grit sizes of 30 μm, 9 μm, and 1 μm in sequence. The polished fiber bundle is then chemically etched in a 0.025 M hydrochloric acid solution for a specific duration. The etching process is stopped by immersing the bundle in water. After etching, the fiber bundle is ultrasonicated in pure water, rinsed, and vacuum-dried. The inner glass of the optical fibers etches faster than the outer cladding, leading to the formation of a microwell array with approximately 4.5 μm diameter wells on the inner glass layer (Figure 3.3).

Soft lithography of PDMS microwell arrays is fabricated as follows: A clean silicon wafer is spin-coated with negative photoresist (SU-8 series, MicroChem). After soft baking to remove excess solvent, UV exposure is

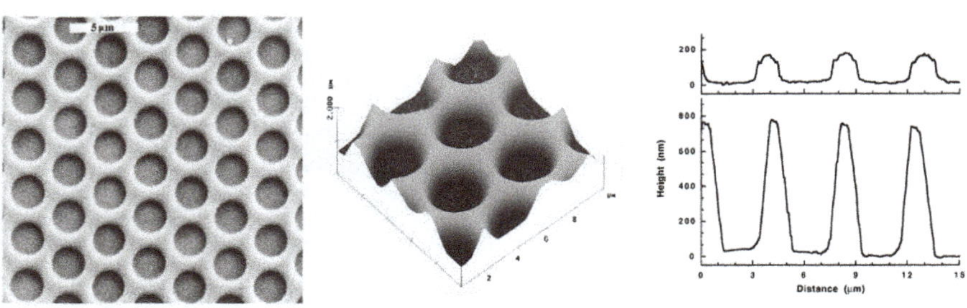

FIGURE 3.3 Optical fiber bundle microwell array.

Note: From left to right, shows: a scanning electron micrograph of the distal face of a polished, chemically etched imaging fiber with a 1,000 μm diameter, an atomic force micrograph of the distal face of a similarly polished and chemically etched imaging fiber (etched for 25 seconds), and representative line profiles of imaging fibers etched for 15 seconds (top) and 60 seconds (bottom).

Source: Reproduced from Pantano, P.; Walt, D.R. Ordered nanowell arrays, *Chemistry of Materials*, 1996, 8, 2832–2835 (ref 21). Copyright 2020 American Chemical Society.

applied through a chromium photomask featuring a circular dot array, with each dot having a diameter of a few micrometers. Post-exposure baking fixes the pattern, and the unexposed photoresist is removed during development, leaving an SU-8 mold with circular pillars. A pre-mixed PDMS precursor is then spin-coated onto the SU-8 mold and cured at 80°C for 0.5–1 hour. The top surface of the PDMS is bonded to a clean glass substrate after plasma treatment, and the PDMS layer is peeled off from the SU-8 mold to obtain the PDMS microwell array. The size of the microwells is determined by the chrome mask pattern, while their depth depends on the photoresist's viscosity, spin speed, and spin time. PDMS microwell arrays are easily combined and integrated with microfluidic chips containing various functional units, enabling fluid manipulation and the integration of immunocapture and signal discretization. Takeuchi et al. designed and fabricated a PDMS chip that integrated perfusion channels with a microwell array (Figure 3.4).[22] The circular wells were approximately 5 μm in diameter, 6 μm in depth, and 120 fL in volume. As shown in Figure 3.4A, β-Galactosidase (β-Gal) and a fluorescent substrate (FDG)

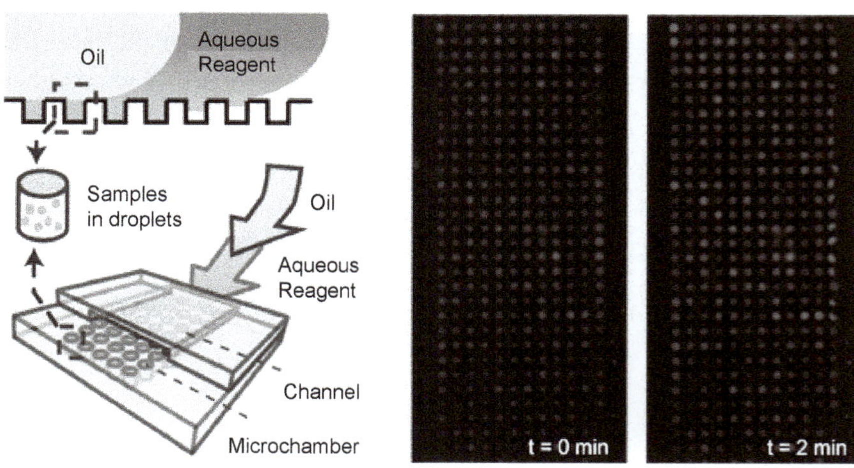

FIGURE 3.4 PDMS microwell arrays for detecting single enzyme activity: (A) assay procedure and (B) typical fluorescent images showing enzymatic activity in the chamber arrays.

Source: Reproduced from Ota, S.; Kitagawa, H.; Takeuchi, S. Generation of femtoliter reactor arrays within a microfluidic channel for biochemical analysis, *Analytical Chemistry*, 2012, 84, 6346–6350 (ref 22). Copyright 2012 American Chemical Society.

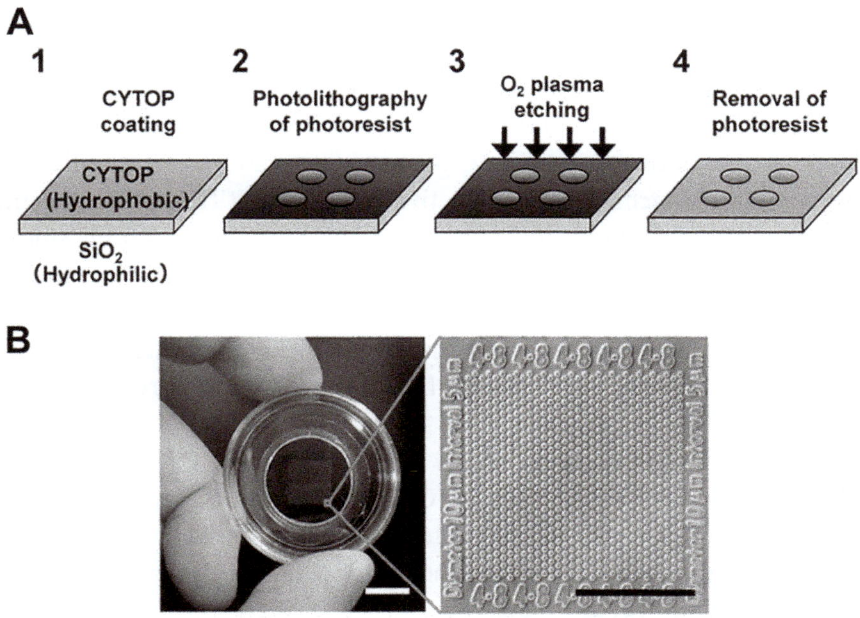

FIGURE 3.5 (A) Fabrication procedure for a hydrophilic-in-hydrophobic microwell array and (B) image of the assembled device and microscopic view of the hydrophilic-in-hydrophobic micropatterned surface.

Note: The scale bars represent 10.0 mm (left) and 0.2 mm (right).

Source: Reprinted with permission from Ryota Iino, Yoshimi Matsumoto, Kunihiko Nishino, et al. Design of a large-scale femtoliter droplet array for single-cell analysis of drug-tolerant and drug-resistant bacteria, *Frontiers in Microbiology*, 2013.

were mixed and introduced into the PDMS microwells via the perfusion channels. An oil phase was then used to seal the wells through the same channels. Inside the microwells, individual β-Gal molecules reacted with the substrate to generate fluorescein, which emitted green fluorescence (Figure 3.4B), showcasing hydrolytic enzyme activity differences at the single-molecule level.[22] However, due to the strong hydrophobicity of PDMS, aqueous solutions had difficulty infiltrating the wells, leading to air bubble formation. Additionally, the solution within the wells could easily be carried away by the oil phase.

The process of dry etching and demolding for creating microwell arrays with hydrophobic walls and hydrophilic bottoms is described as follows in Figure 3.5A:[23] Hydrophobic polymer of carbon-fluorine (CYTOP) is spin-coated onto a glass substrate and baked at 180°C for 1 hour. The spin-coating and baking cycles are repeated multiple times until the CYTOP layer reaches a thickness comparable to the diameter of the discrete microspheres. Positive photoresist (AZP4903) is then spin-coated onto the CYTOP layer, and the mask pattern is transferred to the CYTOP layer via photolithography. The exposed CYTOP layer is removed using oxygen plasma etching with a reactive ion etching (RIE-10NR) system, revealing the hydrophilic silica substrate beneath. The remaining unexposed photoresist is cleaned off with acetone and ethanol, forming a hydrophobic-through-hole array on the hydrophilic glass surface. As shown in Figure 3.5B, the walls of the circular microwells are hydrophobic, while the bottoms are hydrophilic, improving solution retention and minimizing bubble formation.

The process of printing technique for microwell arrays with hydrophobic walls and hydrophilic bottoms is described as follows:[23] A silicon wafer was coated with an adhesion promoter (HMDS), followed by spin-coating with a ~1.2 μm layer of positive photoresist (AZ 6632). After baking at 120°C for 1 minute, the wafer was masked with a chrome mask and exposed to UV light. The exposed regions were developed and removed, followed by a post-bake at 100°C for 1 hour. The exposed silicon areas were then etched into a microwell array using deep reactive ion etching (DRIE). The remaining photoresist was cleaned off using ultrasonic acetone cleaning, followed by rinsing with acetone, isopropanol, and water, resulting in a silicon mold with a microwell array. The mold was cleaned using a sulfuric acid and hydrogen peroxide mixture (3:1), then spin-coated with perfluorosilane to prevent PDMS adhesion. Next, the PDMS prepolymer (monomer and crosslinker in a 5:1 ratio) was mixed, poured onto the silicon mold, and vacuum degassed to ensure the PDMS filled the microwells. The mold was then heated at 60°C for six hours to cure the PDMS. After curing, the PDMS stamp with micropillar arrays was peeled off. Clean glass slides were soaked in a methacryloxy silane solution for 10 minutes, baked at 120°C, and then spin-coated with a 3 μm thick hydrophobic polymer film (OSTE+). The PDMS stamp was pressed into the polymer film. After exposing the assembly to UV light, the PDMS was removed, leaving behind patterned glass slides. The final fL-scale polymer microwell arrays were formed after allowing the patterned slides to react overnight at room temperature.

The procedure of injection molding of thermoplastic microwell arrays is as follows:[25] Photoresist is spin-coated onto a silicon wafer, and laser micromachining is used to create a positive mold with a raised circular dot array. A metal layer is then electroplated onto the mold, and the mold is removed to leave behind a negative mold with a microwell array. Next, a stainless-steel block is machined to create a negative mold for larger fluid channels. A cyclic olefin polymer (COP) is heat-pressed into both negative molds, and then demolded to produce a microwell array plate and a fluid channel plate. The two plates are aligned and bonded together using laser welding, forming a complete circular plastic chip. Duffy et al. used injection molding to fabricate a disk-shaped COP chip featuring 24 radially distributed units. Each unit contained a reservoir, an array of 216,000 microwells, a 500 μm-wide fluid channel, and a vent. The microwells were 4.25 μm in diameter and 3.25 μm deep, with a volume of approximately 40 fL, and spaced 8 μm apart center-to-center. The heat injection molding technique for creating thermoplastic microwells addresses the production capacity and low yield rates associated with chemical etching and photolithography, enabling large-scale, high-throughput manufacturing. This technique has been successfully commercialized and is now used in proprietary counting chips for DI instruments developed by Quanterix.

Discretization Efficiency of Microwell

The process of individual molecules or magnetic beads entering the microwell array is random, meaning many molecules or beads fail to enter the wells. This results in low discretization efficiency and significant digital information loss. Repeated operations, such as washing and separation, make the problem worse, ultimately leading to less than 5% of the target molecules being detected. When the target molecule concentration is extremely low (aM-fM range) and the sample volume is small (10–100 μL), the absolute number of target protein molecules is very low (typically in the range of tens to hundreds). The low discretization efficiency results in very few positive signals being detected, leading to significant Poisson noise ($\sqrt{N} / N$, where N is the number of targets). This noise can severely impact the sensitivity and accuracy of the measurement.

To improve the discretization efficiency of microwell, external magnetic fields, circulating flows, or a combination of both are commonly used to promote the entry of single molecules or magnetic beads into the wells. Lammertyn et al. demonstrated a technique where they pipetted a small volume of magnetic bead suspension from the pipette tip, which formed large droplets on a hydrophobic-hydrophilic microwell array.[24] The pipette tip moved the droplets, and under the influence of an external magnetic field, the magnetic beads were guided into the microwells.

Thanks to the hydrophobic-hydrophilic properties of the microwells, a small volume of solution was retained within the wells, creating femtoliter-sized droplet arrays. After 8 to 12 cycles of loading and dragging, the efficiency of magnetic bead entry into the microwells reached about 96%. However, this method requires microwell arrays with hydrophilic walls and hydrophobic bottoms, which limit the yield. Additionally, the manual manipulation of the liquid across the microwell surface introduces some variability.

Duffy et al. combined magnetic fields with curved meniscus sweeping to enhance the efficiency of magnetic bead entry into microwells—entry rate.[26] In a disk-shaped microwell array chip developed by Quanterix, a magnet was placed at the bottom of the chip, and magnetic bead suspension was introduced into the disk, forming a long liquid column. This column was then pushed forward by air pressure. During this process, the magnetic force pulled the magnetic beads downward, while the circulating flow within the liquid column caused the beads to accumulate at the rear curved meniscus. At the same time, the interface tension of the rear meniscus pushed the beads forward. The combined effect of these three forces significantly increased the efficiency of bead entry into the microwells. For instance, with 120,000 magnetic beads, the entry rate was about 5% under the influence of a single magnetic field. When only the curved meniscus sweeping technique was applied, the entry rate increased to about 15%. However, when both forces acted together, the bead entry rate reached approximately 61%.

In addition to the entry rate, increasing the number of microwells in the array also improves discretization efficiency. In digital single-molecule, enzyme-linked immunoassays, the initial fiber bundle arrays contained around 50,000 microwells, with approximately 20,000–30,000 magnetic beads out of 200,000 being discretized into the microwells. The number of microwells was increased to 216,000 in the disk-shaped COP microwell array chip developed by Quanterix, with about 97,000 out of 500,000 magnetic beads being loaded into the microwells. While the discretization efficiency remained unchanged, the number of beads available for detection was significantly increased. Hiroyuki et al. dry-etched a hydrophilic glass substrate to fabricate ultrahigh-throughput microwell arrays.[27] As shown in Figure 3.6, 8,648 CYTOP microwells form a block, and 120 blocks together create an array of approximately 1 million microwells. After introducing 3×10^6 beads onto the chip and allowing them to settle for 10 minutes, 7×10^5 beads were discretized into the microwells for subsequent

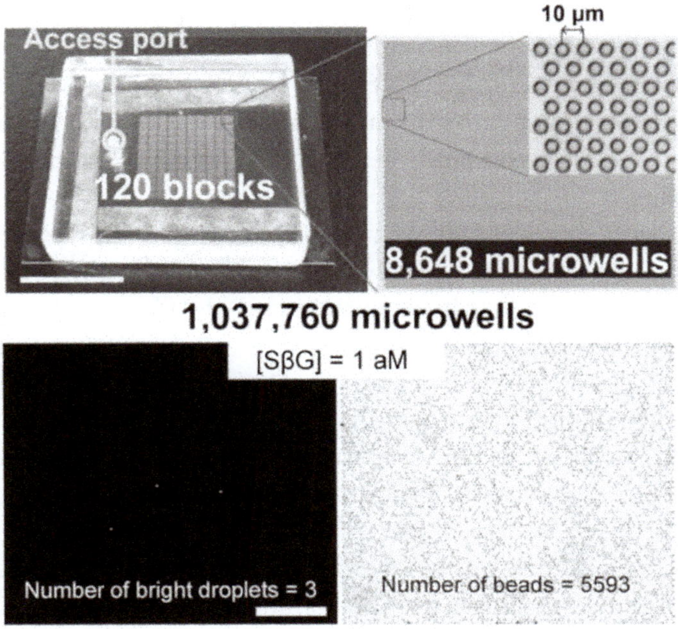

FIGURE 3.6 (A) Ultrahigh-throughput microwell arrays and (B) fluorescence image of the microwells at a β-Gal concentration of 1 aM.

Source: Reproduced from Kim, S.H.; Iwai, S.; Araki, S. et al. *Lab on a Chip* 2012, 12, 4986–4991 (ref 27), with permission of The Royal Society of Chemistry.

detection. Compared to the 50,000 microwells in the fiber bundle array, the million-microwell array greatly increased the number of partitioned single molecules, boosting imaging detection sensitivity by about 20 times. The LOD of streptavidin-labeled β-Gal reached 10 zM.

Microwell Sealing

Sealing the microwells is a crucial step for generating digital signals after single-molecule discretization. The single enzyme molecules carried by the immunocomplexes react with substrates inside the microwells, producing fluorescent or luminescent products. The sealing process confines the fluorescent material within the microwells, significantly increasing its local concentration, which makes single-molecule detection much easier. As a result, conventional fluorescence microscopes and imaging devices can directly detect single molecules. For example, a single enzyme molecule reacts with the substrate in a volume smaller than 50 fL, producing an average of 28,000 fluorophores within 2–3 minutes. The local concentration of fluorescent material can surprisingly reach about 1 μM. Additionally, sealing prevents cross-contamination between wells caused by diffusion and minimizes water evaporation from the wells. Current methods for sealing microwells mainly include PDMS sealing gaskets for mechanical closure, oil-phase sealing, and pressure-sensitive adhesive tape sealing.

In the early stages of digital enzyme-linked immunoassays, fiber bundle microwell arrays were typically sealed mechanically using PDMS sealing gaskets. After the beads were partitioned, the fiber bundle microwell array was fixed onto the microscope stage using custom clamps. The PDMS gasket was placed on a precision mechanical platform, aligning it with the fiber bundle array. A small drop of substrate solution was placed on the PDMS gasket, and the drop was then brought into contact with the microwell array. The mechanical platform was carefully adjusted to ensure that the PDMS gasket made tight contact with the fiber bundle microwell array, forming a separated microreactor space.[28] However, this PDMS gasket sealing process requires manual alignment and platform adjustments, leading to poor repeatability and limited automation, which hinders its industrial application.

Oil-phase sealing is the most commonly used method for sealing microwells. As shown in Figure 3.7, single molecules are partitioned into the microwell array that is integrated with a microfluidic chip. External forces (such as pressure or centrifugal force) are applied to drive the oil phase, which is immiscible with water, through the microwells. Since the oil phase has a wetting effect on the microwell walls, it expels the aqueous

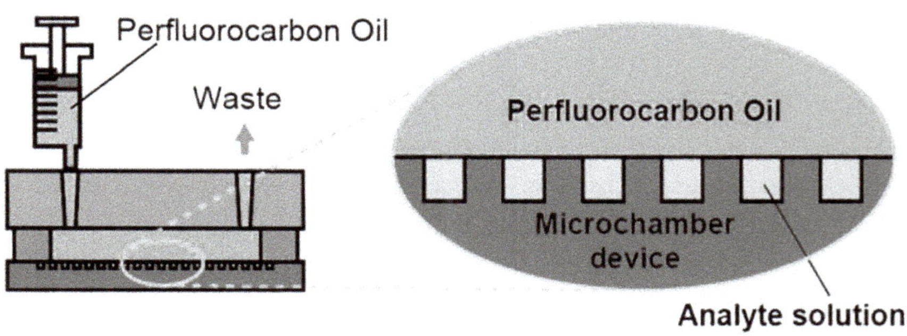

FIGURE 3.7 Scheme of conventional oil sealing.

Source: Reproduced from Yaginuma, H.; Ohtake, K.; Akamatsu, T.; et al. *Lab on a Chip*, 2022, 22, 2001–2010 (ref 29), with permission of The Royal Society of Chemistry.

solution from the wells, creating separate microreactor spaces.[29] However, oil-phase sealing requires attention to three key issues: (1) the extraction of fluorescent products by the oil phase, (2) the gradual replacement of a small amount of aqueous liquid in the microwells by the oil phase over time, and (3) the optical transparency and biocompatibility of the oil phase. Considering these factors, fluorocarbon oils—chemically inert and biocompatible—have become the preferred choice for sealing. By optimizing the viscosity of the oil, the aqueous solution inside the microwells can be stably maintained for several hours while also maintaining good flow properties. Compared to mechanical sealing, oil-phase sealing takes advantage of the fluid control properties of microfluidic chips. It is a simple, reliable method that does not require complex manual operations and can be easily automated. Oil-phase sealing has been successfully used in the commercialization of digital microwell array discretization.

In 2022, Tabata et al. developed sealing microwells using a pressure-sensitive adhesive tape-based sealing method (PITAT) in addition to traditional mechanical and oil-phase sealing techniques.[29] This adhesive tape consists of two layers: an adhesive layer, typically made from a blend of natural or synthetic rubbers, acrylics, siloxane polymers, urethane polymers, and other additives like plasticizers, crosslinkers, and antioxidants, and a support layer. The sealing process involves three main steps, as shown in Figure 3.8 (upper panel): step 1) A small droplet of the reaction mixture is placed onto the microwell array using a pipette. Step 2) The droplet deforms, and some of the liquid is driven into the microwell by the gentle contact with the adhesive tape. Step 3) A roller is then used to press the adhesive tape onto the microwell surface, rolling it back and forth several times to expel any excess liquid, thus sealing the microwells. To verify the effectiveness of the seal, the researchers sealed Cy3 dye within the microwells (Figure 3.8 down panel). After partially bleaching the Cy3 in some microwells, they found that the fluorescence in these bleached wells did not recover even after 25 minutes (Figure 3.8 down panel), indicating that there was no cross-diffusion between the wells and confirming the seal was intact. Since the exact composition of commercial pressure-sensitive adhesive tapes is not publicly disclosed, the researchers tested various materials, including rubber, acrylic, siloxane, and urethane-based tapes. They found that rubber and siloxane-based tapes were effective for sealing microwells. In practical applications, such as the DI of influenza viruses, the sealing method was shown to be comparable to traditional oil-phase sealing after optimizing the composition and sealing conditions.

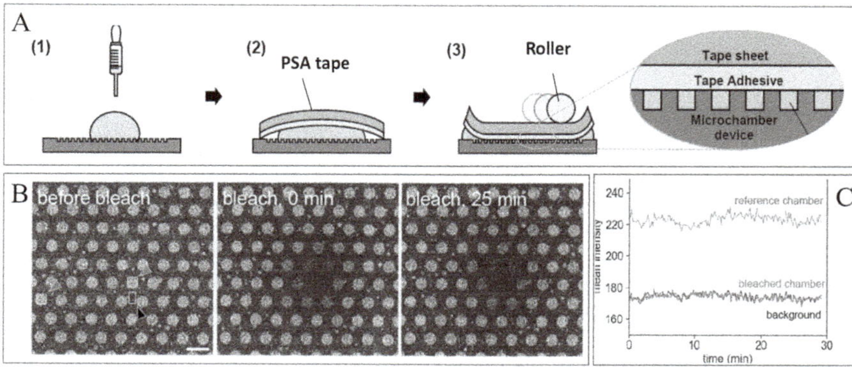

FIGURE 3.8 Microreactor sealing using PITAT and leak testing (upper panel, left to right), and confocal microscopy images of sealed Cy3 dye solution with time-lapse fluorescence intensity changes after photobleaching (lower panel, left to right).

Note: Arrowheads and white rectangles indicate the regions measured during photobleaching. The scale bar represents 10 μm.

Source: Reproduced from Yaginuma, H.; Ohtake, K.; Akamatsu, T.; et al. *Lab on a Chip*, 2022, 22, 2001–2010 (ref 29), with permission of The Royal Society of Chemistry.

The use of pressure-sensitive adhesive tape for microwell sealing offers a low-cost and highly reliable solution for single-molecule detection, making it particularly suitable for use in remote areas or even at home.

Droplet Discretization

Microdroplets produced using microfluidic chip technology have several key features, including small volumes (pL–fL), uniform sizes (with a coefficient of variation under 3%), high throughput (millions of droplets), a stable internal environment, and high mass and heat transfer efficiency. By encapsulating a single immunocomplex within a microdroplet, it becomes possible to generate a detectable optical signal, allowing for the physical isolation of digital signals. This makes it a promising platform for DIs, aligning microfluidic droplet technology with the need for digital signal dispersion and offering strong potential for commercial applications.

Currently, the main techniques for generating microdroplets include microfluidic and electrowetting technologies. Electrowetting technology is known for its precise control of individual droplets but is less suited for high-throughput or ultrahigh-throughput droplet preparation. This chapter will focus on microfluidic chip droplet technology. Based on the used droplet techniques, DI are divided into two categories: continuous moving droplets formed through oil-water phase shear, and stationary droplet arrays.

Continuous Moving Droplet Preparation

Based on the mechanisms of droplet formation, the technology for generating continuous moving droplets within chip channels is categorized into T-channel methods, flow-focusing methods, co-axial focusing methods, and step emulsification methods. In a chip with T-shaped channels, the oil and water phases flow through horizontal and vertical channels, respectively, and meet at the T-junction, where they create an oil/water interface. Under external forces, the aqueous phase continuously flows into the T-structure, and due to the shear force from the oil phase, it moves forward along with the oil. When the tension at the oil-water interface is not strong enough to maintain the movement of the aqueous phase as a whole, the aqueous solution breaks up, forming independent droplets surrounded by the oil phase, as shown in Figure 3.9. At relatively low flow rates, droplets are easily formed. However, the droplet formation rate is slow, and it is challenging to generate droplets with diameters smaller than 10 micrometers.[30]

In a chip with "flow-focusing" channels, the aqueous phase flows through the middle channel, while the oil phase flows through two side channels, as shown in Figure 3.10. A narrow gap is designed where the two

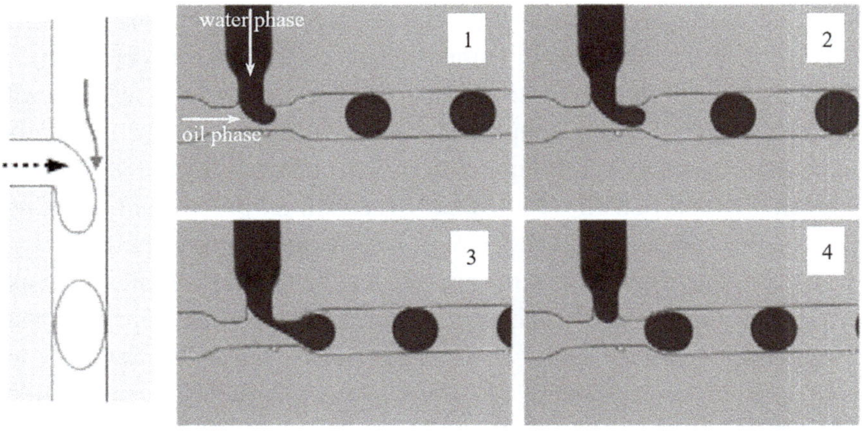

FIGURE 3.9 Droplet formation process using T-shaped channels.

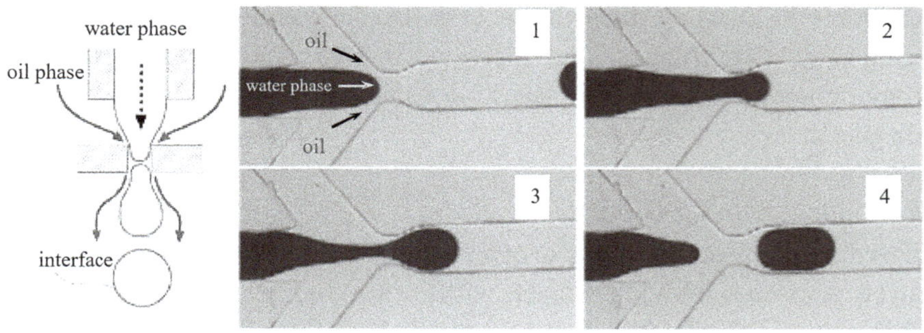

FIGURE 3.10 Droplet formation process using a flow-focusing microfluidic chip.

phases meet, creating an oil-phase "clamp" around the aqueous phase, known as the "focusing" effect. Under the shear force from the oil phase on both sides, the aqueous phase forms a thin stream and eventually breaks into independent droplets. Compared to the T-shaped structure, the flow-focusing structure requires higher flow rates to generate droplets, which leads to faster droplet formation. This method is particularly suitable for systems with higher phase viscosities, and it allows for better control over droplet size across a wider range.[30]

Although droplets formed using the flow-focusing method are smaller than those produced by the T-channel method, it remains challenging to create droplets with diameters smaller than 10 microns using conventional chips (with channel sizes as small as 20 μm). This makes it difficult to generate droplets with diameters as small as 1–2 μm, let alone submicron droplets. Larger droplet sizes or volumes can directly affect the concentration of target molecules inside the droplet, potentially leading to lower target concentrations and higher background signals—both of which are undesirable for single-molecule discretization and detection. To reduce droplet volume, we used flow-focusing channels to form parent droplets containing two components: one volatile (component 1) and one nonvolatile (component 2; Figure 3.11). The parent droplets are encapsulated in fluorocarbon oil (FC40). Due to the high gas permeability of FC40, the volatile component in the parent droplet diffuses into the FC40 and evaporates into the surrounding air, causing the droplet volume to shrink. Once the volatile component has completely evaporated, single-component droplets are formed.[31] Using ethanol and water droplets as examples, we conducted an in-depth study of solvent evaporation rates, size reduction, and other related factors. Under optimal conditions, using a PDMS chip with a minimum channel width of 30 μm, the smallest aqueous droplets

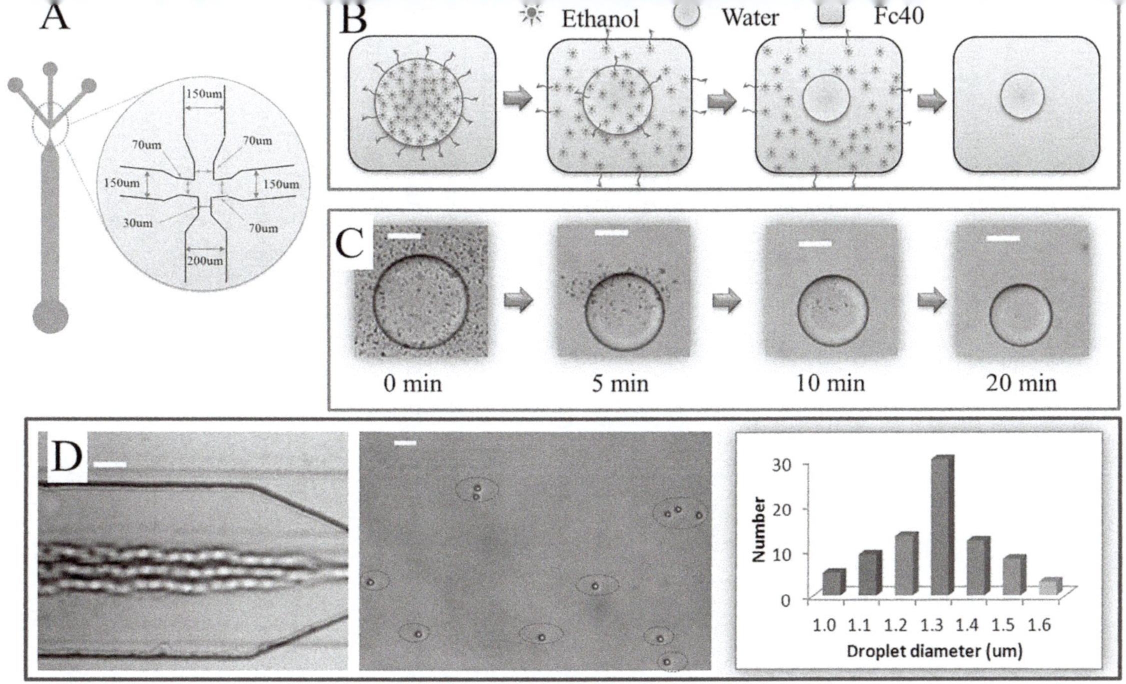

FIGURE 3.11 Generation of ultrasmall droplets via solvent evaporation: (A) schematic of the microfluidic chip, (B) principle of binary droplet size reduction through component evaporation, (C) time-lapse images of ethanol evaporation within binary droplets, and (D) ultrasmall droplet formation achieved by adjusting flow rate and volume ratios, including images of ethanol–water binary droplet generation (left), final water droplets (middle), and their size distribution (right).

Note: The scale bars represent 40 μm in both (C) and the left panel of (D), and 4 μm in the middle panel of (D).

Source: Reproduced from Zhang, Q.Q.; Liu, X.J.; Liu, D.Y. et al. *Lab on a Chip*, 2014, 14, 1395-1400 (ref 31), with permission of The Royal Society of Chemistry.

could be controlled to approximately 1.3 μm, while the smallest organic phase droplets could reach 1.6 μm.

Static Droplet Array Preparation

There are two main techniques for creating static droplet arrays: hydrophilic-hydrophobic patterning and droplet printing. The basic principle of the hydrophilic-hydrophobic patterning technique is as follows: a hydrophilic glass substrate is spin-coated with a layer of positive photoresist. After ultraviolet photolithography and development, a micropatterned photoresist array is formed on the substrate. The entire surface of the substrate is then treated with perfluorosilane to make it hydrophobic. The photoresist pattern is removed using a release agent, leaving behind a hydrophilic micro-pattern array surrounded by the hydrophobic surface. When a solution flows over the hydrophilic pattern, droplets form in the hydrophilic areas due to changes in surface tension and local wetting properties. The droplet volume is determined by the size of the hydrophilic pattern and the contact angle difference between the hydrophobic and hydrophilic surfaces. Zheng et al. developed a nonphotolithographic hydrophilic-hydrophobic patterning.[32] First, a layer of CYTOP thin film was spin-coated onto a clean glass substrate. Polydopamine was then allowed to react with the perfluoropolymer, forming a polydopamine layer on the substrate. A pre-fabricated PDMS stamp containing a 2 μm circular micropore array was treated with oxygen plasma to make its surface hydrophilic. The PDMS stamp was then brought into contact with the polydopamine layer for about 6 minutes. After removing the PDMS stamp, the polydopamine in the contact region was removed, leaving behind a polydopamine dot array on the perfluoropolymer surface. The perfluoropolymer layer was highly hydrophobic, while the polydopamine dots were highly hydrophilic. Next, a PDMS layer with microchannels was bonded to the substrate with the polydopamine dot array. Reagents were injected into the microchannels, and air was used to quickly expel the reagents from the main channel. The hydrophilic nature of the polydopamine dots retained some of the liquid, forming fL-scale microdroplet arrays. This method can rapidly create fL-scale droplets with high throughput, but the monodispersity of the droplets is slightly worse. Moreover, since the droplets are directly exposed to the air, the effect of solvent evaporation cannot be ignored.

Levkin et al. developed a method for preparing static microdroplet arrays with controllable shapes and sizes.[33] The process began by fabricating a superhydrophilic nanoporous polymer film (a copolymer of hydroxyethyl methacrylate and ethylene glycol dimethacrylate) on the surface of a glass substrate, as shown in Figure 3.12. This superhydrophilic film was then immersed in 2′,2′,3′,3′,3′-pentafluoropropyl methacrylate (a superhydrophobic monomer), and local UV polymerization was performed using a quartz-chromium mask. The superhydrophilic polymer film in the irradiated areas grafted the superhydrophobic monomer, transforming those regions into superhydrophobic areas. This technique effectively segmented the superhydrophilic porous film, creating an array of superhydrophilic structures with adjustable shapes and sizes. When a solution was applied to the superhydrophilic structural array, some of the liquid was retained on the porous hydrophilic microstructures, forming a static droplet array. They were able to prepare 85,000 droplets, with sizes that could be flexibly adjusted between 700 pL and 3 μL. These droplets were applied in cell-related fields. However, the droplet size was limited by photolithography, resulting in relatively larger droplets and fewer droplets overall, which presents challenges for DIs.

Recently, Doyle et al. established an embedded droplet printer.[34] A high-yield stress fluid was added to a large-volume container, and the

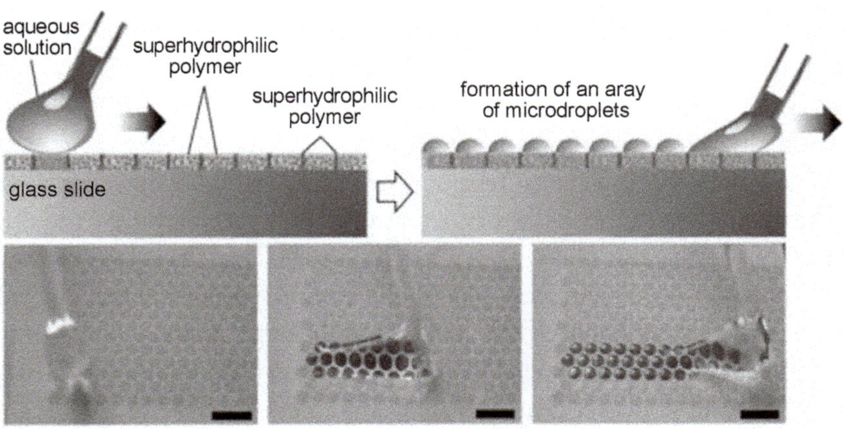

FIGURE 3.12 Formation of a droplet microarray using the rolling droplet method.

Note: The scale bars are 3 mm.

Source: Reproduced from Ueda, E.; Geyer, F.L.; Nedashkivska, V. et al. *Lab on a Chip*, 2012, 12, 5218–5224 (ref 33), with permission of The Royal Society of Chemistry.

printing nozzle was inserted into the fluid. An immiscible fluid flowed out as a line of spherical droplets at the nozzle tip. As the nozzle moved, the immiscible fluid broke into droplets. Yield stress fluids are materials that exhibit complex fluid behaviors. Depending on the pressure, these fluids can behave like either solids or liquids. By exploiting the solid-like behavior of the yield stress fluids, the embedded droplet printer could print droplets layer by layer in three-dimensional space, without the need to consider solid boundaries or convection effects. This feature allowed droplets to be manipulated freely and arranged as needed in a completely stationary state, which prevented droplet motion, fusion, and cross-contamination, and eliminated the need for surfactants. However, the downside is that the printed droplet volume is relatively large (~2 μL), which does not meet the requirements for DIs and necessitates further reduction in droplet size.

Application Examples of Microdroplet Discretization

Microdroplet technology is used to encapsulate individual target proteins or their complexes, which generate optically detectable signals within each droplet. By using static imaging or flow-based detection to count the droplets containing the target, this approach overcomes the limitations of microarray quantity, enabling more accurate measurements with a wider dynamic response range. The key feature of microdroplet technology for digital signal discretization is its high throughput, while the main challenge lies in controlling droplet volume and generation rate. The high-speed formation of fL-scale droplets requires both high fluid shear force and low oil-water interfacial tension. Increasing the fluid shear force can help reduce the channel size or increase the flow rate; however, the flow rate can only be adjusted within a limited range. Decreasing the channel size increases fluid resistance, potentially compromising stability. To balance the trade-off between channel size and fluid resistance, Shim et al. developed an ultrahigh-speed droplet generation chip with a specialized flow-focusing structure (Figure 3.13).[35] The inlet channels for the oil and water phases were sized at 100 × 25 μm (width×depth), while the flow-focusing region was reduced to 10 × 5 μm. This simultaneous reduction in both channel width and depth significantly increased the local fluid flow rate and shear force, which helped reduce droplet size. Additionally, shallow and narrow channels were used only in the flow-focusing region to avoid a significant increase in flow resistance. The use of low-viscosity fluorocarbon oil and fluorosurfactants further reduced the oil-water

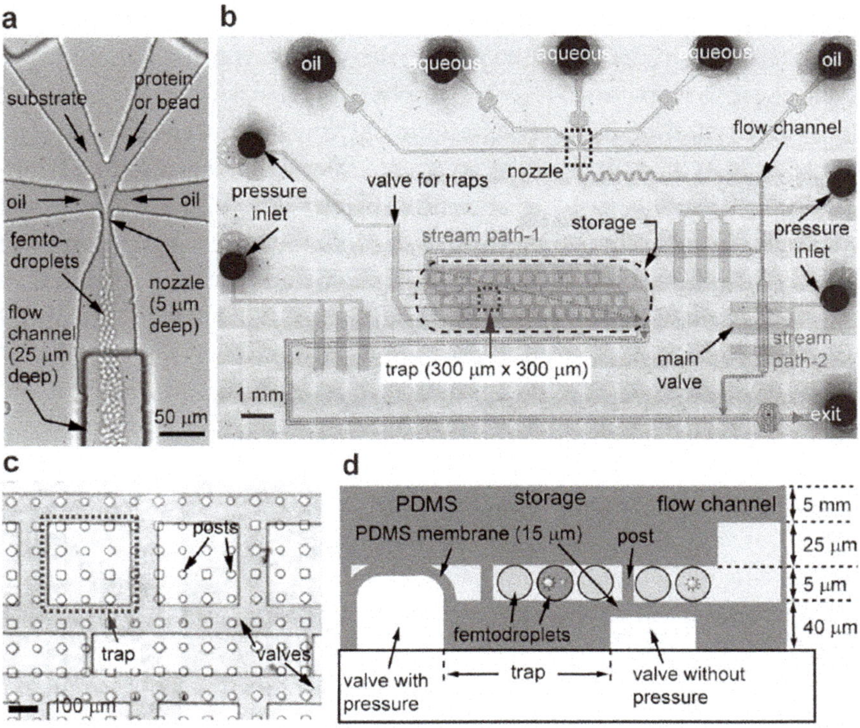

FIGURE 3.13 Microfluidic device for femtodroplet generation and manipulation: (A) formation of femtodroplets at the nozzle of the microfluidic device, (B) photograph of the complete multilayer PDMS device, (C) image of the trap structures used for femtodroplet storage and isolation, and (D) cross-sectional schematic of the storage structure.

Source: Reproduced from Shim, J.U.; Ranasinghe, R.T.; Smith, C.A. et al. Ultrarapid generation of femtoliter microfluidic droplets for single molecule-counting immunoassays, *ACS Nano*, 2013, 7, 5955–5964 (ref 35). Copyright 20213 American Chemical Society.

interfacial tension. This unique design and the specific synergy between the continuous phases enabled ultrahigh-speed generation of fL-scale droplets. The smallest droplet volume reached 1.1 fL, with the highest formation speed reaching 8.9 MHz. Using tumor markers (PSA) as the target, capture antibodies were modified onto 1 μm polystyrene beads to form capture spheres. After incubating with the antigen, biotinylated detection antibody, and enzyme-labeled streptavidin, enzyme-labeled immunocomplexes were formed on the capture beads. A mixture of a microbead suspension (containing both capture beads and enzyme-labeled capture beads) and a fluorescent substrate (fluorescein di-β-D-galactopyranoside, FDG) was used as the dispersed phase, with fluorocarbon oil as the continuous phase to form monodisperse droplets. The droplets were categorized into three types: (i) droplets containing only the fluorescent substrate (no fluorescence), (ii) droplets containing the fluorescent substrate and capture beads (red fluorescence), and (iii) droplets containing the fluorescent substrate and immunocomplexes (green fluorescence). The number of type (iii) droplets and their percentage relative to the total beads showed a positive correlation with PSA concentration. The LOD for PSA in this DI was 46 fM.

Tay et al. used microdroplets as discrete units and combined nucleic acid amplification technology (PCR) with proximity ligation assay (PLA) to develop a highly sensitive digital proximity ligation assay (digital PLA) for single-cell protein digital immunoassay.[36] In PLA, two oligonucleotide sequences were used to separately label the capture and detection antibodies, creating two PLA probes. When the target protein was present, the two PLA probes bound to the same target protein, bringing the probes close together and generating a proximity effect. Catalyzed by a ligase, the DNAs from the two PLA probes formed complementary dsDNA fragments, joined by connector DNA (a short oligonucleotide whose ends were complementary to the DNA of each PLA probe). These new dsDNA fragments were then amplified by fluorescence PCR, and the quantity of the DNA fragment was used to determine the target protein concentration. PLA converts protein quantification into nucleic acid quantification, leveraging the signal amplification effect of nucleic acid amplification, which provides high detection sensitivity. The PLA strategy was used for immunorecognition and capture in single mammalian cell lysates. Probe 1, the antigen, and Probe 2 formed an immunocomplex, and the connector DNA hybridized with the adjacent probe DNAs. Once again, catalyzed by a ligase, dsDNA was formed, and protein detection was shifted to detecting the formed double-stranded DNA. When a commercial microfluidic chip was used to generate monodisperse droplets, the distribution of the dsDNA within the droplets followed the Poisson distribution. Fluorescence PCR was performed within the droplets, and the number of fluorescent droplets was imaged and used to determine protein concentration.

The performance of PLA for quantifying three protein standards (CD147, GFP, and ICAM-1) is shown in Table 3.1. The linear range for CD147 span three orders of magnitude, with an LOD as low as 0.14 fM (the value corresponds to a pre-dilution sample concentration of 16.2 fM).

TABLE 3.1 Detection Parameters of Three Proteins Using Digital Proximity Ligation Assay

Protein	Assay LOD (fM)	Sensitivity (Proteins per DNA)	Sample LOD (fM)
CD147	0.14	70	16.2
GFP	0.18	95	20.1
ICAM-1	0.34	565	38.3

The digital PLA method was also used to quantify two proteins (CD147 and GFP-p65) in a single human embryonic kidney (HEK293T) cells. The LOD for CD147 was found to be 29,300 molecules. Out of 186 cells sorted, 138 cells showed CD147 expression levels above the LOD. The lowest CD147 count in a single cell was 35,900 ± 2,900, while the highest reached 636,000 ± 36,000, showing an 18-fold variation in expression levels and highlighting the heterogeneity of protein expression at the single-cell level. The average CD147 level in single HEK cells was 218,000, which was consistent with the estimated value of 258,978 in single mouse fibroblasts. As the number of single cells counted increased from 1 to 100, the variation in CD147 levels narrowed. For 100 cells, the average CD147 level ranged from 67,200 ± 1,000 to 229,000 ± 3,000. The trend for GFP-p65 followed a similar pattern to CD147, with slight differences in the variation.

The process of encapsulating single molecules, cells, or particles in droplets follows a Poisson distribution. Typically, the limitation of encapsulation efficiency is constrained by the randomness inherent in the Poisson distribution. In DIs, the number of microbeads used is usually in the hundreds of thousands. A 10% encapsulation efficiency for single molecules or particles means that the actual sample size to be detected is small, leading to significant information loss. This limits the ability to accurately measure rare proteins in small-volume samples. Two potential strategies to improve sample size and increase the positive detection rate are (1) overcoming the Poisson distribution limit to enhance the efficiency of encapsulating single molecules in droplets and (2) reducing the number of capture microbeads while maintaining immunocapture efficiency.

To address the limitations posed by the Poisson distribution, Toner et al. proposed a method in 2008 to control the movement of particles or cells in microchannels, enabling them to self-sort using hydrodynamic control.[37] This approach ensured that particles or cells form evenly spaced sequences. If the frequency at which single particles or cells reached the droplet formation region matched the droplet generation frequency, the encapsulation efficiency of single particles or cells could be significantly improved. HL60 cells or 9.9 μm polystyrene microbeads, at high density, were forced to rapidly flow into a high aspect ratio square microchannel (27 μm wide, 52 μm high, and 6 cm long). Since the size of the cells or microbeads was large relative to the channel width, they were focused by the fluid flow and arranged into evenly spaced, unidirectional linear arrays or alternating bidirectional arrays along the flow direction. These cells or microbeads then entered the oil-water junction to form droplets encapsulating single cells or single beads. By synchronizing the entry frequency of the microbeads into the oil-water junction with the droplet formation frequency, this method overcame the random limitations of the Poisson distribution, significantly increasing the encapsulation efficiency of single cells. When $\lambda = 0.89$ (meaning, according to Poisson statistics, each droplet encapsulates 0.89 particles), the ratio of droplets encapsulating single particles exceeded 80%, with empty droplets and multi-particle droplets accounting for less than 20%. This method roughly doubled the encapsulation efficiency compared to the Poisson distribution.

In a similar approach, Xu et al. developed a chip called the "bead-ordered droplet (BOAD) system".[38] This system achieved the ordered alignment of microbeads through a combination of channel design, sheath flow, Dean flow, and compressive flow within a short channel, resulting in improved encapsulation efficiency for single microbeads. The BOAD chip was composed of a Y-shaped injection area, a curved pre-focus area, a particle alignment section, and a droplet formation zone. The microbead suspension and sheath fluid were introduced through their respective Y-shaped inlets and into a curved channel (50 μm wide, 35 μm high, with a curvature radius of 55 μm). The sheath flow pushed the microbeads toward the inner side of the curved channel, while the Dean vortices focused the microbeads along the Z-axis. As the microbeads reached an equilibrium position near the inner exit of the curved channel, they were funneled into a single-sided contracted channel (from 50 μm to 30 μm), which focused the microbead cluster along the X-axis and directed them into the microbead alignment channel (30 μm wide, 35 μm high). In this section, inertial uplift caused the microbeads to rapidly move toward the upper wall of the channel, forming an evenly spaced sequence. These microbeads then entered a cross-channel section containing narrow slits (20 μm wide, 35 μm high, 30 μm long), where they formed droplets. The clever integration of these forces achieved a maximum encapsulation efficiency of 86% for single microbeads. The BOAD system was applied to a cytokine digital immunoassay, achieving an LOD of 0.14 pg/mL.

In DIs, the number of capture microbeads plays a crucial role in determining the proportion of target beads within the total bead population. For example, in a protein sample with a concentration of 10 aM and a volume of 100 μL (which contains approximately 600 protein molecules), if the number of microbeads is set at 1 million, 500,000, or 100,000, the theoretical proportion of target microbeads would be 0.0006, 0.0012, and 0.0060, respectively. As the number of microbeads decreases, the proportion of target beads increases. When dealing with very low molecule concentrations, a higher proportion of microbeads sampled results in more target signals being detected, which enhances accuracy. For instance, with 600 protein molecules in the sample, assuming no loss during immunocapture and washing, 100% sampling would yield 600 positive signals, while only 10% sampling would produce just 60 positive signals. Therefore, increasing the proportion of microbeads sampled boosts detection sensitivity and improves measurement accuracy (Figure 3.14). However, due to the large number of microbeads and limitations in detection techniques and measurement time, fully detecting all microbeads is often impractical. In these cases, reducing the number of microbeads is an effective strategy to improve detection sensitivity when full detection is not possible. Walt et al., after considering the theoretical aspects of DIs, conducted experiments using 100,000 capture magnetic beads and employed fL-scale droplet encapsulation for immunocomplexes.[39] This approach significantly improved sampling efficiency and increased the statistical likelihood of detecting target proteins. Compared to microarray-based DIs, this droplet-based one achieved a 25-fold increase in detection sensitivity and was able to reach LODs in the fL range.

Other Discretization Methods

In addition to microwells and microdroplets, bubbles can also serve as discrete units for single-molecule immunoassay analysis. Recently, Qin et al. developed a plasmon resonance nanoparticle bubble-based DI that has proven successful for fast and highly sensitive detection of nanoparticles or viruses.[40] The principle behind plasmon resonance nanoparticle bubble generation works as follows: a gold nanoparticle suspension is injected into a capillary tube via a syringe pump and flows into the detection area. An ultrafast pulsed laser (28 ps, 532 nm) then irradiates the

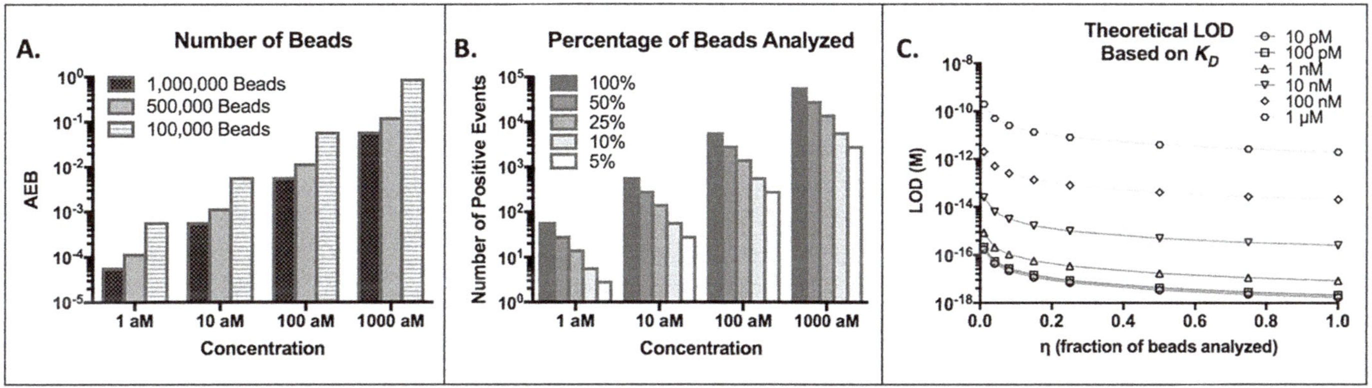

FIGURE 3.14 Theoretical analysis of assay performance: (A) calculated average enzymes per bead (AEB) across varying concentrations and bead numbers, (B) number of positive events at different concentrations and detection percentages, and (C) LOD as a function of dissociation constants.

Source: Reproduced from Cohen, L.; Cui, N.; Cai, Y.M. et al. Single molecule protein detection with attomolar sensitivity using droplet digital enzyme-linked immunosorbent assay, *ACS Nano*, 2020, 14, 9491–9501 (ref 39). Copyright 2020 American Chemical Society.

gold nanoparticles, generating nanoparticle bubbles through plasmon resonance. These nanoparticle bubbles isolate the gold nanoparticles from the surrounding solution, amplifying their absorption and causing the intensity of the simultaneously irradiated probe laser to decrease. Since plasmon resonance nanoparticle bubbles exist for only a very short time (on the order of nanoseconds), time-resolved imaging of a single gold nanoparticle produces a distinct pulse dip signal. The larger the gold nanoparticle, the larger the nanoparticle bubble and the stronger the pulse signal amplitude. Using respiratory syncytial virus (RSV) as a model, antibody-modified gold nanoparticles were used to form virus-gold particle assemblies. When introduced into the capillary tube, plasmon resonance nanoparticle bubbles were generated in the detection area, where they were detected by a sensor. The virus-gold particle assemblies, which contained multiple gold nanoparticles, produced nanoparticle bubbles significantly larger than those generated by a single gold nanoparticle. By measuring the pulse amplitude, the signal of the target virus could be identified, and interference from single gold nanoparticles was eliminated. This plasmon resonance nanoparticle bubble-based DI achieved an LOD of approximately 100 PFU/mL for RSV in nasal swabs, which corresponded to about one RNA copy per mL. This sensitivity is comparable to that of digital PCR for viral detection. This method is promising for the rapid screening and diagnosis of virus-caused diseases.

REFERENCES

1. Todd J, Freese B, Lu A, et al. Ultrasensitive flow-based immunoassays using single-molecule counting, *Clinical Chemistry*, 2007, 53, 1990–1995.
2. Agrawal A, Zhang CY, Byassee T, et al. Counting single native biomolecules and intact viruses with color-coded nanoparticles, *Analytical Chemistry*, 2006, 78, 1061–1070.
3. Akama K, Shirai K, Suzuki S. Droplet-free digital enzyme-linked immunosorbent assay based on a tyramide signal amplification system, *Analytical Chemistry*, 2016, 88, 7123–7129.
4. Han Y, Wu HY, Liu F, et al. Label-free biomarker assay in a microresistive pulse sensor via immunoaggregation, *Analytical Chemistry*, 2014, 86, 9717–9722.
5. Mok J, Mindrinos MN, Davis DW, Javanmard M. Digital microfluidic assay for protein detection, *Proceedings of the National Academy of Sciences of the United States of America*, 2014, 111, 2110–2115.
6. Arima A, Tsutsui M, Yoshida T, et al. Digital pathology platform for respiratory tract infection diagnosis via multiplex single-particle detections, *ACS Sensors*, 2020, 5, 3398–3403.

7. He LQ, Tessier DR, Briggs K, et al. Digital immunoassay for biomarker concentration quantification using solid-state nanopores, *Nature Communications*, 2021, 12, 5348.
8. Krainer G, Saar KL, Arter WE, et al. Direct digital sensing of protein biomarkers in solution, *Nature Communications*, 2023, 14, 653.
9. Löscher F, Böhme S, Martin J, et al. Counting of single protein molecules at interfaces and application of this technique in early-stage diagnosis, *Analytical Chemistry*, 1998, 70, 3202–3205.
10. Zhang Q, Li J, Pan X, et al. Low-numerical aperture microscope objective boosted by liquid-immersed dielectric microspheres for quantum dot-based digital immunoassays, *Analytical Chemistry*, 2021, 93, 12848–12853.
11. Farka Z, Mickert MJ, Hlavacek A, et al. Single molecule upconversion-linked immunosorbent assay with extended dynamic range for the sensitive detection of diagnostic biomarkers, *Analytical Chemistry*, 2017, 89, 11825–11830.
12. Li GH, Zhu L, Wu ZJ, et al. Digital concentration readout of DNA by absolute quantification of optically countable gold nanorods, *Analytical Chemistry*, 2016, 88, 10994–11000.
13. Zhu L, Li G, Sun S, et al. Digital immunoassay of a prostate-specific antigen using gold nanorods and magnetic nanoparticles, *RSC Advances*, 2017, 7, 27595–27602.
14. Liu XJ, Huang CH, Zong CH, et al. A single-molecule homogeneous immunoassay by counting spatially "Overlapping" two-color quantum dots with wide-field fluorescence microscopy, *ACS Sensors*, 2018, 3, 2644–2650.
15. Liu XJ, Zhang Y, Liang A, et al. Plasmonic resonance energy transfer from a Au nanosphere to quantum dots at a single particle level and its homogenous immunoassay, *Chemical Communications*, 2019, 55, 11442–11445.
16. Liu XJ, Huang CH, Dong XL, et al. Asynchrony of spectral blue-shifts of quantum dot based digital homogeneous immunoassay, *Chemical Communications*, 2018, 54, 13103–13106.
17. Liu XJ, Lin XY, Pan X, et al. Multiplexed homogeneous immunoassay based on counting single immunocomplexes together with dark-field and fluorescence microscopy, *Analytical Chemistry*, 2022, 94, 5830–5837.
18. Zhang QQ, Zhang XB, Li JJ, et al. Non-stochastic protein counting analysis for precision biomarker detection: suppressing Poisson noise at ultralow concentration, *Analytical Chemistry*, 2020, 92, 654–658.
19. Hennequin Y, Allier CP, McLeod E, et al. Optical detection and sizing of single nanoparticles using continuous wetting films, *ACS Nano*, 2013, 7, 7601–7609.
20. Zhang QQ, Chai WW, Pan XY, Gai HW, Amplification-free digital immunoassay down to attomolar level by synergistic sedimentation of Brownian motion suppression and dehydration transfer, *Analytical Chemistry*, 2024, 96, 3054–3061.
21. Pantano P, Walt DR. Ordered nanowell arrays, *Chemistry of Materials*, 1996, 8, 2832–2835.

22. Ota S, Kitagawa H, Takeuchi S. Generation of femtoliter reactor arrays within a microfluidic channel for biochemical analysis, *Analytical Chemistry*, 2012, 84, 6346–6350.
23. Iino R, Matsumoto Y, Nishino K, et al. Design of a large-scale femtoliter droplet array for single-cell analysis of drug-tolerant and drug-resistant bacteria, *Frontiers in Microbiology*, 2013, 4, 300.
24. Decrop D, Pardon G, Brancato L, et al. Single-step imprinting of femtoliter microwell arrays allows digital bioassays with attomolar limit of detection, *ACS Applied Materials & Interfaces*, 2017, 9, 10418–10426.
25. Kan CW, Rivnak AJ, Campbell TG, et al. Isolation and detection of single molecules on paramagnetic beads using sequential fluid flows in microfabricated polymer array assemblies, *Lab on a Chip*, 2012, 12, 977–985.
26. Kan CW, Tobos CI, Rissin DM, et al. Digital enzyme-linked immunosorbent assays with sub-attomolar detection limits based on low numbers of capture beads combined with high efficiency bead analysis, *Lab on a Chip*, 2020, 20, 2122–2135.
27. Kim SH, Iwai S, Araki S, et al. Large-scale femtoliter droplet array for digital counting of single biomolecules, *Lab on a Chip*, 2012, 12, 4986–4991.
28. Zhang H, Nie S, Etson CM, et al. Oil-sealed femtoliter fiber-optic arrays for single molecule analysis, *Lab on a Chip*, 2012, 12, 2229–2239.
29. Yaginuma H, Ohtake K, Akamatsu T, et al. A microreactor sealing method using adhesive tape for digital bioassays, *Lab on a Chip*, 2022, 22, 2001–2010.
30. Zhang QQ. *Droplet Microfluidic System and Its Applications in Material Synthesis*, Liaoning: Dalian Institute of Chemical Physics, Chinese Academy of Sciences, 2011.
31. Zhang Q, Liu X, Liu D, et al. Ultra-small droplet generation via volatile component evaporation, *Lab on a Chip*, 2014, 14, 1395–1400.
32. Qian SQ, Wu H, Huang B, et al. Bead-free digital immunoassays on polydopamine patterned perfluorinated surfaces, *Sensors and Actuators: B. Chemical*, 2021, 345, 130341.
33. Erica U, Florian LG, Victoria N, et al. Droplet Microarray: facile formation of arrays of microdroplets and hydrogel micropads for cell screening applications, *Lab on a Chip*, 2012, 12, 5218–5224.
34. Arif ZN, Binu K, Wong WK, et al. Embedded droplet printing in yield-stress fluids, *Proceedings of the National Academy of Sciences of the United States of America*, 2020, 117, 5671–5679.
35. Shim JU, Ranainghe RT, Smith CA, et al. Ultrarapid generation of femtoliter microfluidic droplets for single-molecule-counting immunoassays, *ACS Nano*, 2013, 7, 5955–5964.
36. Cem A, Christian AJ, Christoph Z, et al. Digital quantification of proteins and mRNA in single mammalian cells, *Molecular Cell*, 2016, 61, 914–924.
37. Jon FE, Dino DC, Katherine JH, et al. Controlled encapsulation of single-cells into monodisperse picolitre drops, *Lab on a Chip*, 2008, 8, 1262–1264.

38. Yue XY, Fang XX, Sun T, et al. Breaking through the poisson distribution: a compact high-efficiency droplet microfluidic system for single-bead encapsulation and digital immunoassay detection, *Biosensors and Bioelectronics*, 2022, 211, 114384.
39. Cohen L, Cui NW, Cai YM, et al. Single molecule protein detection with attomolar sensitivity using droplet digital enzyme-linked immunosorbent assay, *ACS Nano*, 2020, 14, 9491–9501.
40. Liu YN, Ye HH, HoangDinh H, et al. Digital plasmonic nanobubble detection for rapid and ultrasensitive virus diagnostics, *Nature Communications*, 2022, 13, 1687.

CHAPTER 4

Heterogeneous Digital Immunoassays

Heterogeneous digital immunoassays (DIs) refer to assays in which a separation step is required to remove unreacted, free antibodies and tags and effectively purify the immunocomplexes. This separation step is followed by partitioning, detection, and quantification of the immunocomplexes. The separation process serves as a double-edged sword: on one hand, it reduces interference from complex background matrices, minimizes nonspecific adsorption, and enhances sensitivity, reproducibility, and lowers the detection limit (LOD). On the other hand, immunocomplexes with weak affinity are more susceptible to dissociation during the separation, leading to a reduction in the number of detectable immunocomplexes, which can hinder the assay's performance. The success of heterogeneous DIs depends on the ability to detect and identify signals from individual, dispersed immunocomplexes. Thus, the selection of appropriate tags and signal amplification strategies is crucial to ensure accurate detection and quantification of individual immunocomplexes in these assays. Chapter 2 outlines the process of immunocomplex formation, where capture and detection antibodies are labeled with functional α and β tags (Figure 2.2b). These α and β tags can serve as signal reporters or as solid-phase surfaces. The chapter further introduces heterogeneous DIs, categorized according to the various combinations of α and β tags.

DOI: 10.1201/9781003729211-4

HETEROGENEOUS DIGITAL IMMUNOASSAYS ON SOLID SUBSTRATE SURFACES

When a two-dimensional solid matrix surface is used as a β-tag for labeling the capture antibody, the process of heterogeneous digital immunoanalysis is illustrated in Figure 4.1. The surface of the substrate is sequentially coated with the capture antibody and blocked to minimize nonspecific binding using a blocking solution. The sample is then introduced onto the blocked matrix, where target antigen molecules diffuse from the solution to the substrate surface and are captured by the capture antibody. Subsequently, any free or nonspecifically bound target antigens are removed by washing with a buffer solution. The α tag-labeled detection antibody is added for incubation, forming a sandwich immunocomplex on the matrix surface. After washing with buffer to eliminate excess detection antibodies, the immunocomplexes are imaged in regions, and the quantity of immunocomplexes is quantified to determine the amount of target antigen present. In immunoassays, the immobilization of antibodies on a surface increases the local concentration of the antibody, thereby enhancing antigen recognition and capture efficiency. However, the chemical and physical properties of the substrate surface can vary significantly depending on the material, which in turn affects both the binding strength and biological activity of the antibodies immobilized on it. Consequently, the choice of substrate and the method of antibody immobilization should be made based on the characteristics of the analyte and the intended detection purpose (see section on heterogeneous immunoassay in Chapter 1 of this book). Furthermore, the α tag attached to the detection antibody plays a crucial role in generating the digital signal, directly influencing both the recognition and reading methods of the signal. Various types of α

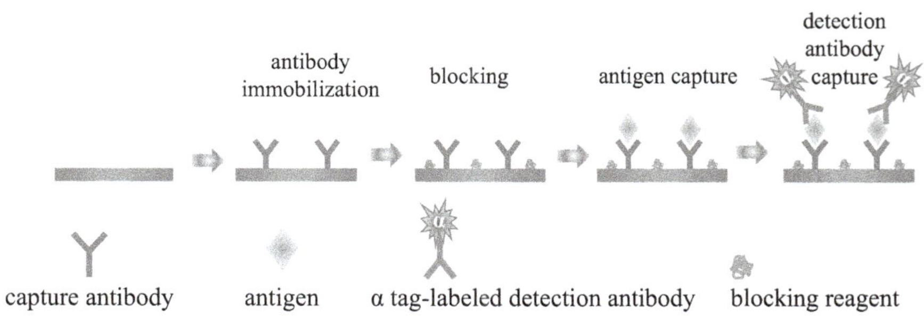

FIGURE 4.1 Schematic of a heterogeneous DI performed on a flat substrate surface.

tags have been reported, including fluorescent dyes, fluorescent nanoparticles, plasmonic nanoparticles, and nucleic acid sequences.

Organic Fluorescent Dye Labeling

Since organic dye molecules are small, their spatial steric hindrance is minimal when conjugated with antibodies, resulting in high labeling efficiency. However, the signal from a single fluorescent molecule is relatively weak and susceptible to photobleaching. Commonly used fluorescent dyes include Sulfo-Cyanine 3 (Cy3), Sulfo-Cyanine 5 (Cy5), fluorescein (FAM), and SYBR Green I. Cy3 belongs to the cyanine family of dyes, with a molecular weight of 767. The excitation and emission peaks of Cy3 are approximately 550 nm and 570 nm, respectively. As a pH-insensitive dye, Cy3 can be excited using laser wavelengths of 488 nm or 532 nm. Optimal imaging quality is achieved with the tetramethylrhodamine (TRITC) filter. The excitation peak of Cy5, with a molecular weight of 792, is around 650 nm, and its emission peak is around 670 nm. Cy5 can be excited with 633 nm or 647 nm laser beams, with the Cy5 filter set providing the best imaging quality. Both Cy3 and Cy5 contain carboxyl groups that form stable amide bonds with the primary amine groups of antibodies, including the N-terminus and lysine side chains. While the fluorescence of Cy5 is slightly brighter than that of Cy3, its stability is weaker in buffer solutions but more stable than Cy3 in anti-photobleaching solutions.

Mitra et al. fixed Cy3-labeled target proteins onto a glass substrate pretreated with bovine serum albumin (BSA), and incubated Cy5-labeled antibodies with the target proteins.[1] Unbound antibodies were subsequently washed away, and total internal reflection fluorescence (TIRF) microscopy was employed to quantify the bound antibodies. Each target protein molecule, polyclonal sheep immunoglobulin (IgG), labeled with an average of approximately 8 Cy3 molecules, appeared as green spots in the fluorescence images. Cy5-labeled polyclonal anti-sheep antibodies, which bound to the sheep IgG, appeared as red spots. The overlapping green and red spots resulted in yellow spots. The proportion of yellow spots was used to calculate the efficiency of specific immunocapture, revealing that approximately 70% of the sheep IgG was successfully recognized and bound by the antibodies. The assay demonstrated a linear detection range for sheep IgG from 55 to 1,676 molecules/1,000 μm^2, with a high linear correlation ($R^2 = 0.98$) and a CV between 1% and 5%. The LOD was found to be 55 molecules/1000 μm^2, corresponding to a target protein concentration of 100 pM.

Single-molecule multicolor detection is achieved by labeling different detection antibodies with Cy3 and FAM. Wang et al. simultaneously detected two cytokines, Interferon (IFN)-γ and TNF-α, using a dual-color barcode encoding and single-molecule counting technique.[2] Briefly, capture antibodies specific to IFN-γ and TNF-α were immobilized on an epoxy-coated glass slide. The cytokines were then incubated and captured by the fixed antibodies. Magnetic nanoparticles, modified with detection antibodies and primary barcode strands, served as dual-function magnetic nanoprobes, where the antibodies were responsible for molecular recognition and the primary barcode DNA strands provided encoding. These nanoprobes subsequently reacted with the antigens, forming two types of immunocomplexes. The primary barcode strands initiated a multibranch hybridization chain reaction (mHCR), generating multiple branched double-stranded polymers with gaps (secondary barcode strands). These secondary barcode strands then hybridized with two types of fluorescent probes, FAM-labeled and Cy3-labeled, producing enhanced fluorescence signals. Under different filter combinations, FAM and Cy3 were imaged separately, allowing the quantification of the target concentration by counting the number of fluorescent spots. This method employed a three-level signal amplification strategy: the dual-function magnetic nanoprobe constituted the first level of amplification, converting single molecules into multiple primary barcode strands; the multibranch hybridization chain reaction of the primary barcode strands represented the second amplification step, generating multiple arms from a single primary barcode strand; and the cross-shaped fluorescent probes formed the third level of amplification, with each fluorescent probe labeling three fluorescent dye molecules, thereby amplifying a single arm of the secondary barcode strand into three fluorescent molecules. This three-level amplification significantly boosted fluorescence intensity and enhanced detection sensitivity. The linear detection range for IFN-γ and TNF-α spans from 5 fM to 120 fM, with an LOD of 5 fM.

SYBR Green I binds to the minor groove of DNA double helices. As a fluorescent dye for double-stranded DNA, it has an excitation wavelength of approximately 497 nm and an emission wavelength of around 520 nm. Free SYBR Green I exhibits weak fluorescence, but once bound to double-stranded DNA (dsDNA), its fluorescence is enhanced by 800–1000 times. That is why SYBR Green I can be used to detect the quantity of dsDNA. Jiang et al. immobilized human IgG onto a silanized glass slide and then used biotinylated mouse anti-human antibodies to bind to the fixed IgG, forming an immunocomplex.[3] Avidin-modified magnetic nanoparticles (350 nm) were conjugated to the biotinylated antibodies in a 1:1 ratio. A 30-base pair dsDNA with biotin at one end was then reacted with the avidin on the magnetic nanoparticles. This reaction led to the accumulation of a large amount of dsDNA on the magnetic beads. SYBR Green I efficiently stained the dsDNA, and fluorescence microscopy was used to capture images of the SYBR Green I bound to the magnetic particles. Through the biotin-avidin interaction, a connection between human IgG and dsDNA was established. The two-step fluorescence signal amplification method, using magnetic particles and dsDNA, significantly increased the detection sensitivity. This allowed for high-sensitivity fluorescence detection of individual protein molecules, with a linear detection range from 3.0 fM to 50 fM.

Inorganic Fluorescent Nanoparticle Labeling

Quantum Dot

Quantum dots (QDs) are inorganic semiconductor nanomaterials with sizes ranging from 2 to 20 nm. Compared to organic fluorescent dyes, QDs exhibit several superior optical properties:[4] (1) The emission spectrum is narrow and symmetric, with a full-width at half-maximum of 25–35 nm. The emission wavelength can be precisely tuned by adjusting the size of the QDs. (2) The absorption spectrum spans a broad range, extending from the near-infrared to the near-ultraviolet region, coupled with a high molar absorption coefficient. (3) QDs possess a high quantum yield, typically ranging from 0.2 to 0.9 at room temperature. (4) They demonstrate exceptional photostability and are resistant to photobleaching. (5) The fluorescence emission of individual QDs displays a periodic on-off blinking pattern. These unique optical properties make QDs highly effective for single-molecule fluorescence labeling. Jiang et al. immobilized biotinylated human IgG antibodies onto a glass substrate, allowing avidin-labeled QDs to specifically bind to the biotin on the substrate surface.[5] To perform single-molecule fluorescence imaging and antibody counting, both TIRFM and wide-field epi-fluorescence microscopy were employed. These microscopies were equipped with high-magnification, high numerical aperture objectives (60×, 1.45 numerical aperture (NA)) and an electron-multiplying charge-coupled device (EMCCD) camera. The linear detection range for human IgG antibody was found to be 8.0×10^{-14} to 5.0×10^{-12} M for TIRFM and 5.0×10^{-14} to 3.0×10^{-12} M for wide-field epi-fluorescence microscopy.

To maximize light collection efficiency and enhance imaging resolution, imaging individual QDs necessitates the use of high-magnification oil immersion objectives with a NA greater than 1.2. However, this increases experimental costs and raises the threshold for QD imaging-based assays, presenting challenges for the broader application of QD counting technology beyond research laboratories and hindering the miniaturization of imaging instruments. As a result, the development of techniques for imaging individual QDs with low-NA air objectives has become a pressing need. We enhanced the emission and collection efficiency of QDs by utilizing microspheres with a high refractive index, thereby developing a low-cost imaging approach for single QDs using low-NA air lenses. 6 As shown in Figure 4.2A, a layer of QDs is deposited on the surface of a glass substrate, with micrometer-sized barium titanate ($BaTiO_3$) microspheres (refractive index ~2.2) dispersed on top of the QDs. A cover glass seals the QDs and $BaTiO_3$ microspheres, and the slides are observed using 10×/0.30 NA and 20×/0.4 NA air objectives, with images captured by a

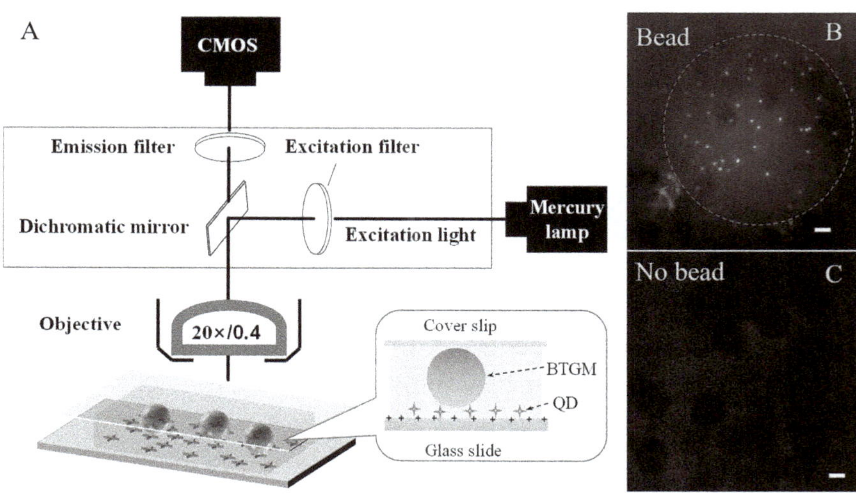

FIGURE 4.2 Scheme of the single-QD imaging setup (A), and fluorescent images of QDs (emission, 655 nm) in the (B) presence or (C) absence of BTGM.

Note: The scale bars represent 11 μm

Source: Reproduced from Zhang, Q.Q.; Li, J.J.; Pan, X.Y. et al. Low-numerical aperture microscope objective boosted by liquid-immersed dielectric microspheres for quantum dot-based digital immunoassays, *Analytical Chemistry*, 2021, 93, 12848–12853 (ref 6). Copyright 2021 American Chemical Society.

CMOS camera. In the presence of $BaTiO_3$ microspheres, randomly distributed bright spots are observed in the fluorescence images, most of which exhibit the "blinking" characteristics typical of QDs. The intensity of 100 fluorescence spots follows a Gaussian distribution, and the photobleaching curve of individual spots matches that of single QDs, confirming that the fluorescence spots in the image correspond to the emission of individual QDs (Figure 4.2B). In the control group without $BaTiO_3$ microspheres, no bright spots appear in the fluorescence image (Figure 4.2C). As miniatured spherical lenses, we believe that the $BaTiO_3$ microspheres serve three key functions in the imaging process: (1) they focus the light emitted from the mercury lamp, thereby enhancing the excitation efficiency of the QDs; (2) they alter the direction of the emitted light and enlarge the collection angle, improving the efficiency of light collection; and (3) they magnify the real image, thus enhancing the imaging resolution. After optimizing the material and size of the microspheres, the signal-to-background ratio of single QDs observed with the 20×/0.4 NA objective is nearly equivalent to that obtained using the 100×/1.4 NA objective.

We conducted a DI for detecting prostate-specific antigen (PSA) based on $BaTiO_3$ microspheres and a 20×/0.4 NA air objective. The process and the performance of this assay are outlined in Figure 4.3. The capture antibody against PSA was covalently immobilized on the glass slide surface using glutaraldehyde as a bridge. PSA and QD 655-labeled detection antibodies sequentially reacted with the fixed capture antibody to form QD-labeled immunocomplexes. Unbound detection antibodies were washed away, and $BaTiO_3$ microspheres were deposited onto the top layer of the immunocomplexes. The slide was then sealed with a cover glass, and the immunocomplexes containing QDs were imaged using the 20×/0.4 NA air objective and a CMOS camera. The number of QDs was counted to determine the PSA concentration. The linear detection range for PSA was 0.5–50 ng/mL, with an LOD as low as 0.17 ng/mL (Figure 4.3). The five interfering proteins, Alpha-fetoprotein (AFP), BSA, carcinoembryonic antigen (CEA), neuron-specific enolase (NSE), squamous cell carcinoma antigen (SCC), at concentrations five times that of PSA did not affect PSA detection, confirming the high specificity of this method. The concentrations of PSA in clinical serum samples measured using our approach were in good agreement with those measured by hospital labs. In terms of linear range, selectivity, and LOD, the performance of the combination of the 20×/0.4 NA air objective and $BaTiO_3$ microspheres was comparable to that

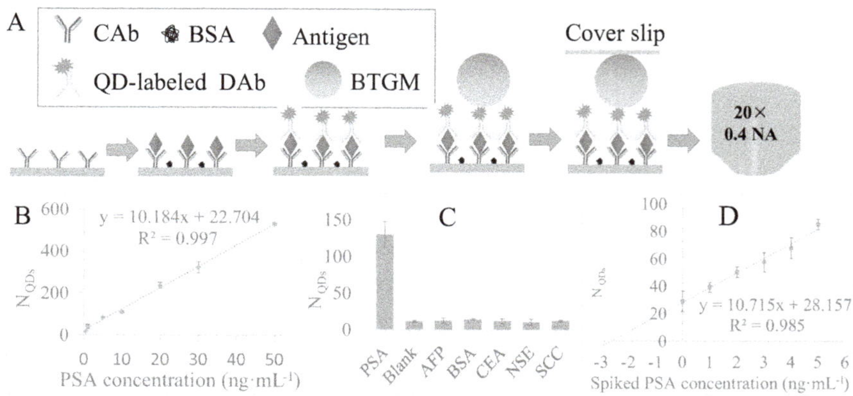

FIGURE 4.3 DI of PSA by using a 20×/0.4 NA objective with 150 μm BTGMs: (A) immunoassay procedure, (B) standard curve, (C) specificity evaluation, and (D) standard addition curve.

Source: Reproduced from Zhang, Q.Q.; Li, J.J.; Pan, X.Y. et al. Low-numerical aperture microscope objective boosted by liquid-immersed dielectric microspheres for quantum dot-based digital immunoassays, *Analytical Chemistry*, 2021, 93, 12848–12853 (ref 6). Copyright 2021 American Chemical Society.

of conventional 100×/1.4 NA objectives. The QD counting technique based on microsphere lenses eliminates the need for high-NA objectives, reducing the threshold for observing individual QDs while avoiding added complexity in sample preparation. This approach makes it more suitable for the widespread application of single-particle level QD imaging and counting techniques.

Upconversion Nanoparticle

Upconversion nanoparticles (UCNPs) are a class of nanomaterials that convert low-energy light into high-energy light through a nonlinear anti-Stokes process.[7,8] Compared to conventional luminescent probes, UCNPs offer several advantages: (1) Their near-infrared excitation wavelengths (980/808 nm) fall within the "optical window" of biological tissues (700–1,100 nm), effectively reducing light scattering, enhancing tissue penetration depth, and eliminating interference from endogenous fluorescence background. (2) UCNPs exhibit superior chemical and photostability, without blinking. Their photoluminescence decay time spans the microsecond range, significantly longer than that of organic dyes and QDs (0.1–20 ns). When used with time-resolved detectors, UCNPs can effectively filter out the short-lived fluorescence of biological sample autofluorescence. (3) UCNPs exhibit sharp, multi-peak emission spectra, with their emission peaks being tunable by adjusting the host lattice and the doping concentration of activators. (4) Lanthanide-doped UCNPs possess a relatively low extinction coefficient and low quantum efficiency. These properties make UCNPs highly suitable for applications in biological imaging and photodynamic therapy.

Gorris et al. immobilized capture antibodies against PSA on the bottom of a 96-well plate.[9] PSA and UCNPs (β-$NaYF_4$: 18 mol % Yb^{3+}, 2 mol % Er^{3+})-labeled detection antibodies were used to form immunocomplexes. A 980 nm laser was employed to excite the UCNPs within the immunocomplexes, and two quantification modes were utilized: (1) A custom microplate luminescence reader was used to measure the upconversion luminescence intensity, with a linear range from 100 pg/mL to 10 ng/mL and an LOD of 20.3 pg/mL. (2) Wide-field microscopy, coupled with a CMOS camera, was used for single-particle imaging and counting, with a linear range from 10 pg/mL to 1.0 ng/mL and an LOD of 1.2 pg/mL. The combination of these two detection modes extended the linear range by three orders of magnitude and enhanced sensitivity by 10 to 100 times compared to traditional enzyme-linked immunosorbent assay (ELISA). However, nonspecific adsorption on the well surface and spatial hindrance limited further improvements in sensitivity.

To address the spatial hindrance of UCNPs, Gorris et al. developed a new strategy for immunorecognition and capture.[10] In their study, anti-mouse antibodies were immobilized on the bottom of a well plate, to which mouse anti-PSA antibodies were subsequently bound. After PSA in the sample was captured onto the substrate, biotinylated PSA antibodies were added to form an immunocomplex. Avidin-coated UCNPs were then used to label the immunocomplex, and fluorescence imaging and counting were performed to determine the PSA concentration. Hydrophilic PEG chains were employed to modify the UCNPs, improving the dispersion of the UCNPs in aqueous solutions and reducing particle aggregation. These modifications also introduced steric and repulsive effects against interfering proteins in serum, decreasing nonspecific adsorption and minimizing the interaction between the UCNPs and the well plate, thereby reducing spatial hindrance. The stepwise reaction of the detection antibody and UCNPs allowed for the use of higher concentrations of detection antibodies, ensuring that all immobilized PSA on the surface was bound to the detection antibody. Furthermore, the high affinity between biotin and avidin allowed for a reduced concentration of UCNPs, minimizing nonspecific adsorption. These improvements resulted in a 50-fold reduction in the LOD for PSA, achieving a limit of 23 fg/mL (800 aM).

Plasmon Resonance Nanoparticle Labeling

Surface plasmons refer to the collective oscillations of free electrons on a metal surface when light (electromagnetic waves) is incident on the interface between the metal and the surrounding medium. When surface plasmons are confined to the surface of metal nanoparticles with diameters much smaller than the wavelength of the incident light, localized surface plasmon resonance (LSPR) occurs.[11] Noble metal nanoparticles, such as gold, silver, and gold-silver alloy particles, exhibit strong LSPR effects, which are influenced by factors including particle size, morphology, composition, surrounding medium, and interparticle distance. Currently, techniques for detecting the scattering properties of noble metal nanoparticles at the single-particle level for accurate particle counting include interferometric reflection imaging, dark-field microscopy, and surface plasmon resonance (SPR) technology.

Ünlü et al. designed and fabricated an interferometric reflection imaging sensor (IRIS) capable of imaging individual gold nanorods.[12] Specific nucleotide probes were immobilized on the surface of the IRIS chip to create a nucleic acid microarray. Complementary strands of these probes, fixed on the gold nanorods, hybridized with the immobilized probes. A dynamic tracking technique precisely measured the binding and dissociation processes, as well as the equilibrium time for each hybridization event. This approach allowed for the differentiation of specific nucleic acid hybridization reactions from nonspecific surface adsorption, effectively eliminating background signals and accurately counting the number of specific hybridization points. The sensor achieved highly sensitive detection of 12 synthetic nucleic acid sequences, with a LOD as low as 19 fM.

They further applied the IRIS for single-molecule protein immunoassays.[13] In their DI, capture antibodies specific for β-casein were immobilized on the surface of the IRIS chip, where β-casein was captured and subsequently hybridized with detection antibodies labeled with gold nanoparticles. The number of gold nanoparticle-labeled immunocomplexes was quantified using IRIS. The LOD for β-casein in undiluted serum samples was as low as 60 aM, with an upper concentration limit of approximately 100 pM, and a linear range spanning ten orders of magnitude. In whole blood samples, the LOD for β-casein was around 500 aM, with a linear range extending over five orders of magnitude. Additionally, they applied this approach to detect Hepatitis B surface antigen (HBsAg), achieving a LOD of 3.2 pg/mL.

Dark-field microscopy is a well-established and widely used imaging technique for gold and silver nanoparticles that exhibit LSPR. In dark-field images, the intensity and color of the scattered light produced by nanoparticles vary with their composition and size, making it easy to identify and encode multiple components. Huang Chengzhi's research group leveraged the color differences in dark-field images of gold and silver nanoparticles to achieve a color-coded dual-component DI.[14] Capture antibodies specific to CEA and alpha-fetoprotein (AFP) were immobilized on the surface of a glass substrate. Each antibody captured its respective target antigen from the sample. AFP antibodies labeled with gold nanoparticles and CEA antibodies labeled with silver nanoparticles were used to bind to the corresponding captured antigens, forming sandwich-structured immunocomplexes. Unreacted detection antibodies were washed away, and dark-field images were acquired. The CEA immunocomplexes, labeled with silver nanoparticles, appeared as blue light spots, while AFP immunocomplexes, labeled with gold nanoparticles, appeared as green light spots. The number of blue and green light spots correlated positively with the respective antibody concentrations. By counting the blue and green light spots separately, the concentrations of CEA and AFP could be simultaneously determined. The simultaneous measurements for the two proteins within the range of 0.5–10 ng/mL were consistent with those obtained from independent assays, with negligible cross-interference. When a 5 ng/mL standard solution of CEA and AFP was added to serum samples, the measured values closely matched the actual added amounts. The recovery rates for CEA ranged from 107.4% to 109.4%, while those for AFP ranged from 90.0% to 96.6%, demonstrating that this platform is suitable for the dual-component testing of clinical samples.

In addition to conventional dark-field techniques, SPRM is also employed for imaging nanoparticles. Unlike dark-field imaging, which uses glass substrates, SPRM requires the deposition of a gold nanoparticle thin film on a regular substrate. The object to be observed (such as molecules, nanoparticles, or cells) induces SPR on the gold surface. By using a P-polarized filter, a CCD camera, and a high-NA oil immersion objective, SPRM captures plasmon resonance images of the object. When P-polarized light illuminates the object deposited or bound to the gold-coated substrate, bright light spots with a V-shaped diffraction pattern are captured on the CCD. This V-shaped diffraction pattern results from the scattering of surface resonance waves by the nanoparticles. Based on the intensity of the signal, different object sizes can be distinguished at the nanoscale. The research group led by Tao Nongjian employed SPRM for the detection of influenza viruses.[15] First, a 47 nm thick gold film was deposited on a cover glass, and polyethylene glycol (PEG)/PEG-COOH was self-assembled on the gold surface to introduce carboxyl groups. Antibodies against the influenza A virus were covalently immobilized on the gold film through an amide coupling reaction, enabling the capture of influenza A viruses in solution. Correspondingly, typical light spots with V-shaped diffraction patterns appeared in the images, facilitating the detection of 0.8 ± 0.35 fg of influenza virus (H1N1 Influenza A/PR/8/34) and 6.5 ± 0.8 fg of human cytomegalovirus. The imaging area per unit reached 0.2 fg/mm^2. This platform is compatible with DIs and only requires nanoparticles to label detection antibodies. Nanoparticle-labeled immunocomplexes can be formed on the gold surface, enabling the imaging and counting of nanoparticle numbers, as well as quantitative protein analysis at the single-molecule level.

Surface-Enhanced Raman Scattering Labeling

When analyte molecules adsorb onto specific rough metal surfaces, the Raman scattering intensity can increase by a factor of 10^4 to 10^6. This phenomenon is known as surface-enhanced Raman scattering (SERS). The spectra generated by SERS allow for rapid qualitative and quantitative analysis of substances. Trau et al. developed a single-molecule nanopillar SERS platform for predicting and detecting immune toxicity in immunotherapy.[16] The fabrication process for the nanopillar SERS substrate is as follows in Figure 4.4A-C: electron-beam lithography was first employed to create an array pattern on a photosensitive material, followed by physical vapor deposition of a gold film to form pillars with gold-coated tips. Selective reactive ion etching was then used to reveal the nanopillar structure. The final chip consisted of 250,000 nanopillars, each 1,000 nm in width with a 1,000 nm spacing. This design aligned with the lateral resolution of Raman microscopy (~1,000 nm) and adhered to the scattering separation principle, ensuring that the SERS spectra from adjacent nanopillars did not overlap. Using succinimidyl propionate as a coupling reagent, capture antibodies were immobilized onto the tips of nanopillars via gold-sulfur bonds to form a capture nanopillar array. Detection antibodies and Raman reporter molecules were co-modified

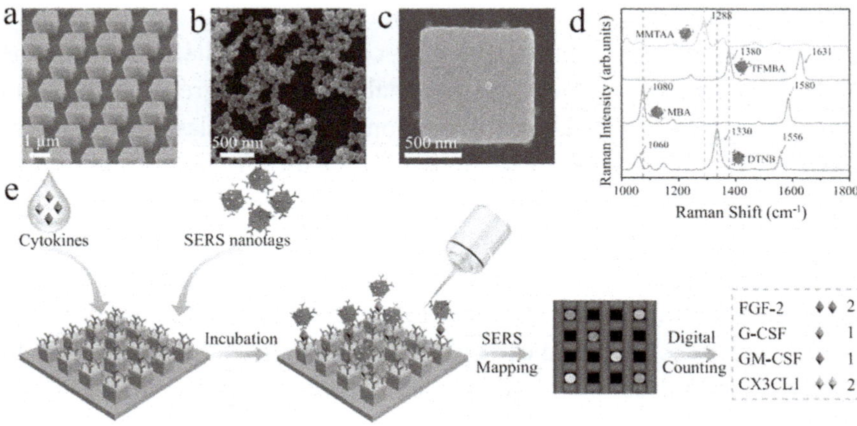

FIGURE 4.4 DIs for simultaneous detection of four cytokines using single-molecule nanopillar SERS: (A) SEM image of the nanopillar array from a side view, (B) SEM image of nanoboxes, (C) SEM image of a single nanobox positioned on top of a pillar, (D) SERS spectra of nanoboxes conjugated with four distinct Raman reporters, and (E) workflow for multiplexed counting of four cytokines.

Source: Reprinted by permission from Springer Nature Limited, Li, J.R.; Wuethrich, A.; Sina, A.I.; Cheng, H.H.; Wang, Y.L.; Behren, A.; Mainwaring, P.N.; Trau, M., *Nature Communications*, 2021, 12, 1087 (ref 16). Copyright 2021.

onto gold-silver alloy nanoboxes (~80 nm) to create bifunctional detection probes (Figure 4.4D). The detection antibody in the probe specifically bound to the target protein, while the reporter molecules generated distinct Raman signals. The target protein, along with the detection probe, formed an immunocomplex at the gold-coated nanopillar tips, producing SERS signals detectable by Raman microscopy. This platform exhibited an LOD at the attomolar (aM) level and was well-suited for multicolor analysis (Figure 4.4E). To quantify cytokines, the target proteins FGF-2, G-CSF, GM-CSF, and CX3CL1 were analyzed using four distinct SERS reporter molecules: 5,5-dithiobis(2-nitrobenzoic acid) (DTNB, Raman signal at 1,330 cm^{-1}), 4-mercaptobenzoic acid (MBA, Raman signal at 1,080 cm^{-1}), 2,3,5,6-tetrafluoro-4-mercaptobenzoic acid (TFMBA, Raman signal at 1,380 cm^{-1}), and 2-mercapto-4-methyl-5-thiazoleacetic acid (MMTAA, Raman signal at 1,288 cm^{-1}). The enhancement factors for DTNB, MBA, TFMBA, and MMTAA were 8.14×10^6, 1.46×10^7, 4.01×10^7, and 3.26×10^7, respectively, all surpassing those reported for gold nanoparticles and silver nanoprisms. In the color-enhanced Raman images, the nanopillar array appeared as a blue-black grid (blue representing the silicon substrate and black representing the gold-coated nanopillars). Within the black squares, four distinct Raman signal colors (red, green, purple, and cyan) were randomly produced, corresponding to fibroblast growth factor 2 (FGF-2), granulocyte colony-stimulating factor (G-CSF), granulocyte-macrophage colony-stimulating factor (GM-CSF), and chemokine (c-x3-c motif) ligand 1 (CX3CL1), thereby enabling multicomponent identification (Figure 4.4E). Blood serum samples from ten melanoma patients undergoing immunosuppressive therapy were analyzed to quantify the four cytokines as described earlier. The results showed that an increase in cytokine concentration correlated with a rising risk of developing severe immune toxicity in patients. This confirmed that the single-molecule nanopillar SERS platform was a viable tool for dynamic monitoring of immune-related adverse events during clinical immunotherapy.

Nonspectral Particle Labeling

Nonspectral particles refer to micron- or nanoscale materials that do not produce significant spectral signals, such as fluorescence, absorption, or scattering, and instead rely on particle volume to generate impedance or ion flow pulse signals. The principle of impedance or ion flow pulse detection is illustrated in Figure 4.5. Miniature electrodes are positioned at both ends of the sensing slit in the detection area, forming an electrical circuit. Nonspectral particle solutions are pressure-driven into the detection area, causing a change in the circuit's resistance. When no particles are present in the sensing slit, the circuit resistance is relatively low, allowing for higher ion flow, with no noticeable fluctuation, appearing as a smooth baseline on the impedance-time curve. However, when particles pass through the sensing slit, they displace the ionic solution, significantly increasing the circuit's resistance and reducing ion flow. This results in a pulse-like increase in impedance (or a decrease in ion flow) that is visible

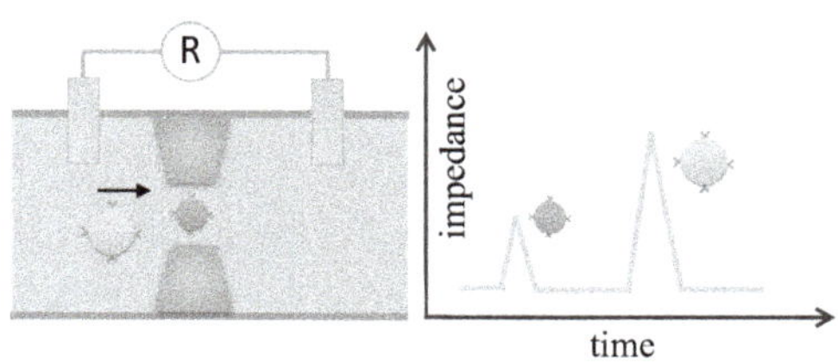

FIGURE 4.5 Principle of particle detection based on impedance or ion flow pulses.

on the impedance-time curve. The magnitude of the impedance pulse is positively correlated with particle size, while the frequency of pulses is directly proportional to the number of particles.

Javanmard et al. utilized micron-sized magnetic beads (2.8 μm) to label detection antibodies for quantifying cytokine, interleukin-6 (IL-6) concentrations, and tyrosine kinase activity.[17] They developed a microfluidic chip that integrates an immunoassay region with an impedance pulse sensor. The chip consisted of three primary zones: an immunoreaction zone, a fluid control zone, and an impedance detection zone. Capture antibodies were immobilized on the bottom of the microfluidic channel in the immunoreaction zone, while detection antibodies were conjugated to the micron-sized magnetic beads. In the immunoreaction zone, steps such as antigen capture, sandwich complex formation, and washing were performed. Alkaline buffer (1M NaOH) was then used to elute the specifically bound magnetic beads. The released magnetic beads sequentially passed through the sensing slit, generating impedance pulse signals. The concentration of the target protein was determined by the number of impedance pulses detected. This DI platform effectively integrated immunoreaction and signal counting, requiring minimal sample volume (less than 5 microliters per measurement). However, the LOD for proteins remained relatively high; for IL-6, the LOD was approximately 50 pM. The primary limitation for sensitivity arose from the large steric hindrance of the micron-sized magnetic beads, which reduced the binding efficiency in the immunoassay. Reducing the particle size to the nanoscale can alleviate this steric hindrance, improving binding efficiency. However, this reduction also leads to a smaller impedance change when nanoparticles pass through the sensing slit, making the signal more challenging to detect.

Nanopore sensors are designed to enhance the impedance signals generated by nanoparticles. These sensors detect ion flow pulses that occur when individual molecules or particles pass through the nanopore. Due to the matching size between the nanopore and the molecules or particles, nanopores produce stronger signal pulses and higher signal-to-noise ratios compared to micron-sized slits. However, several challenges remain in nanopore sensing. One issue is low perforation efficiency when dealing with low-concentration samples, which leads to extended testing periods. Another challenge is the high background interference caused by nontarget substances that generate additional pulses during perforation. Gooding et al. addressed some of these issues by using 50 nm magnetic nanoparticles to label detection antibodies, which then blocked the nanopore.[18] This nanopore-based approach enabled the first-ever single-protein counting, achieving an LOD of 0.8 fM. In this method, the walls of the nanopore were modified with capture antibodies specific to PSA, while the surface of magnetic nanoparticles was modified with PSA detection antibodies (anti-PSA magnetic beads). These anti-PSA magnetic beads bound to PSA in the sample, and after magnetic separation, background interference was minimized. The magnetic beads that carried PSA were considered target beads, while those without PSA were empty beads. A mixture of both types of beads was introduced into the flow cell of the nanopore. Under the influence of an external magnetic field, both types of beads entered the nanopore. The target magnetic beads specifically interacted with the PSA capture antibodies on the nanopore wall, resulting in a blockage of the nanopore. When the external magnetic field was reversed, the empty magnetic beads were expelled from the nanopore, thereby eliminating nonspecific blockages. The ion flow reduction caused by the blockage of the nanopore by the target magnetic beads was detected and counted by the ion flow detector.

Counting single immunocomplex molecules formed on the surface of a solid substrate represents an early form of DI, typically without the use of signal amplification strategies or discrete techniques. This approach is straightforward and relies primarily on single-molecule imaging technology. However, several challenges limit its effectiveness. The distance between target proteins in the bulk solution and the substrate reduces the likelihood of collisions between protein molecules and the antibodies on the substrate due to diffusion dynamics, resulting in low capture efficiency for the target proteins. Additionally, immunocomplexes are often labeled with highly luminescent nanoprobe tags (such as QDs, gold nanoparticles, or UCNPs). The spatial steric hindrance between the probes, the substrate, and neighboring probes further diminishes the affinity efficiency, hindering the formation of immunocomplexes. Diffusion dynamics and spatial steric effects directly influence the efficiency of immune reactions, thereby limiting the detection sensitivity and quantitative range. The number of capture sites available on a two-dimensional surface is finite and can become easily saturated by high concentrations of target proteins, restricting the upper limits of quantitative analysis. As a result, the linear range of DIs performed on substrate surfaces is typically narrow, with LODs often confined to the sub-pM level. Current research is focused on improving the luminescence efficiency of probes, reducing the technical complexity and cost of single-molecule imaging, and facilitating portable, real-time detection capabilities.

HETEROGENEOUS DIGITAL IMMUNOASSAYS ON MICROBEAD SURFACES

Due to the significant influence of diffusion dynamics on immunoreactions occurring on substrate surfaces, which result in prolonged reaction times and low efficiency, microbead surfaces have become increasingly preferred for immunoreactions. Commonly used microbeads include magnetic beads and fluorescent microspheres. Capture antibodies are immobilized on the surfaces of these microbeads, where they interact with detection antibodies and antigens. Immunocomplexes form on some of the microbeads and are then separated from unreacted detection antibodies using magnetic or centrifugal force. The microbeads are subsequently dispersed, and the number of immunocomplexes is quantified based on the properties of the tags associated with the detection antibodies.

Enzyme Labeling

Enzyme labeling refers to the assays in which enzymes are used to mark immunocomplexes, and enzymatic reactions occur within confined, small spaces. The signal from the immunocomplex is converted into a detectable signal from the enzyme substrate, making enzyme labeling a widely used method for signal recognition and amplification in heterogeneous DIs. The principle of enzyme catalysis is illustrated in Figure 4.6. In this process, magnetic beads modified with capture antibodies, antigens, and detection antibodies form immunocomplexes. Enzyme labels, either single or multiple, are then applied to generate enzyme-labeled immunocomplexes. The mixture of enzyme-labeled immunocomplexes and empty

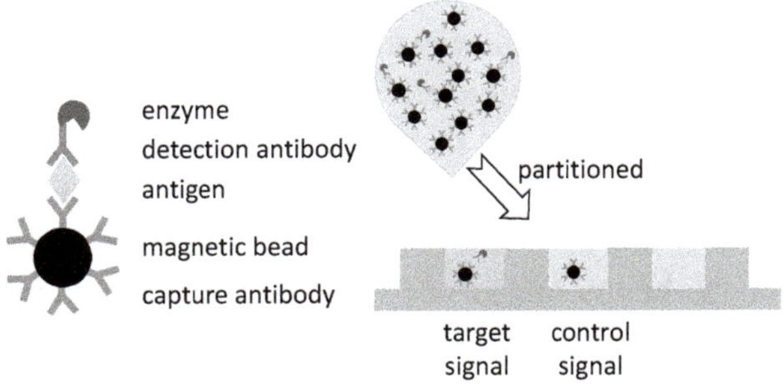

FIGURE 4.6 Enzyme-catalyzed DIs.

capture magnetic beads is partitioned into discrete microstructures or microdroplets, where enzyme-catalyzed substrate reactions occur in isolated small spaces. The enzyme-labeled immunocomplexes generate target signals, such as fluorescence, chemiluminescence, or color changes, while the empty capture beads provide blank signals. By statistically comparing the target signals to the blank signals, the concentration of the antigen can be accurately quantified.

Enzyme catalysis in DIs typically operates via a "turn-on" mechanism. For effective enzymatic reactions, specific requirements must be fulfilled for both the enzymes and the substrates. In DIs, the reaction typically occurs at the femtoliter (fL) scale, and the catalytic efficiency of individual enzyme molecules must be sufficiently high. Together, these factors ensure that a single enzyme molecule can generate a detectable signal within a limited timeframe. Additionally, enzymes, being proteins, are sensitive to their external environment, so they must exhibit good stability to maintain both their biological and catalytic activities. Substrates, on the other hand, should exhibit little to no fluorescence or luminescence prior to enzyme catalysis. The enzymatic reaction itself must efficiently convert substrates into fluorescent or luminescent products, enabling detection. Common enzymes used in DIs include β-galactosidase (β-Gal), alkaline phosphatase, and horseradish peroxidase (HRP).

β-Gal

β-Gal and its fluorescent substrate, resorufin-β-D-galactopyranoside (RGP), represent the earliest systems employed in enzyme catalysis. In this setup, the surface of magnetic beads is modified with capture antibodies to form capture beads, while biotin and β-Gal are used to label detection antibodies (biotin-detection antibodies) and streptavidin (enzyme-streptavidin), respectively. The capture beads, antigen, and biotin-detection antibodies are incubated and reacted sequentially to form a biotinylated immunocomplex. The enzyme-streptavidin complex then binds with biotin to form enzyme-labeled magnetic beads. When the antigen concentration in the sample is very low, the ratio of antigen molecules to capture beads is less than 1:1, and the distribution of enzyme-labeled complexes on the magnetic beads follows a Poisson distribution. In this case, only a small fraction of the microbeads carry enzyme-labeled complexes, while the majority are unoccupied. For instance, in a 100 μL sample with an antigen concentration of 50 aM, approximately 3,000 target proteins are captured by 200,000 microbeads. As a result, only 1.5% of the microbeads carry a single enzyme-labeled complex, while around 98.5% of the beads remain empty. These magnetic beads, both enzyme-labeled and empty, are randomly distributed into a picoliter-scale array of microholes. RGP is introduced into the microholes, where it reacts. In the microholes containing β-galactosidase, RGP is catalytically broken down to produce a fluorescent product, resorufin. Distinct fluorescence spots corresponding to these reactions are observed under fluorescence imaging, which are considered as target signals. The number of fluorescence spots correlates with the target signal, while the number of spots in bright-field imaging represents the total number of magnetic beads. By comparing the ratio of fluorescence spots to bright-field spots, the concentration of the target protein can be determined.[19]

The sensitivity of a microwell array-based DI platform for detecting PSA and tumor necrosis factor-α (TNF-α) in blood samples was investigated. Human PSA and TNF-α standards were added to bovine serum, and the mixture was then diluted four folds to minimize the matrix effect. In a 25% bovine serum matrix, the LOD for PSA was found to be 50 aM (1.5 fg/mL), which was equivalent to 200 aM (6 fg/mL) in whole serum. For TNF-α, the LOD was 150 aM (2.5 fg/mL), corresponding to 600 aM (10 fg/mL) in whole serum. These LODs for PSA and TNF-α were considerably lower than those of commercial ELISA, which had LODs of 0.1 ng/mL for PSA and 0.34 pg/mL for TNF-α. PSA expression levels in serum were important for prostate cancer screening and monitoring disease recurrence after prostatectomy. After prostatectomy, most patients exhibited undetectable PSA levels in their serum, indicating that their concentrations fell below the LODs of conventional clinical immunoassays (typically around 3 pM or 0.1 ng/mL). While a significant rise in PSA levels was used to monitor disease recurrence, it often took several years for PSA concentrations to increase sufficiently after surgery, resulting in missed opportunities for early intervention. The microwell array DI platform offered the advantage of accurately quantifying PSA concentrations in serum from post-prostatectomy patients, even at levels below 300 fM (10 pg/mL), which was valuable for early detection of disease recurrence. To further assess the platform's utility, serum samples were collected from 30 patients (aged 60–89) six weeks after prostatectomy, and these samples were diluted four folds for detection. The results revealed that PSA concentrations in the serum samples ranged from 14 fg/mL to 9.4 pg/mL, with an average concentration of 1.5 pg/mL. This was significantly lower than the LODs of traditional clinical immunoassays, demonstrating the potential of the microwell array-based DI for early detection and monitoring of prostate cancer recurrence.

When the target concentration is very low, the affinity between the antigen and antibody becomes particularly crucial. The affinity constant (K_D) for typical antigen-antibody interactions typically ranges from 10^{-6} to 10^{-10} M. The binding between biotin and streptavidin is highly stable, specific, and minimally affected by external interference. It has been demonstrated to be the strongest known noncovalent interaction, with affinity constants ranging from 10^{-12} to 10^{-14} M. This affinity is significantly higher than that of antigen-antibody interactions, making it the preferred choice for enzyme labeling of immunocomplexes. Additionally, secondary antibody-based indirect labeling is also commonly employed for enzyme labeling of immunocomplexes.[20] The secondary antibody, labeled with β-Gal, binds to the detection antibody rather than the capture antibody in the immunocomplex, resulting in the formation of an enzyme-labeled immunocomplex. This complex is spatially dispersed within array microwell that have hydrophilic bottoms and hydrophobic walls. Fluorescein-di-β-D-galactopyranoside is introduced into the microwell as a fluorescent substrate, where it undergoes an enzyme-catalyzed chemical reaction to produce fluorescence. The affinity between the primary and secondary antibodies plays a significant role in determining the enzyme labeling efficiency of the immunocomplex, and this efficiency can vary depending on the origin of the antibodies. Researchers utilized SPR technology to determine the binding kinetics of various antibodies and their affinity constants. Based on these measurements, they selected antibodies with higher affinity for detecting the influenza A virus nucleoprotein. The LODs in buffer and throat swab samples were found to be 4 ± 1 fM and 10 ± 2 fM, respectively, which are comparable to the LODs of assays using the biotin-streptavidin system.

Alkaline Phosphatase

Alkaline phosphatase (ALP) is a widely used enzyme in commercial enzyme-linked immunoassays. Its substrates differ from those of β-Gal. For ALP, a common fluorescent substrate is a coumarin derivative, such as 4-methylumbelliferyl phosphate. This compound is a phosphorylated coumarin derivative that does not emit fluorescence on its own. However, when catalyzed by ALP, it is hydrolyzed to produce 4-methylumbelliferone and inorganic phosphate. 4-methylumbelliferone emits fluorescence, with a maximum excitation wavelength of 372 nm and a maximum emission wavelength of 445 nm. Hiroyuki Noji et al. developed a hydrophilic-hydrophobic microwell array, with each microwell having a volume of 44 fL. They mixed ALP with 4-methylumbelliferyl phosphate and introduced the mixture into the microhole array. The array was subsequently sealed with fluorocarbon oil and subjected to fluorescence imaging after an appropriate incubation period.[21] When the average number of enzyme molecules (λ) in a single microwell was much lower than 1, the fluorescence imaging of the microhole array revealed intermittent fluorescence spots. The distribution of microwell without fluorescence, with weak fluorescence, and with strong fluorescence followed a Poisson distribution. For different values of λ, the proportion of fluorescent microholes was positively correlated with the concentration of ALP. Moreover, the fluorescence emission wavelength of the ALP and 4-methylumbelliferyl phosphate system (445 nm) did not overlap with that of the fluorescein system (585 nm), allowing the two systems to be used together in a multicolor DI. This provided an alternative wavelength for encoding, expanding the capabilities of the assay. Unfortunately, coumarin-based compounds were prone to leaking into the fluorocarbon oil, leading to signal cross-interference between adjacent microwell. While 4-methylumbelliferone, with its additional hydroxyl group, improved water solubility, it required deprotonation to minimize molecular leakage. Consequently, the use of 4-methylumbelliferone derivatives was limited under alkaline conditions. Another challenge with using coumarin-based fluorescent substrates in DIs was the high fluorescence background signal. Even in the absence of ALP, approximately 0.03%–0.04% of the microholes still emitted detectable fluorescence, resulting in elevated blank signals and complicating the achievement of fM LODs in immunoassays.

In addition to fluorescence imaging, ALP catalyzes the hydrolysis of phosphate ester bonds, producing electroactive substances that enable electrochemical detection of single molecules. One commonly used substrate for electrochemical sensing is p-aminophenyl phosphate monohydrate, which is hydrolyzed by ALP to release the highly electroactive p-aminophenol (p-AP). This p-AP can then be directly oxidized to generate an electrochemical signal. However, the amount of p-AP produced by a single ALP molecule is limited, and the electrochemical signal generated by the direct oxidation of p-AP remains weak, posing a challenge for sensitive detection. A team led by Zhang Zhiling developed an enzyme-induced metallization approach to enhance the electrochemical signal.[22]

ALP was mixed with p-aminophenyl phosphate and silver nitrate solution, then introduced into polydimethylsiloxane (PDMS) microholes with a diameter of 1 mm. The bottoms of the microholes were equipped with gold electrodes that are 100 μm wide. When enzyme molecules were present in the PDMS microwell, p-aminophenyl phosphate was hydrolyzed, releasing p-aminophenol. The released p-AP reduced silver ions to zero-valent metallic silver, which subsequently deposited onto the surface of the gold electrode, producing a distinct signal in the linear voltammogram. Compared to the direct oxidation signal of p-AP, the electrochemical signal generated by enzyme-induced metallization was enhanced by approximately two orders of magnitude.

To enhance detection throughput, Zhang et al. next bonded a PDMS block with 100 through-holes to a 10×10 microelectrode array substrate, thereby fabricating a single-molecule electrochemical array. In the blank microwell, the signal generated by the microelectrode array was approximately 1.8 nA, while in microwell containing a single enzyme molecule, the signal increased to approximately 8.0 nA. Within the 1–10 aM range, the distribution of enzyme molecules detected by electrochemistry closely followed the theoretical Poisson distribution, validating the feasibility and accuracy of the design. Liver cancer cells (Hep G2) and breast cancer cells (MCF-7) were used as real samples. After cell lysis and ALP extraction, the cell extracts were analyzed using the single-molecule electrochemical array described earlier. The results revealed that the average concentration of ALP in a single liver cancer cell was approximately 12.1 aM, further confirming that the design is robust against complex backgrounds and offers excellent accuracy.

HRP

HRP is commonly used in clinical testing due to its high specific activity, stability, low preparation cost, and ease of storage. HRP is widely employed in colorimetric analysis, chemiluminescent immunoassays, and fluorescent immunoassays. In DIs, HRP, in combination with the fluorescent substrate 10-acetyl-3,7-dihydroxyphenoxazine (ADHP), is well-suited for fluorescence imaging detection. The fluorescence chemical reaction is depicted in Figure 4.7. ADHP itself does not emit fluorescence; however, when catalyzed by HRP, it generates a highly fluorescent compound called resorufin. Resorufin is soluble and exhibits maximum excitation and emission wavelengths at 570 nm and 585 nm, respectively. The HRP and ADHP system is well-established, with commercial kits available. For

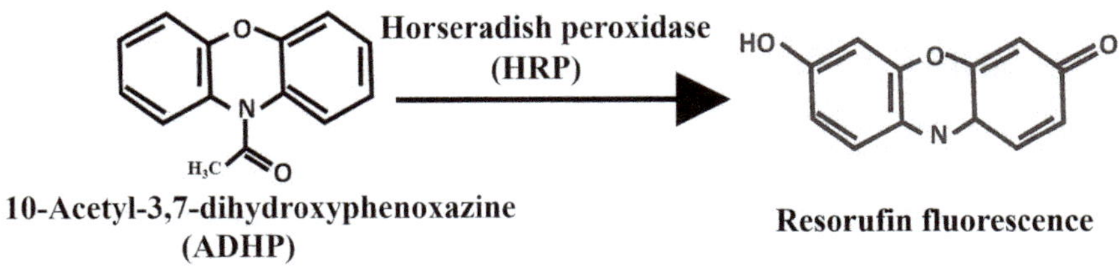

FIGURE 4.7 HRP-catalyzed conversion of ADHP to fluorescent resorufin.

instance, the QuantaRed™ Enhanced Chemiluminescent HRP Substrate Kit (Thermo Fisher) is frequently used in DIs.

Issadore and colleagues at the University of Pennsylvania developed a multicolor digital droplet immunoassay using a smartphone in conjunction with the HRP and ADHP system.[23] The microfluidic chip consisted of several components: a microbead processing unit, a reagent mixing unit, a droplet generation unit, a droplet incubation unit, and a droplet fluorescence detection unit. In the microbead processing unit, microbeads modified with capture antibodies captured the target antigen from a serum sample and immobilized it onto the micropores on a semipermeable membrane. After washing with buffer, HRP-labeled detection antibodies were introduced, forming enzyme-labeled immunocomplexes. Unbound enzyme-labeled antibodies were removed via buffer washing, and the microbeads were released through reverse fluid driving. In the reagent mixing unit, the microbeads were mixed with the fluorescent substrate and flowed through a 14 mm-long fishbone-structured micromixer, ensuring rapid reagent mixing. This fast mixing minimized enzyme reaction time before droplet encapsulation, which helped reduce background signals. In the droplet generation unit, a gradient emulsification technique was employed, with a 12:1 height difference in the wedge-shaped structure. The system included 100 droplet generators working in parallel, capable of generating up to 100,000 droplets per second. By adjusting the microbead concentration, individual droplets could encapsulate either one microbead or none. The droplet incubation unit featured an enlarged channel length and reduced fluid flow speed to increase the incubation time. The channel dimensions were expanded from the micrometer scale to the millimeter scale (width 1.8 mm, height 1.5 mm), significantly increasing the cross-sectional area and slowing the flow speed. Additionally, four layers of spiral channels were stacked, and interlayer through-holes connected them, further extending the channel length. With a fluid flow rate of 67 mL/h, the droplets took approximately 3.2 minutes to pass through the incubation channel, ensuring sufficient time for the enzyme-catalyzed fluorescence reaction. In the droplet detection unit, an optofluidic device with time-domain encoded smartphone was used to overcome the challenge that smartphones could not resolve high-speed moving droplets in close proximity. This innovation increased the droplet detection rate by a factor of 100, enabling multicolor, ultrahigh-throughput analysis.

This optofluidic platform was successfully applied to conduct both single-component and dual-component assays for GM-CSF and IL-6 in bovine serum. For the single-component assay, the LODs were 0.0045 pg/mL (320 aM) for GM-CSF and 0.0070 pg/mL (350 aM) for IL-6. In the dual-component assay, the LOD was approximately 0.004 pg/mL. The levels of endogenous GM-CSF and IL-6 in human serum measured using this assay were consistent with those obtained using commercial digital enzyme-linked immunoassays. By combining microfluidic droplet technology with a smartphone, this platform overcame the frame rate limitations of conventional microscopy, enabling a detection throughput of up to 1 million droplets per second. This innovation increased the droplet detection rate more than 100-fold, significantly reducing measurement time. Additionally, the use of smartphones eliminated the need for expensive optical components, lowering the detection threshold and promoting the widespread adoption of digital droplet immunoassay technology.

In addition to catalyzing chemiluminescent reactions, HRP also functions as a key enzyme in tyrosine signal amplification, generating locally amplified fluorescent signals. Walt et al. incorporated tyrosine signal amplification into hydrogel-immobilized microbeads to develop a simplified version of the digital ELISA.[24] In this approach, target proteins were captured on the surface of magnetic beads, followed by binding of biotinylated detection antibodies to the captured proteins. Single immunocomplexes on the magnetic beads were then conjugated with an avidin-conjugated polymer that carries multiple HRP molecules, thus completing the HRP labeling of the immunocomplexes. After the immunoreaction, the mixture of magnetic beads—comprising both target-bound and empty magnetic beads—was suspended in a solution containing hydrogen peroxide and tyramine-Alexa Fluor 488. In the presence of hydrogen peroxide, HRP catalyzed the formation of a radical intermediate from tyramine-Alexa Fluor 488, which covalently bound to aromatic rings near the HRP (e.g., tyrosine residues on the surface proteins of the magnetic beads). This process resulted in a significant labeling of the magnetic beads carrying the immunocomplexes with fluorescent dye, thereby completing the tyrosine signal amplification process. Due to the short lifespan of the free radicals, their diffusion into the outer regions of the microbeads in the diluted solution was highly unlikely, minimizing cross-contamination between beads. As a result, empty magnetic beads carried only a small amount of fluorescent dye and remained undetectable. After the HRP-catalyzed tyrosine reaction, the fluorescent dye was primarily labeled on the target magnetic beads' surface. Consequently, there was no need to disperse the microbeads in microplates. Instead, the microbead suspension was directly dropped onto a substrate surface, followed by the addition of a hydrogel precursor solution. The solution quickly formed a hydrogel onto the beads. Bright-field and fluorescence imaging were then employed to count both the total magnetic beads and the fluorescently labeled beads. This assay enabled highly sensitive detection of IL-6, with a linear range and LOD comparable to those of commercial digital ELISA. Furthermore, tests of saliva samples were consistent with clinical detection results.

Organic Fluorescent Dye Labeling

Magnetic beads serve as the immunoreaction carrier, and fluorescent dye molecules are used as the labeling signals for detection antibodies. By employing heterogeneous immunoassays, the unique property of magnetic beads—easy separation—can be utilized to isolate the immunocomplexes from complex background matrices, thereby reducing the interference of background signals on the fluorescence emission of dye molecules and improving detection sensitivity. There are two types of detection methods for the fluorescent dye in the immunocomplexes: single-molecule fluorescence flow cytometry and single-molecule fluorescence imaging.

The single-molecule fluorescence flow cytometry detection strategy is described as follows:[25] Commercialized avidin-modified magnetic beads react with biotinylated capture antibodies to form capture magnetic beads. The capture magnetic beads then successively react with target proteins and fluorescent dye-labeled detection antibodies, forming immunocomplex-labeled magnetic beads. The elution solution strips the fluorescent detection antibodies from the surface of the immunocomplex-labeled magnetic beads, resulting in an elute containing the fluorescent antibodies (the number of fluorescent antibody molecules is directly proportional to the target protein). The capillary draws the eluate, and the fluorescent antibodies flow through the detection window, where a photon counter detects and records the signal of individual fluorescent antibodies. The target protein concentration is determined by counting the number of fluorescence pulse signals. This strategy has LODs for multiple protein markers below pg/mL. For example, the LOD for cardiac troponin I can reach 0.11 pg/mL.

Single-molecule fluorescence imaging detection technology follows the same immunocomplex formation and elution process as the single-molecule fluorescence flow cytometry detection, except that the fluorescent dye-labeled antibodies eluted from the immunocomplex are deposited onto the substrate surface. Confocal fluorescence imaging or TIRF imaging is then used to detect the fluorescent dye-labeled antibodies. Zhang Chunyang's research group developed a multicolor single-molecule fluorescence imaging detection method for histone-modifying enzymes using a magnetic bead elution strategy.[26] A synthetic peptide substrate with a biotin-labeled N-terminal (biotin-ART KQT ARK STGGKA PRK QLA) corresponded to the first 21 amino acid residues of histone H3. In the presence of acetyl donor (acetyl-CoA) and methyl donor (S-adenosyl-L-methionine), histone acetyltransferase (GcN5) and histone methyltransferase (G9a) transferred acetyl and methyl groups to the corresponding lysine residues on the peptide substrate. The acetylation sites of the peptide substrate bound to Alexa Fluor 488-conjugated anti-acetyl lysine antibody, and the methylation sites bound to Alexa Fluor 647-conjugated anti-methyl lysine antibody, thus converting the enzyme information into distinguishable fluorescence signals. The fluorescent dye-labeled peptide substrates specifically bound to the magnetic beads through biotin-avidin interaction, and the unbound fluorescent antibodies were removed by magnetic separation. The fluorescently labeled peptides captured by the magnetic beads were then eluted and deposited onto the substrate surface. TIRF technology was used for single-molecule imaging and counting. This method allowed for highly sensitive and specific detection of multiple histone-modifying enzymes, with LODs for GcN5 and G9a at 21 pM and 12 pM, respectively. It can also be used for screening inhibitors of histone acetyltransferase.

Inorganic Fluorescent Particle Labeling

Fluorescent particles used in magnetic bead based heterogeneous DIs on the surface of magnetic beads include QDs, fluorescent polystyrene (PS) microspheres, etc. The detection methods are generally divided into flow cytometry-based point detection and fluorescence imaging-detection. Fluorescent

PS microspheres are a type of polymer microsphere that encapsulates dyes with different emission wavelengths within the PS shell. Their size can be adjusted from tens of nanometers to several micrometers, and the emission wavelength can be tuned from 390 nm to 700 nm. The synthesis technology of fluorescent PS microspheres is relatively mature, and commercial products (such as those from Bangslab) are available. These microspheres can utilize fluorescence color coding and are widely used in multiplex analysis of biomarkers.

Xu Hong's research group proposed a "fluorescent PS microsphere labeling of magnetic beads" strategy.[27] This strategy enabled DI for CEA using flow cytometry. Magnetic beads (3.2 μm) were surface modified with capture antibodies to form capture magnetic beads. The capture magnetic beads, target antigens, and biotinylated detection antibodies then formed immunocomplex beads (target magnetic beads). Avidin-conjugated HRP was bound to the surface of the target magnetic beads, and biotinylated tyramine, under the catalysis of HRP, bound to the proteins on the surface of the target magnetic beads, amplifying the number of biotin molecules and providing a sufficient number of functional sites (Ligand A). Avidin-conjugated fluorescent PS microspheres (1.2 μm) were used to label the target magnetic beads with a signal. Magnetic separation removed excess fluorescent beads, and the flow cytometer (NovoCyte2040R) detected the number of magnetic beads and fluorescent PS microspheres. The protein concentration was determined based on the ratio of the two.

Due to the significant steric hindrance between micrometer-scale PS microspheres, the efficiency of immunorecognition and binding is reduced. Li's research group used nanoscale fluorescent PS nanospheres to reduce the steric hindrance effect in immune binding.[28] Nanoscale PS nanospheres with different fluorescent emission wavelengths were used for target signal encoding, achieving multiplex, attomoles, and amplification-free analysis of DNA and miRNA. Green (λex/em: 480/520 nm), blue (λex/em: 360/450 nm), and red (λex/em: 360/450 nm) fluorescent PS nanospheres (200 nm) were modified with capture DNAs for three target DNA sequences, while magnetic beads (1 μm) were modified with capture DNAs for the target DNA. The fluorescent PS nanospheres, target DNA, and magnetic beads react to form sandwich complexes. The three-color fluorescent PS nanospheres were then eluted from the complex surface and deposited onto a glass substrate. Fluorescence microscopy was used to count the three-color nanospheres separately, thereby determining the target DNA concentration. The linear range for analyzing 40-nt

HIV DNA, TB, and HBV sequences was 20–10,000 pM, with LODs ranging from 3–52 pM (15–260 attomoles). The linear range for detecting three human cancer-related microRNAs (miR-141, miR-21, and let-7c) was 20–20,000 pM, with LODs of 3 pM (miR-141), 1 pM (miR-21), and 2 pM (let-7c), respectively. This analysis platform showed no significant discrimination effect for DNA and miRNA and allowed for highly sensitive multiplex analysis. Although it had not been used for DI of proteins, it was evident that replacing the capture nucleic acid sequence with an antibody and the target nucleic acid with a target protein, using the same strategy, would enable highly sensitive DIs.

To further reduce the LOD for biomolecules, Zhang Chunyang and his colleagues used QD-encapsulated liposomes as the labeling signal for target molecules.[29] Magnetic beads, target DNA, and liposomes formed sandwich complexes, and the liposomes bound to the magnetic beads were eluted. The eluted liposomes were then lysed to release QDs, amplifying a single target molecule into multiple QDs. Using single-particle counting detection of the QDs greatly improved the sensitivity of the analysis. The LODs for two HIV-related DNAs reached 1 aM (HIV1) and 2.5 aM (HIV2), which was five orders of magnitude lower than the LODs of fluorescent microsphere-labeled nanosensors and three orders of magnitude lower than the LODs of single quantum dot-labeled nanosensors.

Plasmon Resonance Particle Labeling

Plasmon resonance nanoparticles have been used as labeling signals in magnetic bead based heterogeneous DIs. In one study conducted by Wu et al., avidin-functionalized magnetic beads bound with biotinylated capture antibodies to form capture magnetic beads.[30] The capture magnetic beads, antigens, and detection antibodies then formed immunocomplexes (magnetic beads-antigen-detection antibody). The second antibody labeled with gold nanoparticles bound to the detection antibody in the immunocomplex, forming a gold nanoparticle-labeled complex (magnetic beads-antigen-detection antibody-gold nanoparticles). The gold nanoparticles in the complex were eluted, and riched by magnetic separation. Then, the eluted gold nanoparticles were quantitatively transferred to the substrate surface. Dark-field microscopy was then used to image and count the gold nanoparticles. The number of gold nanoparticles was directly proportional to the concentration of the antigen. Using a standard curve based on the number of gold nanoparticles, the concentration of specific protein biomarkers in biological samples can be calibrated. The linear ranges of PSA, CEA, and AFP were found to be 1–20 ng/mL, 5–30 ng/mL, and 1–25 ng/mL, with the corresponding LODs of 1 ng/mL, 5 ng/mL, and 1 ng/mL, respectively. Compared with existing single-particle counting assays, the LODs of this method were relatively higher, which might be attributed to two factors: (1) loss of gold nanoparticles during the elution, transfer, and deposition processes and (2) the limited dark-field imaging area, which could lead to missed gold nanoparticles.

To enhance detection sensitivity, He and his colleagues replaced gold nanospheres with gold nanorods and adopted a strategy for elution and separation for magnetic bead–gold nanoparticle complex.[31] They improved analytical sensitivity from two aspects: the deposition efficiency of gold nanorods and the imaging area. The surface of the gold nanorods was simultaneously modified with antibodies and single-stranded DNA. The single-stranded DNA gave the eluted gold nanorods more negative charges, which in turn accelerated their deposition onto the positively charged substrate surface. Compared to natural sedimentation, the deposition efficiency of the gold nanorods was significantly improved. Low-magnification objectives (10X) were used to repeatedly image the fixed gold nanorods, and algorithms were developed to eliminate overlapping regions in the images, allowing for accurate identification of the deposition area. High-magnification objectives (40X) were then used for high-resolution imaging of the gold nanorods in the deposition area to count their number, thus improving the statistical accuracy and minimizing missed nanorods. This method achieved a linear detection range spanning three orders of magnitude for PSA (0.01-10 pg/mL) with an LOD of 8 fg/mL (226 aM), and a LOD as low as 6.5 aM for human papillomavirus-related DNA sequences, greatly enhancing the potential of this method for detecting low-abundance biomarkers in clinical samples.

Catalytic Particle Labeling

Fluorescent or plasmonic particles are typically detected using fluorescence or dark-field imaging, which requires high-resolution, complex microscopic imaging systems. Mobile devices, such as smartphones, generally only support bright-field imaging, making it challenging to detect fluorescent or plasmonic particles, and posing significant challenges for high-sensitivity, real-time on-site detection. Precious metal particles like platinum (Pt) and palladium (Pd) have excellent catalytic properties and can efficiently catalyze various chemical reactions, making them commonly used catalysts in chemical synthesis. By replacing biological enzyme molecules with precious metal catalysts, the chemical reactions catalyzed by these metals can amplify single-molecule signals, thus replacing the enzyme-catalyzed signal amplification process in DIs. This approach not only addresses the activity differences between different enzyme molecules but also avoids issues related to enzyme stability. Wang et al. used catalytically active Pt nanoparticles to label single-molecule immunocomplexes and employed the gas bubbles generated by Pt-catalyzed hydrogen peroxide decomposition as the single-molecule signal.[32] Using a mobile phone's bright-field imaging as the detector, they developed a highly sensitive and portable DI. In their study, capture magnetic beads, target proteins, and biotinylated detection antibodies formed a sandwich immunocomplex. The Pt nanoparticles modified with anti-biotin antibodies reacted with the biotin in the immunocomplex to complete the Pt particle labeling. The immunocomplex was then mixed with H_2O_2 and introduced into a microbubble generation chip with a square PDMS microwell array (microwell size: 14 μm × 14 μm, depth 7 μm) under the influence of an external magnet. When the ratio of immunocomplexes to microwells was less than 1, the loading of immunocomplexes into the microwells followed a Poisson distribution. In microwells containing target proteins, Pt nanoparticles catalyzed the decomposition of H_2O_2, generating oxygen gas, which accumulated to form bubbles detectable by the mobile phone microscope. In microwells lacking target proteins, Pt nanoparticles were absent, and minimal oxygen was produced by the natural decomposition of H_2O_2, resulting in no detectable bubbles. Therefore, the number of microbubbles was directly proportional to the concentration of the target protein. The time for bubble formation and the size of the bubbles varied with target proteins. Conventional microscope observations showed that after 8–9 minutes of reaction, the number of detectable bubbles stabilized. Researchers used a smartphone (iPhone 6s) in combination with a commercial mobile microscope (magnification up to 9×) to capture bright-field images of microbubbles on the bubble-generation chip. They applied a convolutional neural network to identify and count the microbubbles, significantly improving the reliability and accuracy of microbubble image processing algorithms. After training the algorithm with 493 frames of images, the program successfully identified the edges of the microarray and counted the microbubbles within one minute. The counting results from 22 frames of test images matched those obtained from manual with ImageJ, achieving automatic and accurate counting through machine learning.

The researchers used PSA, a marker protein monitored after prostate cancer surgery, and beta-subunit human chorionic gonadotropin (β-hCG), a marker protein for early pregnancy detection, as examples to validate the performance of the microbubble counting DI in clinical samples. In the range of 0.06–1 pg/mL, the number of microbubbles showed a linear relationship with PSA concentration. When the number of microbubbles reached 500, fusion of adjacent microbubbles occurred, meaning that the bubble density reached saturation. The LOD for PSA was as low as 2.1 fM (0.060 pg/mL), which was 167 times lower than the LOD of electrochemiluminescence assays used in clinical central laboratories (Roche Elecsys Cobas Total PSA assay, 0.01 ng/mL). The average coefficient of variation for PSA analysis was 16.5%, which was slightly higher, and might be further reduced through microfluidic automation for sample preparation, reaction mixing, washing, and other steps. The blind serum test results from 13 prostate cancer patients were largely consistent with the clinical electrochemiluminescence detection results. For β-hCG, in the concentration range of 3.75–30 pg/mL, the number of microbubbles was linearly related to the concentration, with an LOD as low as 0.034 mIU/mL (2.84 pg/mL), showing significantly better sensitivity than existing central laboratory techniques (Beckman Coulter chemiluminescent immunoassay, 0.5 mIU/mL) and point-of-care testing technologies (Abbott i-STAT Total b-hCG Test, 5 mIU/mL).

The microbubble counting-based immunoassay platform was also used for the high-sensitivity detection of SARS-CoV-2 antigens in nasopharyngeal swabs.[33] This platform helped reduce false-negative results and could quantitatively track the antigen load in respiratory samples from different groups. In nasopharyngeal swabs, SARS-CoV-2 released nucleocapsid protein (N protein) upon cleavage. An immune reaction occurred in solution, generating microbubbles on a bubble-generation chip. These microbubbles were then imaged and automatically counted. The LOD reached 0.5 pg/mL (10.6 fmol/L), corresponding to 4,000 copies/mL of inactivated SARS-CoV-2. Analysis of 372 clinical nasopharyngeal swab samples showed that the positive and negative percentage agreement between the microbubble counting immunoassay and RT-PCR assay was 97%. For individuals with persistent positive nucleic acid swab test and strong immune responses, the antigen positivity rate decreased as the number of days since symptom onset increased. For patients with malignant blood diseases that impair the immune system, antigen testing could be performed over a prolonged and variable period. As a quantitative indicator of antigen load, the total volume of microbubbles was inversely proportional to the cycle threshold and the number of days since symptom onset. Viral sequence mutations could also be detected in patients who maintained high antigen loads for a prolonged period. Therefore, the microbubble counting-based viral antigen detection platform can be used to screen patients undergoing high-activity viral replication and closely monitor viral sequence changes.

Microparticle Labeling

In heterogeneous DIs using magnetic beads as the immunoreaction carriers, subsequent deposition and spectral imaging of the labeled particles are required. Whether through natural sedimentation or charge attraction, existing deposition techniques cannot achieve 100% deposition and ensure uniform distribution. Particle aggregation, which causes signal overlap and missed detection, is inevitable, and random imaging analysis increases the uncertainty of low-concentration sample analysis. To address this, we proposed a strategy that involves solvent evaporation-induced particle deposition, grid-assisted multiframe imaging, and microparticle signal enhancement.[34] This strategy has preliminarily achieved nonstochastic counting quantification of proteins, effectively suppressing Poisson noise in DI of ultralow concentration samples. The principle of the method is as follows: Capture antibodies were modified onto magnetic beads, and detection antibodies were modified onto PS spheres. The capture magnetic beads successively incubated with the antigen and the PS spheres to form immunocomplexes (magnetic beads-antigen-PS spheres). Magnetic separation was used to remove unbound PS spheres. The PS spheres in the immunocomplexes were then eluted and concentrated in a centrifuge tube. A PDMS sheet with 3 mm diameter holes (1 mm thickness) was made and reversibly sealed with a hemocytometer to form a PS sphere counting chip. A certain volume of the eluted PS sphere solution from the immunocomplexes was added to the microwells of the counting chip. The solution was heated to evaporate and remove water. This liquid-addition-evaporation process was repeated until all eluted PS spheres were fixed to the bottom of the microwells. The grid lines at the bottom of the microwells assisted in determining the positions of the spheres. A microscope (Olympus IX71) was used to take multiframe bright-field images of all the PS spheres deposited on the bottom of the microwells. The number of PS spheres in all microscopic images was counted, and the antigen concentration was calibrated based on a custom proportional factor.

To achieve 100% counting of all labeled particles, we made improvements to the single-particle imaging and counting technique from three aspects: (1) nanoparticles with micron-sized PS spheres to enhance the signal resolution of individual particles and ensure accurate recognition of particle number even with aggregates; (2) employing water evaporation-induced particle deposition guarantees high-density deposition of all labeled particles in a limited area, achieving complete deposition of the labeled particles; (3) using a micro-fabricated auxiliary grid allows precise localization of each frame image, eliminating issues of imaging overlap or region omission, ensuring that all deposited particles are accurately counted. Thus, we transformed the existing random deposition, random imaging, and random statistical approach of single-particle counting analysis into a new way of full deposition, full imaging, and full statistics, which enhanced the sensitivity and accuracy of DIs. Using 5 aM CEA as a positive control and PBS buffer as a negative control, we optimized parameters such as the amount of capture magnetic beads, antigen incubation time, blocking time, the amount of PS microspheres, and incubation time. Under optimal reaction conditions, the linear range for nonrandom counting analysis of CEA in PBS buffer reached three orders of magnitude (5×10^{-18} –5×10^{-16} M), with a relative standard deviation between 2.2% and 7.7% (measured three times). The LOD reached as low as 4.9 aM. In plasma samples without CEA, 20 aM and 40 aM CEA were added, and the relative standard deviation of the measured CEA concentration was less than 8%, with recovery rates of 92.5% (for 20 aM) and 98.8% (for 40 aM), confirming the specificity and accuracy of this method for blood sample analysis. We collected two clinical plasma samples, diluted them 105 times, and used the standard addition method to quantify CEA. The obtained CEA concentrations were 0.53 ng/mL and 0.46 ng/mL, respectively, which were close to the chemiluminescent immunoassay results provided by the hospital (0.54 ng/mL and 0.78 ng/mL). This method suppressed the Poisson noise from random counting analysis, reduced the LOD, and improved the accuracy of the results. It is suitable for the detection of rare proteins in clinical samples and shows potential for early screening of major diseases.

Nucleic Acid Sequence Labeling

Nucleic acid molecule labeling involves using specific transformation technologies to link the target protein with nucleic acid probes, converting the counting of protein molecules into that of target nucleic acid probes. Nucleic acid signal amplification technologies are then employed to amplify the of individual target nucleic acid probes, followed by imaging or flow cytometry to determine the number of target nucleic acid probes, and thus, quantify the concentration of the target protein. Based on differences in nucleic acid molecule detections, nucleic acid-based heterogeneous DI can be divided into two categories: (1) PCR technology to amplify nucleic acid molecule signals, followed by fluorescence imaging detection; (2) nanopore sensor flow cytometry for counting nucleic acid molecules.

Walt's research group used magnetic beads as immunocapture carriers and DNA molecules as labeling signals for target proteins.[35] They employed rolling circle amplification (RCA) technology to amplify the signal of DNA molecules, developing a highly sensitive, non-microhole, heterogeneous dropcast single-molecule assays (dSimoa), as shown in Figure 4.8. The process involves the following steps: Capture magnetic beads, target proteins, and biotinylated antibodies formed an immunocomplex with

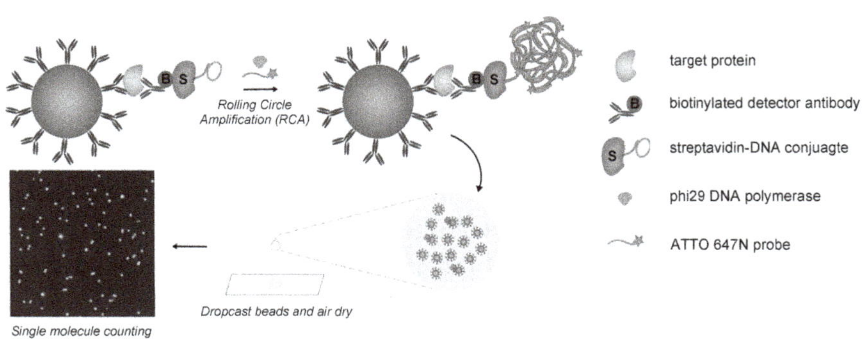

FIGURE 4.8 Scheme shows the principle of dropcast single-molecule assays.

Reproduced from Wu, C.; Garden, P. M.; Walt, D. R. Ultrasensitive Detection of Attomolar Protein Concentrations by Dropcast Single Molecule Assays. *Journal of the American Chemical Society* 2020, 142 (28), 12314–12323 (ref 35). Copyright 2020 American Chemical Society.

magnetic beads (target magnetic beads). Pre-annealed RCA primers and circular DNA templates formed conjugates, which were then labeled with streptavidin and bound to biotin on the target magnetic beads to form DNA-labeled magnetic beads. Under the action of DNA polymerase, RCA occurred on the target magnetic bead surface, forming DNA multimers. Fluorescently labeled DNA probes underwent in situ hybridization reactions, accumulating a large number of fluorescent molecules on the target magnetic bead surface. After cleaning and concentrating the empty and target magnetic beads, they were transferred to a glass slide, and the water was evaporated to form a monolayer film of fixed magnetic beads. Fluorescence imaging detected the number of fluorescent microbeads, and bright-field imaging counted the total number of beads. The ratio of both determined the target protein concentration. The RCA process on the surface of the microbeads not only amplified the signal but also confined the signal near the beads, eliminating the need for microwell-limited fluorescence signals. This avoided the low bead loading efficiency issue and improved the sample loading efficiency to 40%–50%, thereby increasing detection sensitivity. On the other hand, the initial bead usage was reduced from 500,000 to 100,000. This improved the sampling efficiency for rare proteins, increased the ratio of target beads to blank beads, and reduced Poisson noise. For IL-10 and IL-1β, the LOD of dSimoa reached the aM level, with sensitivity 15 times and 25 times higher than Simoa based on microwell arrays and enzyme signal amplification, respectively. The quantification limit (LOQ) was improved by one order of magnitude. The recovery rate of different concentrations of human recombinant IL-10 added to human saliva samples ranged from 76% to 122%, confirming the reliability of the method for protein detection in human saliva samples. Gradient dilution experiments of IL-10 in human saliva showed that the matrix interference from saliva was negligible.

To develop the clinical diagnostic value of dSimoa, they conducted a test of the mouse short-tail mutant phenotype (Brachyury) in plasma. Brachyury is a transcription factor protein product of the T-box gene, which regulates notochord-derived tissues and tumor development. It is a sensitive and specific marker for chordoma (a type of primary bone cancer in the spine or cranial base). To date, there have been no reports of detecting Brachyury in the serum of chordoma patients. The LODs of Brachyury using dSimoa and Simoa were 244.6 aM and 841.4 aM, respectively. The sensitivity of dSimoa was approximately three times higher, though the increase was not substantial due to a reduction in the amount of capture magnetic beads, which lowered the immunocapture efficiency, offsetting the signal-to-noise ratio improvement from the increased sample loading efficiency. Spiked recovery experiments in serum and plasma showed that the recovery rate for Brachyury spiking reached at least 65%–70%, with a relative error of less than 10%. Using plasma samples from six chordoma patients, six healthy donor serum or plasma samples, and one plasma sample from a chondrosarcoma patient, the following results were observed: (1) Endogenous Brachyury in the plasma of all six chordoma patients was below the LOD of conventional Simoa and was not detected by Simoa. However, dSimoa detected Brachyury expression levels in all samples. (2) The plasma from one chondrosarcoma patient was similarly not detected by Simoa but was detected by dSimoa. (3) In six healthy donor serum or plasma samples, one was detected by Simoa and three were detected by dSimoa. This demonstrated that even a small increase in sensitivity could sometimes meet the clinical measurement needs of important biomarkers. Compared with conventional Simoa, dSimoa's efficiency in quantifying low-concentration ranges in plasma or serum samples is attributed to several factors:

(1) Low LODs and high sample efficiency increase the accuracy of measurements in low-concentration ranges. (2) The reduction of microbead usage by a factor of five reduces matrix effects, improves signal-to-noise ratio and recovery, ensuring the accuracy of serum or plasma sample analysis. (3) Compared to large biomolecules (e.g., enzymes), nonspecific adsorption of oligonucleotide-labeled probes is smaller. (4) The increased washing volume further reduces interference from serum or plasma components.

The solid-state nanopore sensor for quantitatively detecting specific proteins in serum samples requires precise and reliable recognition of the electrical signal generated by individual target protein molecules. Tabard-Cossa et al. designed a pair of star-shaped dsDNA nanostructures as probes (P1 and P2) and connecting DNA.[36] Half of the connecting DNA was complementary with P1 and P2, respectively. Therefore, P1 and P2 were then connected to form a dumbbell-shaped DNA nanostructure (DB) by the connecting DNA. In brief, P1 and P2 each consisted of 12-armed star-shaped dsDNA, with 11 of the arms being 25 bp long, and the remaining arm extended to form 175 bp of linear dsDNA in P1, or 150 bp in P2. The extended arms ended with a 25-nt single-stranded DNA region, which was complementary to half of the connecting DNA sequence. In other words, these single-stranded DNA regions served as the binding sites for P1 and P2 to the connecting DNA. Using these specially designed probes and the solid-state nanopore sensor technology, the researchers developed a DI for protein biomarkers. Capture magnetic beads extracted target proteins from complex matrices, and streptavidin-labeled detection antibodies bound to the target protein to form a sandwich immunocomplex with the magnetic beads. Biotinylated single-stranded connecting DNA (50 nt) bound with the detection antibody, labeling the target protein. This process converted the quantification of target protein into that of connecting DNA. Ultraviolet light was used to detach the connecting DNA from the bead. After centrifugal recovery, the connecting DNA solution was incubated with probes of known concentration, followed by high-concentration salt solution for nanopore detection. When unbound probes (P1 or P2) passed through the nanopore, they generated a single electrical peak, viewed as a digital signal "0". When the dumbbell-shaped complex (P1+P2) passed through the nanopore, it generated a double electrical peak, viewed as a digital signal "1", thereby converting the nanopore electrical signal into a digital signal for easy counting. Under a fixed probe concentration, the proportion of dumbbell-shaped complexes was used to calibrate the concentration of the target protein in biological samples.

Under optimized experimental conditions, the LOD for recombinant thyroid-stimulating hormone (rTSH) was approximately 20 pM, with a recovery rate of 121%±5% for serum samples spiked with 0.48 nM rTSH, which is at the edge of the acceptable recovery range (80%–120%). The slightly higher recovery rate may be due to nonspecific adsorption and matrix effects. To improve sensitivity, the team used 30 nm bifunctional gold nanoparticles (with detection antibodies and connecting DNA modified on the particle surface) instead of biotinylated connecting DNA. This approach converts the detection of a single-protein molecule into that of several hundred connecting DNA molecules, increasing the dumbbell-shaped complex by about 100 times, and the LOD was reduced to 385 fM. Compared to existing nanopore immunosensing techniques, this method has a linear range from high fM to low nM, and the sensitivity was improved by approximately 1000 times, which meets the detection needs for rare biomarkers in serum.

MULTIPLE-COMPONENT HETEROGENEOUS DIGITAL IMMUNOASSAYS

Many biological functions require multiple proteins to work together in signaling pathways. Simultaneous detection of multiple proteins is of great significance for monitoring biological dysfunction and screening for major diseases. Due to differences in protein expression levels, protein concentrations within the same biological sample can span several orders of magnitude. Even for the same protein, its expression levels can vary significantly in different samples, making the simultaneous detection of multiple proteins a great challenge. Multiplex detection requires two types of signals for quantitative measurement of multiple proteins. The signal that identifies the protein's identity is called the qualitative signal, and the signal that determines the protein's concentration is called the quantitative signal. Qualitative and quantitative signals can exist independently, or they can be combined into a single signal. This section introduces the heterogeneous multiplex DIs categorized by how their in qualitative and quantitative signals and implemented.

Qualitative Signal Independent of Quantitative Signal

In the mode where qualitative and quantitative signals are independent, a multiple-component DI requires assigning a unique identity signal to each protein being detected during immunocomplex formation. In other words, each protein is encoded with its own distinct signal, meaning that for N proteins, N unique encoding signals are needed. A common signal is needed to count the immunocomplexes. Therefore, to detect N proteins, at least N+1 signals are needed. Currently, the encoding is primarily achieved by optical encoding of the microbeads linked to the antibody. Microbeads are encoded with dyes of different wavelengths and intensities, and then the encoded microbeads are modified with the capture antibodies. Proteins are immunocaptured and bound to the beads, converting the optical encoding of the microbeads into that of the proteins. Walt's research team used optical fiber bundle microwell to contain encoded microbeads, targeting six respiratory disease-related proteins to conduct multiple-component DI.[37] Carboxylated PS microbeads were swollen in an organic solvent containing two luminescent substances (C30 and Eu-TTA), allowing the luminescent substances entered into the microbeads. Then the PS bead were immersed into deionized water, causing the beads to contract and physically trapping the luminescent substances inside the microbeads. Finally, the luminescent microbeads were encoded. By adjusting the concentrations of C30 and Eu-TTA, eight types of encoded microbeads were prepared, with six used as specific protein capture beads and one used as a negative control. Target proteins such as VEGF, IP-10, IL-10, EGF, MMP-9, and IL-1β were analyzed. The LODs for saliva sample reached the sub-picomolar level. A total of 291 human saliva samples were collected and divided into three groups: an asthma patient group (164 individuals), a cystic fibrosis patient group (71 individuals), and a healthy control group (56 individuals). The measurement results showed that the expression levels of VEGF, IP-10, IL-8, and EGF were significantly elevated in the cystic fibrosis patient group, and VEGF, IL-8, and EGF were significantly increased in the asthma patient group ($p < 0.01$), which aided in the diagnosis of respiratory diseases. However, the limited number of optical fiber bundle microwells, along with the increased number and types of microbeads, greatly reduced the sample loading efficiency for each component, and the LOD for multiplex immunoassay was significantly higher than for single-component analysis.

Duffy et al. optically encoded magnetic beads by coupling Cy5 and Hilyte 750 on the bead surface and used a disc-shaped microwell array chip to improve sample loading efficiency.[38] They attempted the simultaneous quantification of four cytokines (IL-6, IL-1α, IL-1β, and TNF-α) spiked to bovine serum samples, and the LODs for multiple-component immunoassay were improved to the femtomolar level. However, the detection sensitivity of the assay remained insufficient for detecting many rare proteins, and it required special instrumentation and lengthy procedures, limiting its wider application. In the microwell array combined with fluorescence imaging detector, as the number of components increases, the sample loading efficiency for each individual decreases, making it challenging to increase the number of components analyzed further.

To address the issue of increasing components and sample loading efficiency, Walt et al. developed a surface signal amplification technique for microbeads and adopted a flow cytometry strategy for detecting both qualitative and quantitative signals, and thereby achieved 8-plex protein biomarker measurement.[39] Eight commercially available magnetic fluorescent-encoded microbeads were selected, with each bead surface modified with a capture antibody for one analyte, totally forming eight types of capture microbeads. These eight capture microbeads sequentially reacted with the sample and biotinylated detection antibodies, forming eight biotinylated immunocomplexes. Avidin-labeled DNA fragments bound to the biotin on the immunocomplexes, thereby completing the nucleic acid labeling of the immunocomplexes. The magnetic beads carrying the immunocomplexes underwent RCA, and fluorescent dye molecules aggregated on the bead surface. These beads generating quantitative signals were detected using the flow fluorescence detector. The pure liquid-phase detection mode allowed the sample loading rate of the encoded microbeads to reach 50%–60%, an increase of approximately one order of magnitude compared to the conventional microwell array imaging detection mode, resulting in a 3–12 times increase in sensitivity. Its LOD was down to the attomolar level. The capability of multiplex detection increased to eight proteins (IL-6, IL-1, IL-10, IFN-γ, IL-12p70, IL-5, IL-18, and VEGF). Recovery experiments of diluted 16-fold plasma samples and 8-fold diluted saliva samples showed that the recovery rates of all analytes were within the acceptable range, with consistent dilution linearity and minimal cross-reactivity in the measured concentration range, demonstrating high clinical sample applicability.

Xu et al. developed a two-color encoded microbead using mesoporous microspheres (3 μm in diameter, with a 10 nm pore size) as the carrier beads and two fluorescent dyes for encoding signals.[40] They adsorbed polyetherimide (PEI) polymers labeled with Cy5 and FITC into the mesoporous spheres through electrostatic interactions. Magnetic nanoparticles were then assembled on the surface of the spheres. The entire assembly was coated with silica, modified with polyacrylic acid, and functionalized with carboxyl groups on the surface to provide protein binding sites. The content of Cy5 and FITC inside the spheres was regulated by the ratio of PEI, PEI-Cy5, and PEI-FITC, forming magnetic microbeads with different luminescent properties. Multicolor-encoded capture microbeads, multiple antigens, and various biotinylated detection antibodies underwent immune reactions, forming multiple biotinylated immunocomplexes. Strepavidin-labeled polyHRP (SA-polyHRP) bound to the biotin groups on the immunocomplex, completing the enzyme labeling of the immunocomplexes. The multiple enzyme-labeled immunocomplexes and fluorescent substrates were mixed in a microfluidic chip, forming droplets. By controlling the flow rate, each droplet contained only one encoded microbead. Fluorescence microscopy was used to image and detect the droplets in the storage region. The fluorescence emitted by the encoded microbeads served as the qualitative signal for the protein, and the fluorescence emitted by the enzyme reaction products in the droplets (582–636 nm) served as the quantitative signal for the protein. This enabled high-sensitivity digital analysis of five cytokines (TNF-α, IL-6, IFN-γ, IL-17A, and IL-10), with LODs reaching 75.15 fg/mL (TNF-α), 16.78 fg/mL (IL-6), 112.96 fg/mL (IFN-γ), 7.57 fg/mL (IL-17A), and 1.63 fg/mL (IL-10). The multiple enzymes labeling strategy improved the local enzyme concentration, increased the reaction rate, and allowed detectable fluorescent signals to be generated in picoliter droplets, avoiding the formation and manipulation of femtoliter droplets.

Qualitative Signal the Same as Quantitative Signal

Integrated qualitative and quantitative signals refer to an immunoassay where a target protein, through immune reactions and signal generation, forms a single signal that serves both as the recognition signal (qualitative) and the counting signal (quantitative) for the target protein. To measure N proteins, N signals need to be generated. In the integrated signal multiple modes, the surface of a solid phase (substrate or microbead) is modified with capture antibodies, and multiple target proteins are captured onto the solid-phase surface. The detection antibodies, encoded with dyes or particles, further react, forming an encoded single-molecule immunocomplex, which is then detected using flow cytometry or fluorescence imaging. The encoding process occurs at the section of detection antibody, with codes primarily focusing on emission wavelengths, particle sizes, and so on.

A typical example of this technology using dyes with different fluorescence emission wavelengths is Singulex's Erenna Immunoassay System. By simply upgrading the single-wavelength detection module to a multi-wavelength detection one, this system allows for simultaneous detection of multiple proteins. Gilbert et al. used three fluorescent dyes (Alexa 647, ATTO532, and BV421) to modify detection antibodies for IL-4, IL-6, and IL-10, respectively, enabling the simultaneous detection of the targets using the modified Erenna system.[41] A three-component test on 100 healthy human plasma samples showed that, with 95% confidence, the reference limits for IL-4, IL-6, and IL-10 were less than 0.61 pg/mL, 6.53 pg/mL, and 1.08 pg/mL, respectively. However, to perform multicomponent test with the Erenna system, the excitation light source must be tailored to the number of components, limiting the increase in the number of targets. Cheng-Zhi Huang et al. labeled the detection antibodies for AFP and CEA with gold particles and silver nanoparticles, respectively.[14] Under dark-field imaging, CEA-induced immuncomplexes showed blue light spots, and AFP-induced immunocomplexes showed green light spots. By counting the light spots in the two colors, they achieved the simultaneous quantification of CEA and AFP. However, dark-field imaging is limited by the imaging area and color resolution, making it difficult to extend beyond 2–3 components.

In addition to wavelength, particle size can also serve as a code. Chen et al. used PS beads of varying sizes as codes and employed bright-field microscopy for detection. They then applied artificial intelligence (AI) to decode the bright-field images, enabling the development of a high-sensitivity, multicomponent DI.[42] Magnetic beads (1 μm) were modified with capture antibodies, and PS beads were modified with detection antibodies. The size of the PS beads was used for encoding, with different detection antibodies corresponding to different-sized PS bead. The immune reaction formed sandwich immunocomplexes, and the immunocomplexes were magnetically separated. The number of unreacted PS beads in the supernatant was inversely related to the protein concentration. The supernatant was collected and transferred to the surface of a substrate for bright-field multiframe imaging. AI algorithms were used to automatically recognized the images of each PS bead size, intelligently classified and counted the PS beads, and filtered out interference from the matrix and background signals, completing signal decoding and counting. The quantity of proteins could be accurately quantified based on the number of different-sized PS beads. AI automatically and rapidly analyzes multiple frames of images, significantly improving throughput and efficiency compared to manual counting, reducing the counting process to just a few minutes. They focused on three inflammation-related biomarkers (IL-6, C-reactive protein (CRP), Procalcitonin (PCT), using 2 μm, 3 μm, and 4 μm as detection probes in DI. The linear ranges for CRP, PCT, and IL-6 were 75 pg/mL–1 μg/mL, 50 pg/mL–500 ng/mL, and 20 pg/mL–100 ng/mL, respectively. The LODs were 25.88 pg/mL for CRP, 17.30 pg/mL for PCT, and 3.66 pg/mL for IL-6. Spike recovery experiments with 10% fetal bovine serum showed recovery rates of 95.2%–115.3%, 106%–98.7%, and 113.7%–107.5% for CRP, PCT, and IL-6, respectively, all within the acceptable range, confirming the reliability and specificity of the method. This approach also has great potential for expanding to small molecule detection, such as neomycin, kanamycin, and chloramphenicol, demonstrating significant application potential.

One Shared Quantitative Signal and No Qualitative Signal

The absence of qualitative signals means that only a single signal is generated during the immunoassay process, regardless of how many proteins are analyzed. There is only one common quantitative signal, and there is no encoding or decoding process for qualitative signals. In this scenario, protein types cannot be identified through the signal; instead, microfluidic chip technology is used to create different test regions through arrayed microchannels. The corresponding relationship between the pre-set test regions and the components to be analyzed is established, and multicomponent identity recognition is completed based on the spatial positioning of the test regions. When analyzing N proteins, a DI without qualitative signals generates only one signal but requires N immunocapture strips. Whether the formation of the immunocapture strips or the operation of the immunoreaction fluid, multicomponent test without qualitative signals relies on the development of microfluidic chip technology. Chen et al. combined microfluidic chips with AI to develop a machine learning-assisted microfluidic nanoplasmonic resonance DI.[43] The immunocapture strips for six cytokines (IL-1β, IL-2, IL-6, IL-10, TNF-α, and IFN-γ) were prepared on a glass substrate, and silver nanocubes were used to label detection antibodies, forming six detection probes. Parallel arrayed channels were orthogonally sealed with the immunocapture strips, creating square test regions. The sample and the six detection probes were mixed and reacted within the test regions, forming immunocomplexes labeled with silver nanocubes. EMCCD was used to perform dark-field imaging of each test region, capturing digital LSPR images of sliver nanocubes. Based on a machine learning image processing method, the light silver spots in the images were intelligently identified and counted, with the results corresponding to the test regions (i.e., the cytokine types), enabling the accurate quantitative classification of the cytokines. Within the range of 1–10,000 pg/mL, all six cytokines showed clear positive signals, spanning four orders of magnitude. In the 1–200 pg/mL range, IL-1β (R^2 = 0.998), IL-10 (R^2 = 0.995), TNF-α (R^2 = 0.998), and IFN-γ (R^2 = 0.998) showed high-quality linear correlations. In the 1–100 pg/mL range, IL-2 (R^2 = 0.999) and IL-6 (R^2 = 0.999) exhibited very high linear correlations. The LODs for the six cytokines were 0.91 pg/mL (IL-1β), 0.47 pg/mL (IL-2), 0.46 pg/mL (IL-6), 1.36 pg/mL (IL-10), 0.71 pg/mL (TNF-α), and 1.08 pg/mL (IFN-γ). In severe COVID-19 patients' serum, cytokine concentrations spanned multiple orders of magnitude. For example, IL-1β ranged from 0.5–130 pg/mL, IL-2 from 1–18 pg/mL, IL-6 from 1–10,000 pg/mL, IL-10 from 1–20 pg/mL, TNF-α from 1–1,000 pg/mL, and IFN-γ from 4–80 pg/mL. Multicomponent immunoassays generally cannot achieve such a wide concentration range. Researchers used this platform to test serum samples from 40 severe COVID-19 patients. Among 25 untreated COVID-19 patients, all had abnormally high levels of IL-6 in their serum, while 52%, 60%, 64%, 52%, and 80% of patients had high levels of IL-1β,

IL-2, IL-10, TNF-α, and IFN-γ, respectively. The high expression of pro-inflammatory cytokines and anti-inflammatory cytokines was consistent with literature reports. The high expression of serum cytokines indicates that most patients were experiencing a cytokine storm. The test of these serum cytokines confirmed that the DI platform offers high sensitivity, accuracy, a wide dynamic range, and multicomponent detection capabilities, which is beneficial for precise and dynamic monitoring of inflammatory responses. It helps reveal the nature of cytokine storms in COVID-19 patients and their immune status, aiding in the efficient, graded treatment of COVID-19.

REFERENCES

1. Tessler LA, Reifenberger JG, Mitra RD. Protein quantification in complex mixtures by solid phase single-molecule counting, *Analytical Chemistry*, 2009, 81, 7141–7148.
2. Li W, Jiang W, Dai S, et al. Multiplexed detection of cytokines based on dual bar-code strategy and single-molecule counting, *Analytical Chemistry*, 2016, 88, 1578–1584.
3. Xue Q, Jiang D, Wang L, et al. Quantitative detection of single molecules using enhancement of Dye/DNA conjugate-labeled nanoparticles, *Bioconjugate Chemistry*, 2010, 21, 1987–1993.
4. Wegner KD, Hildebrandt N. Quantum dots: bright and versatile in vitro and in vivo fluorescence imaging biosensors, *Chemical Society Reviews*, 2015, 44, 4792–4834.
5. Jiang D, Wang L, Jiang W. Quantitative detection of antibody based on single-molecule counting by total internal reflection fluorescence microscopy with quantum dot labeling, *Analytica Chimica Acta*, 2009, 634, 83–88.
6. Zhang Q, Li J, Pan X, et al. Low-numerical aperture microscope objective boosted by liquid-immersed dielectric microspheres for quantum dot-based digital immunoassays, *Analytical Chemistry*, 2021, 93, 12848–12853.
7. Han S, Deng R, Xie X, et al. Enhancing luminescence in lanthanide-doped upconversion nanoparticles, *Angewandte Chemie International Edition*, 2014, 53, 11702–11715.
8. Wang CL, Li XM, Zhang F. Bioapplications and biotechnologies of upconversion nanoparticle-based nanosensors, *Analyst*, 2016, 141, 3601–3620.
9. Farka Z, Mickert MJ, Hlavacek A, et al. Single molecule upconversion-linked immunosorbent assay with extended dynamic range for the sensitive detection of diagnostic biomarkers, *Analytical Chemistry*, 2017, 89, 11825–11830.
10. Mickert MJ, Farka Z, Kostiv U, et al. Measurement of sub-femtomolar concentrations of prostate-specific antigen through single-molecule counting with an upconversion-linked immunosorbent assay, *Analytical Chemistry*, 2019, 91, 9435–9441.

11. Willets KA, Van Duyne RP. Localized surface plasmon resonance spectroscopy and sensing, *Annual Review of Physical Chemistry*, 2007, 58, 267–297.
12. Sevenler D, Trueb J, Ünlü MS. Beating the reaction limits of biosensor sensitivity with dynamic tracking of single binding events, *Proceedings of the National Academy of Sciences of the United States of America*, 2019, 116, 4129–4134.
13. Monroe MR, Daaboul GG, Tuysuzoglu A, et al. Single nanoparticle detection for multiplexed protein diagnostics with attomolar sensitivity in serum and unprocessed whole blood, *Analytical Chemistry*, 2013, 85, 3698–3706.
14. Ma J, Zhan L, Li RS, et al. Color-encoded assays for the simultaneous quantification of dual cancer biomarkers, *Analytical Chemistry*, 2017, 89, 8484–8489.
15. Wang S, Shan X, Patel U, et al. Label-free imaging, detection, and mass measurement of single viruses by surface plasmon resonance, *Proceedings of the National Academy of Sciences of the United States of America*, 2010, 107, 16028–16032.
16. Li J, Wuethrich A, Sina AAI, et al. A digital single-molecule nanopillar SERS platform for predicting and monitoring immune toxicities in immunotherapy, *Nature Communications*, 2021, 12, 1087.
17. Mok J, Mindrinos MN, Davis RW, et al. Digital microfluidic assay for protein detection, *Proceedings of the National Academy of Sciences of the United States of America*, 2014, 111, 2110–2115.
18. Chuah K, Wu Y, Vivekchand SRC, et al. Nanopore blockade sensors for ultrasensitive detection of proteins in complex biological samples, *Nature Communications*, 2019, 10, 2109.
19. Rissin DM, Kan CW, Campbell TG, et al. Single-molecule enzyme-linked immunosorbent assay detects serum proteins at subfemtomolar concentrations, *Nature Biotechnology*, 2010, 28, 595–599.
20. Leirs K, Tewari Kumar P, Decrop D, et al. Bioassay development for ultrasensitive detection of influenza a nucleoprotein using digital ELISA, *Analytical Chemistry*, 2016, 88, 8450–8458.
21. Obayashi Y, Iino R, Noji H. A single-molecule digital enzyme assay using alkaline phosphatase with a cumarin-based fluorogenic substrate, *Analytical Chemistry*, 2015, 140, 5065–5073.
22. Wu Z, Zhou CH, Pan LJ, et al. Reliable digital single molecule electrochemistry for ultrasensitive alkaline phosphatase detection, *Analytical Chemistry*, 2016, 88, 9166–9172.
23. Yelleswarapu V, Buser JR, Haber M, et al. Mobile platform for rapid sub-picogram-per-milliliter, multiplexed, digital droplet detection of proteins, *Proceedings of the National Academy of Sciences of the United States of America*, 2019, 116, 4489–4495.
24. Maley AM, Garden PM, Walt DR. Simplified digital enzyme-linked immunosorbent assay using tyramide signal amplification and fibrin hydrogels, *ACS Sensors*, 2020, 5, 3037–3042.
25. Todd J, Freese B, Lu A, et al. Ultrasensitive flow-based immunoassays using single-molecule counting, *Clinical Chemistry*, 2007, 53, 1990–1995.

26. Ma F, Liu M, Wang ZY, et al. Multiplex detection of histone-modifying enzymes by total internal reflection fluorescence-based single-molecule detection, *Chemical Communications*, 2016, 52, 1218–1221.
27. Xu S, Wu J, Chen C, et al. A micro-chamber free digital biodetection method via the "sphere-labeled-sphere" strategy, *Sensors and Actuators B: Chemical*, 2021, 337, 129794.
28. Pei XJ, Yin HY, Lai TC, et al. Multiplexed detection of attomoles of nucleic acids using fluorescent nanoparticle counting platform, *Analytical Chemistry*, 2018, 90, 1376–1383.
29. Zhou J, Wang QX, Zhang CY. Liposome-quantum dot complexes enable multiplexed detection of attomolar DNAs without target amplification, *Journal of the American Chemical Society*, 2013, 135, 2056–2059.
30. Wu X, Li T, Tao GY, et al. A universal and enzyme-free immunoassay platform for biomarker detection based on gold nanoparticle enumeration with a dark-field microscope, *Analytical Chemistry*, 2017, 142, 4201–4205.
31. Zhu L, Li GH, Sun SQ, et al. Digital immunoassay of a prostate-specific antigen using gold nanorods and magnetic nanoparticles, *RSC Advances*, 2017, 7, 27595–27602.
32. Chen H, Li Z, Zhang LZ, et al. Quantitation of femtomolar-level protein biomarkers using a simple microbubbling digital assay and bright-field smartphone imaging, *Angewandte Chemie International Edition*, 2019, 131, 14060–14066.
33. Chen H, Li Z, Feng S, et al. Femtomolar SARS-CoV-2 antigen detection using the microbubbling digital assay with smartphone readout enables antigen burden quantitation and tracking, *Clinical Chemistry*, 2022, 68, 230–239.
34. Zhang QQ, Zhang XB, Li JJ, et al. Nonstochastic protein counting analysis for precision biomarker detection: suppressing Poisson noise at ultralow concentration, *Analytical Chemistry*, 2019, 92, 654–658.
35. Wu C, Garden PM, Walt DR. Ultrasensitive detection of attomolar protein concentrations by dropcast single molecule assays, *Journal of the American Chemical Society*, 2020, 142, 12314–12323.
36. He L, Tessier DR, Briggs K, et al. Digital immunoassay for biomarker concentration quantification using solid-state nanopores, *Nature Communications*, 2021, 12, 5348.
37. Nie S, Benito-Peña E, Zhang H, et al. Multiplexed salivary protein profiling for patients with respiratory diseases using fiber-optic bundles and fluorescent antibody-based microarrays, *Analytical Chemistry*, 2013, 85, 9272–9280.
38. Rissin DM, Kan CW, Song L, et al. Multiplexed single molecule immunoassays, *Lab on a Chip*, 2013, 13, 2902–2911.
39. Wu CN, Tyler JD, David RW, et al. High-throughput, High-multiplex digital protein detection with attomolar sensitivity, *ACS Nano*, 2022, 16, 1025–1035.
40. Yi JW, Gao ZH, Guo QS, et al. Multiplexed digital ELISA in picoliter droplets based on enzyme signal amplification block and precisely decoding strategy:

a universal and practical biodetection platform, *Sensors and Actuators: B Chemical*, 2022, 369, 132214.
41. Michele G, Richard L, Jackie F, et al. Multiplex single molecule counting technology used to generate interleukin 4, interleukin 6, and interleukin 10 reference limits, *Analytical Biochemistry*, 2016, 503, 11–20.
42. Zhou Y, Zhao WQ, Feng YZ, et al. Artificial intelligence-assisted digital immunoassay based on a programmable-particle-decoding technique for multitarget ultrasensitive detection, *Analytical Chemistry*, 2023, 95, 1589–1598.
43. Gao ZQ, Song YJ, Te YH, et al. Machine-learning-assisted microfluidic nanoplasmonic digital immunoassay for cytokine storm profiling in COVID-19 patients, *ACS Nano*, 2021, 15, 18023–18036.

CHAPTER 5

Homogeneous Digital Immunoassays

HOMOGENEOUS DIGITAL IMMUNOASSAYS (DIs) are a type of immunoassay in which immunocomplexes are directly partitioned, identified, detected, and quantified without the need to separate free antibodies and matrix molecules. In these assays, a significant number of free antibodies, often labeled with signal tags, coexists with the immunocomplexes. The signals emitted by these free antibodies are typically similar to those from the immunocomplexes, which pose a challenge in distinguishing between the two. If the signals from the immunocomplexes cannot be differentiated from those of the free antibodies, accurately counting the immunocomplexes becomes impractical. Therefore, the critical challenge in homogeneous DIs is to develop methods that can distinguish the signals of immunocomplexes from those of free antibodies. Current homogeneous DI technologies primarily rely on three types of signal differences to achieve this differentiation: optical signals, diffusion, and binding events. Among these, differences in optical signals are the most commonly used, while those based on diffusion and binding events are employed less frequently.

OPTICAL SIGNAL DIFFERENCE-BASED RECOGNITION

Optical signal difference-based recognition can be classified into two primary categories: differences in intensity and differences in spectra. Intensity differences include both intensity enhancement and intensity

DOI: 10.1201/9781003729211-5

reduction, where variations in signal strength are used to differentiate between immunocomplexes and free antibodies. Spectral differences arise from temporal coincidence and spatial overlap of two or more spectra, allowing for the differentiation of signals based on their unique spectral characteristics. The homogeneous DIs that utilize these recognition mechanisms will be discussed in detail in this chapter.

Signal Intensity Difference

The intensity difference between the signals of immunocomplexes and free antibodies is primarily induced through three mechanisms: proximity nucleic acid conversion, plasmonic nanobubble-induced signal enhancement, and resonance energy transfer-induced signal attenuation. The first two mechanisms are associated with intensity enhancement, while the latter involves intensity reduction.

Proximity Nucleic Acid Conversion-Induced Signal

In this approach, two single-stranded nucleic acid (ssDNA) fragments are labeled on the capture and detection antibodies, respectively. When these antibodies form a sandwich-structured immunocomplex, the two ssDNA fragments are brought into close proximity. This proximity enables a series of subsequent reactions, including ligation, amplification, hybridization, and displacement, leading to the conversion of the nucleic acid fragments into a stronger signal. As a result, the analyte is transferred from the immunocomplex into the nucleic acid fragments, which emit a more intense signal. Two strategies are commonly used to establish proximity nucleic acid conversion-induced signal-based recognition: one is proximity ligation assay-based droplet partitioning, and the other is proximity displacement-based microwell partitioning.

Proximity Ligation Assay-Based Droplet Partitioning The proximity ligation assay (PLA) was introduced by Landegren et al. in 2002.[1] Two antibodies were each conjugated with distinct segments of ssDNA. When the immunocomplex formed, the two ssDNA tails came into close proximity. With the assistance of DNA ligase, an excess ssDNA, known as the linker, hybridized with the complementary sequences at the 5' and 3' ends of the adjacent ssDNA strands. This ligation process formed a circular complex comprising the detection antibody, antigen, capture antibody, and ssDNA. The ssDNA portion in the circular complex was then amplified through real-time quantitative polymerase chain reaction (PCR), and the antigen concentration was quantified based on the amount of amplified DNA. This process essentially transformed the detection of the immunocomplex into the detection of DNA (Figure 5.1A). Because cycling PCR amplification was employed, a single immunocomplex was converted into a much larger amount of DNA, significantly enhancing the signal. As a result, the sensitivity of the assay was increased by a factor of 1,000 compared to traditional enzyme-linked immunosorbent assay (ELISA)-based methods.

In 2020, Byrnes et al. conducted PLA in droplets, achieving a simplified DI.[2] They partitioned the PLA solution into a polydisperse droplet array, where immunoreactions and DNA ligation occurred. The solution was then transferred to a centrifuge tube for thermal cycling PCR. The droplets containing immunocomplexes emitted fluorescence after PCR amplification (Figure 5.1B). These fluorescent droplets were imaged using fluorescence microscopy and quantified. The number of glowing droplets increased with antigen concentration (Figure 5.1C). The limit of detection (LOD) for interleukin (IL) IL-8 was 0.793 pM. However, compared to other homogeneous DIs, the performance was suboptimal. This was likely

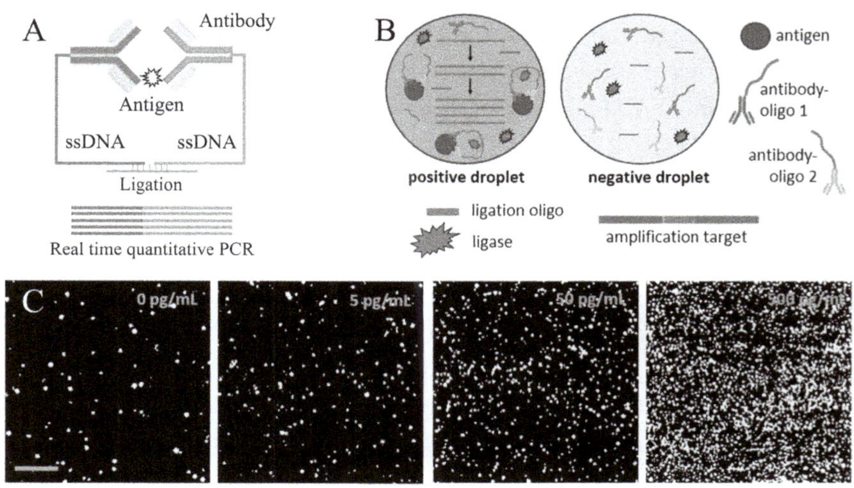

FIGURE 5.1 Schematic of PLA: (A) Conventional PLA, (B) droplet discretization-based PLA, and (C) fluorescence images captured at various antigen concentrations.

Source of Figure B and C: Reproduced from Byrnes, S.A.; Huynh, T.; Chang, T.C.; Anderson, C.E.; McDermott, J.J.; Oncina, C.I.; Bernhard, H.; Weigl, B.H.; Nichols, K.P. Wash-free, Digital Immunoassay in Polydisperse Droplets. *Analytical Chemistry* 2020, 92, 3535–3543 (ref 2). Copyright 2020 American Chemical Society.

due to the high probability of false-positive signals resulting from coincident ligation of the PLA itself. The likelihood of false positives was further exacerbated by reagent enrichment during droplet transfer.

Proximity Displacement-Based Microwell Partitioning PLA requires the usage of enzymes and PCR, and its background noise resulted from nonspecific proximity ligation is high, which reduces sensitivity. Carlo et al. proposed an isothermal, enzyme-free, entropy-driven amplification-based DI.[3] As illustrated in Figure 5.2, the detection antibody was tagged with a ssDNA fragment containing a toehold region (CS1), while the capture antibody was labeled with a different ssDNA fragment (CS2). Upon immunocomplex formation, CS1 and CS2 came into close proximity. This proximity allowed

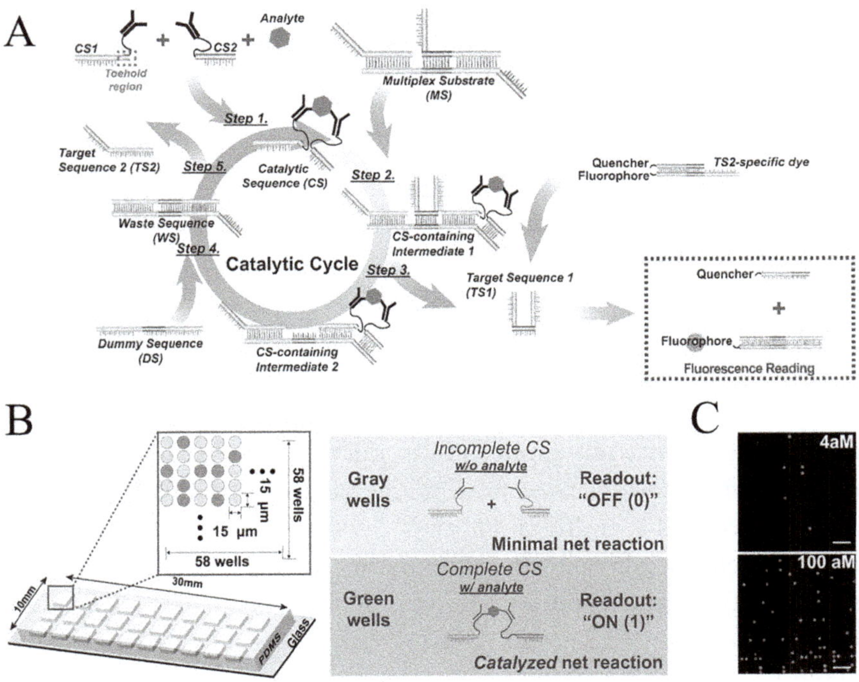

FIGURE 5.2 Schematic illustration of entropy-driven homogeneous DI: (A) overall assay workflow, (B) reactions occurring in "ON" wells (with analyte) and "OFF" wells (without analyte), and (C) representative fluorescence images using influenza A nucleoproteins (NPs) as the target analyte.

Reproduced from Kim, D.; Garner, O.B.; Ozcan, A.; Carlo, D.D. Homogeneous entropy-driven amplified detection of biomolecular interactions. *ACS Nano* 2016, 10, 7467–7475 (ref 3). Copyright 2016 American Chemical Society.

CS1 and CS2 to hybridize with a substrate DNA fragment (MS), triggering a series of displacement reactions. In brief, the hybridization of CS1 and CS2 with MS resulted in CS1 displacing part of MS to form target sequence 1 (TS1). TS1 then hybridized with a signal probe, displacing the quenching sequence and recovering fluorescence. Notably, the displacement reaction only occurred when CS1 and CS2 were in close proximity, i.e., when the immunocomplex was formed. Consequently, the fluorescence signal correlated with antigen concentration. A "dumb" sequence (DS) hybridized with the remaining substrate after TS1 was displaced, releasing the adjacent CS1 and CS2. This indicated that the immunocomplex had detached from MS, and waste sequences and target sequence 2 (TS2) were generated. The detached immunocomplex then reacted with a new substrate, entering the next amplification cycle. In summary, the process required only a single step of mixing CS1, CS2, MS, DS, and the fluorescent probe, without the need for separation or elution. The cycling of CS1 and CS2 within the immunocomplex led to an amplified fluorescence signal through the binding of the displaced TS1 to the fluorescent probe. At the ensemble level, the LOD for influenza nucleoprotein was as low as 10 fM. However, the overall performance was not ideal. This measurement was performed in a microwell array for DI purposes. The microchip, photolithographically fabricated in PDMS, consisted of a 10×3 block array, with each block measuring 2 mm×2 mm. Each block contained 58×58 microwells (a total of 3,364 wells), each with a diameter and depth of 15 microns, corresponding to a volume of approximately 2.65 pL. In the absence of a target, since CS1 and CS2 were not in close proximity, no displacement reaction occurred, and no fluorescence was emitted. These microwells were counted as 0. In contrast, when the target was present, the microwell emitted fluorescence and was counted as 1. The number of fluorescent microwells showed a linear correlation with antigen concentration in the range of 4 aM to 1 fM, significantly enhancing the sensing performance.

Plasmonic Nanobubble-Induced Signal Enhancement

Plasmonic nanobubbles (PNBs) are a type of bubble generated by the vaporization of liquid caused by the heating of noble metal nanoparticles when excited by plasmonic resonance light. Under irradiation, these nanoparticles efficiently convert light energy into heat, resulting in a local temperature increase around the nanoparticles. PNBs exhibit unique photothermal effects and have been applied in various fields, including cell therapy, cell imaging, and drug delivery and release. However, the fundamental theories regarding its generation mechanisms, influencing factors, and dynamics remain incompletely understood. Experimental studies have demonstrated that the scattering intensity of PNBs, generated by gold nanoparticle (GNPs), is influenced by nanoparticle size, structure, aggregation, and laser power intensity. For example, the PNB intensity of 60 nm GNPs has been reported to be approximately ten times greater than that of the gold particles themselves.[4]

Ye et al. developed a PNB scattering enhancement-based DI for virus detection.[5] The assay, which took approximately 40 minutes—comprising 30 minutes of incubation and 10 minutes of detection—achieved an LOD of 100 PFU/mL. Respiratory syncytial virus (RSV) infection is a significant cause of respiratory illness in infants and young children, making early diagnosis crucial for timely and appropriate treatment. In this assay, 15 nm GNPs were cross-linked to RSV antibodies using 3,3'-dithiodipropionic acid disulfonate. When incubated with the virus at room temperature, the antibodies specifically bound to the fusion protein on the virus surface, leading to the adsorption of a substantial, yet undefined number of GNPs onto each virus particle. This resulted in a marked increase in the scattering intensity of the PNBs, allowing for a clear distinction between the virus-bound GNPs and the free, single GNPs. Importantly, the virus detection occurred in a homogeneous manner, without the need for separation steps. PNBs functioned both as an amplified signal and a partitioning tool. The detection setup, shown in Figure 5.3 A-C involves injecting the sample into a square capillary with a 200 μm inner diameter. Under pressure, the sample flows through the detection window at a rate of 6 μL/min. Two laser beams illuminate the detection window: a 532 nm pulsed laser heats the GNPs and generates PNBs, while a 633 nm continuous laser is used to detect PNB scattering. The bubbles have a very short lifetime, lasting only a few nanoseconds. The cylindrical shape, represented by the dashed line in Figure 5.3C, denotes the virtual discretized space. When the virus is encapsulated within the bubble, the PNB intensity significantly increases, indicating a positive detection ("on"). In contrast, free GNPs do not generate PNBs, resulting in low scattering intensity, recorded as "off." Figure 5.3D presents the detection results, which include more than 100 pulses of the virus. Positive and negative signals are distinguished by their amplitude and the area under the curve. The frequency of positive signals is then used to generate a standard curve, as shown in Figure 5.3E.

Although PNB-based detection demonstrates the potential for performing homogeneous DIs for viruses, several aspects require further development. First, detecting proteins, even large ones, remains

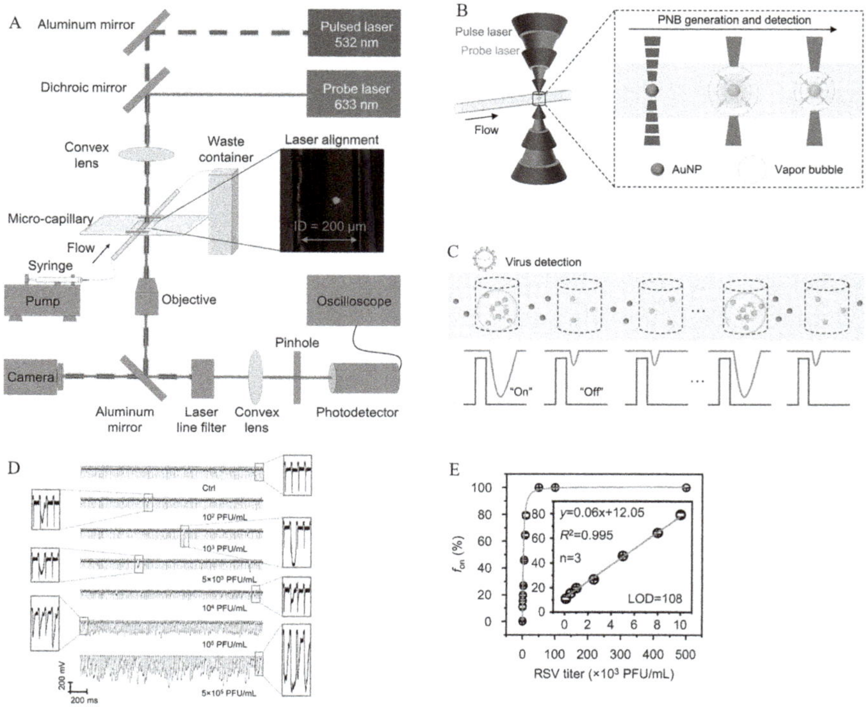

FIGURE 5.3 Schematics and results of plasmonic nanobubble (PNB)-based digital virus detection: (A) equipment setup, (B) generation and detection of PNBs, (C) digital counting strategy for virus detection, (D) representative PNB signal traces from 100 laser pulses, and (E) calibration curve for RSV detection.

Source: Reprinted by permission from Springer Nature Limited, Liu, Y.; Ye, H.H.; Huynh, H.D.; Xie, C.; Kang, P.Y.; Kahn, J.S.; Qin, Z.P., *Nature Communications*, 2022, 13,1687 (ref 5). Copyright 2022.

challenging. While viruses are significantly larger than proteins and can adsorb numerous GNPs, resulting in strong and easily detectable PNB signals, protein molecules typically adsorb only a limited number of GNPs—usually no more than two. It is unclear whether the PNB signals generated by immunocomplexes would be sufficiently strong to distinguish between immunocomplexes and free proteins. Second, the detection efficiency is limited, calculated to be only 10%, based on the size of the laser focal spot (20 μm) and the square capillary (200 μm). Consequently, a large portion of the sample does not pass through the laser focal spot, which compromises the sensitivity and dynamic range of the assay. Third, the system has not yet been demonstrated to detect multiple components simultaneously.

Resonance Energy Transfer-Induced Signal Attenuation

Another example of a difference in optical signal intensity-based homogeneous DI utilizes upconversion nanomaterials (UCNPs) as tags[6]. Polyacrylic acid (PAA)-modified UCNPs (NaYF4:Yb^{3+}, Er^{3+}) with an average diameter of approximately 42 nm were activated with 1-Ethyl-3-(3-dimethylaminopropyl)carbodiimide (EDC) and conjugated with detection antibodies specific to prostate-specific antigen (PSA). Meanwhile, 60 nm GNPs were linked to capture antibodies through thiol groups. In the presence of PSA, the UCNPs and GNPs formed a sandwich-type immunocomplex. The resonance scattering spectrum of the GNPs exhibited a peak at 540 nm, while the fluorescence emission spectrum of the UCNPs peaked at 555 nm. These two spectra overlap significantly, and when the particles were in close proximity—such as within the immunocomplex—resonance energy transfer may occur. Specifically, the UCNPs' fluorescence was absorbed by the GNPs, leading to a reduction or even quenching of the UCNP fluorescence. Figure 5.4 illustrates the immunoassay procedure and the experimental setup. A conventional upright fluorescence microscope was used, equipped with a 980 nm continuous-wave laser as

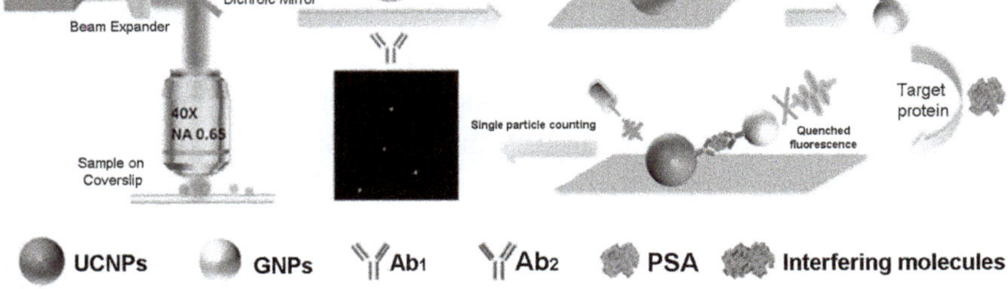

FIGURE 5.4 Principle of a homogeneous DI based on resonance energy transfer between gold nanoparticles and upconversion nanoparticles.

Source: Reproduced from Li, X.; Wei, L.; Pan, L.L.; Yi, Z.Y.; Wang, X.; Ye, Z.J.; Xiao, L.H.; Li, H-W.; Wang, J.F. Homogeneous immunosorbent assay based on single-particle enumeration using upconversion nanoparticles for the sensitive detection of cancer biomarkers, *Analytical Chemistry*, 2018, 90, 4807–4814 (ref 6). Copyright 2018 American Chemical Society.

the excitation source and a 40× objective lens for single-particle imaging. UCNPs were fixed onto a glass slide. In the control experiment (without antigen), the number of UCNPs emitting green light was counted as N_0. Upon the addition of the antigen, resonance energy transfer occurred, reducing the number of UCNPs emitting green fluorescence, which was then counted as N_i. The ratio of the decrease in fluorescence, $(N_0 - N_i) / N_0$, was used to quantify the target PSA. The LOD of this assay was found to be 1.0 pM, with a dynamic range extending up to 60 pM. The relatively high LOD and limited dynamic range were likely attributed to the low efficiency of resonance energy transfer, suboptimal quenching effects, and uneven emission from the UCNPs.

GNPs not only function as energy acceptors that absorb donor energy and quench donor fluorescence but also can act as energy donors themselves, which can be quenched by an acceptor. In 2007, Liu et al. discovered that under dark-field microscopy, GNPs adsorbed with cytochrome c exhibited a quenching dip in their scattering spectrum, with the dip wavelength overlapping the maximum absorption wavelength of cytochrome c.[7] They proposed that cytochrome c absorbed part of the energy from the GNPs through plasmon resonance, resulting in a reduction in scattering intensity. This was the first report on plasmon resonance energy transfer (PRET). Since then, it has become increasingly clear that: (1) PRET is a phenomenon of resonance coupling between noble metal nanoparticles (the donor) and receptor molecules when they are in close proximity, and (2) when the conditions for PRET are met, energy from the donor is non-radiatively transferred to the acceptor molecule. This results in a reduction in the scattering intensity of the donor, which may even exhibit a dip in the spectrum. The prerequisites for PRET are similar to those for fluorescence resonance energy transfer (FRET): the scattering spectrum of the donor must overlap with the absorption spectrum of the acceptor, the dipole moments of both must be aligned, and the donor-acceptor distance must be sufficiently small. However, constructing an efficient PRET system is challenging. Common PRET acceptors include cytochrome c, Cu^{2+} coordination complexes, and I_3^-.

In 2019, we discovered a novel PRET system involving GNPs and QDs, which we utilized to establish a single-particle homogeneous immunoassay.[8] Under dark-field microscopy, the in situ spectra of individual GNPs (70 nm in diameter) were observed both before and after they were conjugated to QD800 via dsDNA. Prior to linking to the QD800, the spectral intensity of individual GNPs remained stable for 80 minutes. However,

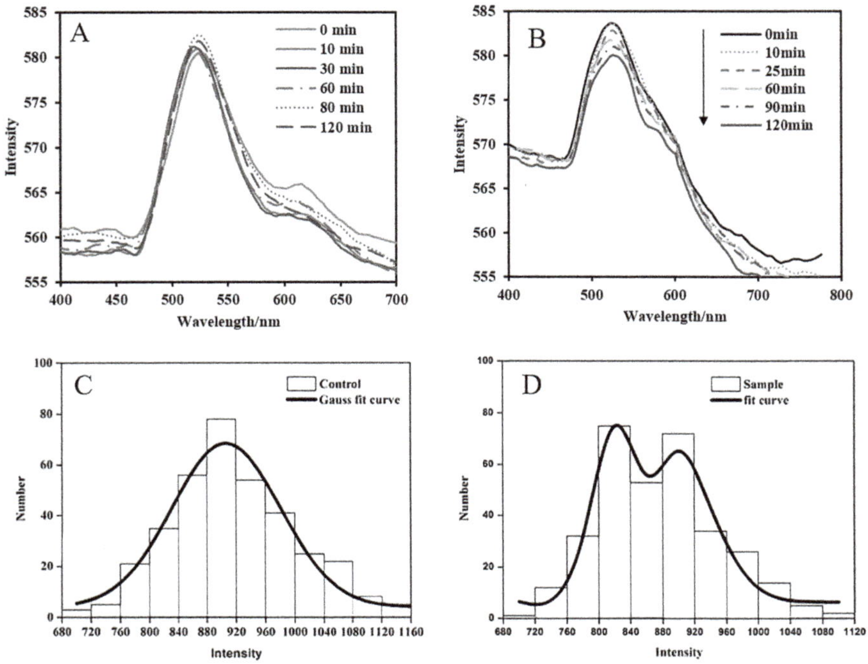

FIGURE 5.5 Scattering spectra and intensity analysis of a single ssDNA-GNP exposed to different probes: (A) Time-dependent scattering spectrum after exposure to ssDNA-QD 800, (B) Time-dependent scattering spectrum after exposure to QD 800, (C) Histogram of scattering intensity in the presence of ssDNA-QD 800, and (D) Histogram of scattering intensity in the presence of QD 800.

Source: Reproduced from Liu, X.J.; Zhang, Y.S.; Liang, A.Y.; Ding, H.W. Gai, H.W. *Chemical Communications*, 2019, 55, 11442–11445 (ref 8), with permission of The Royal Society of Chemistry.

after conjugation to the QDs, a noticeable decrease in intensity is observed (Figure 5.5A, B). Figures 5.5C and D show the distribution histograms of the scattering intensity at the maximum wavelength of the GNPs. Before conjugation, the intensity follows a Gaussian distribution, whereas after conjugation, a bimodal distribution emerges. The efficiency of the PRET process is quantified using the following Equation (5.1):

$$\mathrm{PE} = \frac{\mathrm{P}_a^w}{\mathrm{P}_a^s} - \frac{\mathrm{P}_0^w}{\mathrm{P}_0^s}. \tag{5.1}$$

Here, P_0^w and P_0^s represent the probability of GNPs with an intensity weaker than the expected value in the absence of PRET. Conversely, P_a^w

and P_a^s represent the probability of GNPs with an intensity weaker than the expected value in the presence of PRET. The expected value refers to the average intensity observed without the occurrence of PRET. The parameter PE denotes the degree of distortion in the intensity distribution caused by PRET. In the absence of PRET, no distortion occurs, and PE equals 0. As PRET efficiency increases, a higher proportion of GNPs exhibit intensities weaker than the expected value, resulting in a larger PE value. Furthermore, PE is influenced by the length of the DNA used in the system. When the DNA length ranged from 10 to 27.5 nm, the maximum PE was observed at a DNA length of 15 nm, after which it gradually decreased.

We further adapted the PRET for homogeneous immunoassays, as illustrated in Figure 5.6. In this system, detection antibodies targeting PSA were immobilized onto GNPs through the reaction between their primary amine groups and the carboxyl groups at the 5′ end of dsDNA adsorbed onto the nanoparticles. Corresponding capture antibodies were conjugated to the surface of QDs via a reaction between their primary amine groups and the carboxyl groups on the QD surface. Under dark-field microscopy, the GNPs exhibited strong scattering light. Upon the addition of PSA, specific antigen-antibody binding formed a capture antibody-GNPs@

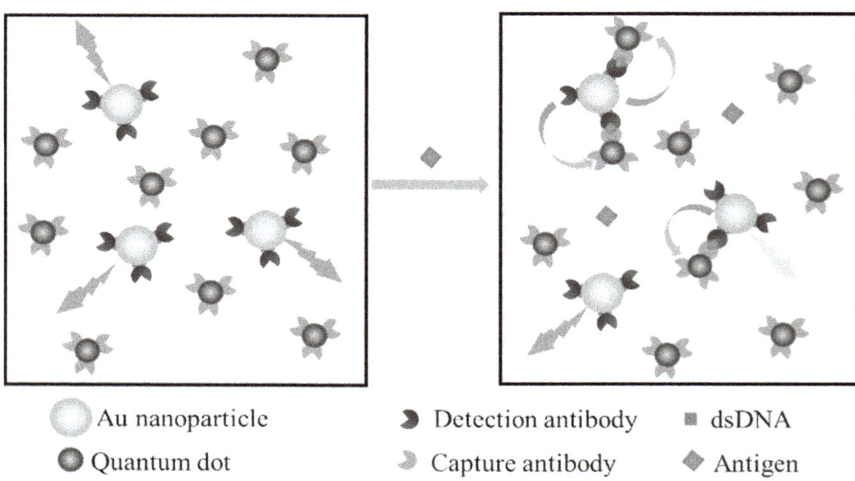

FIGURE 5.6 PRET-based homogeneous immunoassay.

Source: Reproduced from Liu, X.J.; Zhang, Y.S.; Liang, A.Y.; Ding, H.W. Gai, H.W. *Chemical Communications*, 2019, 55, 11442–11445 (ref 8), with permission of The Royal Society of Chemistry.

Antigen@detection antibody-QDs complex. The proximity between the GNPs and the QDs facilitated PRET. This coupling resulted in a reduction in the scattering intensity from the GNPs. Notably, the decrease in intensity correlated with the PSA concentration, with higher target concentrations leading to a more significant reduction. Importantly, no PRET occurred between free GNPs and QDs. Energy transfer efficiency was quantified according to Equation (5.1), and a standard curve was constructed using the response of PRET to different concentrations of PSA standard solutions. At a signal-to-noise ratio of 3σ, the LOD for PSA was as low as 0.2 fM, with a linear range spanning from 0.3 fM to 200.0 fM. The high sensitivity of this assay could be attributed to several factors: the large absorption cross-section of the QDs causing the high PE, the strong and stable scattering light from the GNPs, and the high signal-to-noise ratio achieved through single-particle dark-field imaging.

Strictly speaking, the two previous examples are not considered true DIs. In these assays, the recognition of the immunocomplex is based on the decrease in luminescence intensity caused by resonance energy transfer. The quantitative process relies on statistical counting, such as the proportion of certain single particles, rather than the absolute quantity of the signaling particles. The efficiency of resonance energy transfer plays a critical role in determining the sensitivity of these assays.

Spectral Difference

When the capture antibody and detection antibody are labeled with optical tags of distinct spectra, the free capture or detection antibody generates a single optical signal. In contrast, in the absence of nonspecific adsorption, the immunocomplex produces two optical signals with distinct spectra that either coincide temporally or spatially. This is a straightforward strategy for developing homogeneous DIs. However, it is essential that the optical signals from the labeling tags are sufficiently strong to ensure that the individual signals can effectively surpass background noise in high-throughput detection scenarios.

Temporal Coincidence of Two Tags

In 2006, during the early development of single-molecule counting detection, Nie Shuming et al. labeled the capture antibody and detection antibody specific to tumor necrosis factor-α (TNF-α) with green (40 nm diameter, λ_{ex} = 505 nm, λ_{em} = 515 nm) and red (40 nm diameter, λ_{ex} = 488 nm, λ_{em} = 645 nm) fluorescent microspheres, respectively.[9] The mixture of the

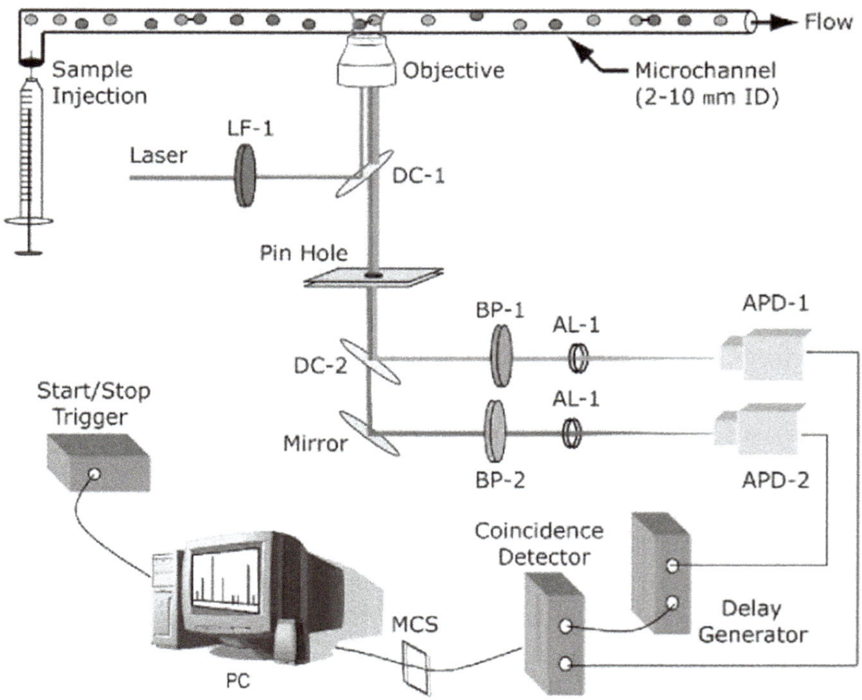

FIGURE 5.7 Instrumental setup for real-time detection of single nanoparticles.

Note: LF-1, laser filter; DC-1, dichroic filter in the microscope filter cube; DC-2, dichroic filter outside the microscope side port; BP-1, band-pass filter (HQ514 M10); BP-2, band-pass filter (D670 M40); AL-1 and AL-2, aspheric focusing lenses; ADP-1 and APD-2, single-photon counting avalanche photodiodes; MCS, multichannel scalar; PC, personal computer.

Source: Reproduced from Agrawal, A.; Zhang, C.Y.; Byassee, T.; Tripp, R.A.; Nie, S.M. Counting single native biomolecules and intact viruses with color-coded nanoparticles, *Analytical Chemistry*, 2006, 78, 1061–1070 (ref 9). Copyright 2021 American Chemical Society.

antigen and the two antibodies was incubated and subsequently injected into a square capillary. The schematic of the instrument is shown in Figure 5.7. A 100x objective lens was used to focus a 488 nm laser onto the detection window, exciting the passing fluorescent particles and collecting the resulting fluorescence. The collected fluorescence passes through the first dichroic mirror to filter out the excitation light. After passing through a 100 μm pinhole, the green fluorescence is reflected by the second dichroic mirror and directed to avalanche photodiode-1, while the red fluorescence passes through the second dichroic mirror, is reflected by a mirror, and then directed to avalanche photodiode-2. Single-photon events from the two detection channels are assessed for coincidence using a coincidence detector. Coincidence occurs when both antibodies are present in the detection window simultaneously. The likelihood of two free antibodies randomly appearing in the detection window at the same time is extremely low. However, the immunocomplex, labeled with two distinct tags, will trigger signals in both detection channels, resulting in a coincident signal. Therefore, the occurrence of coincidence signals serves as an indication of the presence of the immunocomplex. Figure 5.8 illustrates the single-photon signals and corresponding coincidence signals in both detection channels. In the left column, in the absence of antigen, both the green and red detection channels exhibit photon burst signals, but no coincidence signals are observed. In the right column, with the presence of antigen, an immunocomplex forms, and coincident signals are detected by the coincidence detector. The number of coincident signals increases with the antigen concentration, eventually reaching a plateau due to the Hooke effect. The LOD is between 20 and 30 fM, with a detection efficiency of approximately 40%.

Spatial Overlap of Two Tags

When optical tags with distinct spectra, also known as distinctly spectral labels, spatially overlap, their colors will blend. For instance, the spatial overlap of green and red may result in yellow. This color difference is sufficient to distinguish between free antibodies and immunocomplexes. In the case of tags with the same spectral properties (similarly spectral labeling), their spatial overlap must be identified based on their unique optical characteristics, which will be discussed next.

Labeling Using Tags with Distinct Spectra The capture antibody and detection antibody were conjugated to QD 655 (red) and QD 565 (green), respectively. The reaction solution was imaged using an inverted fluorescence microscope equipped with a color CCD camera.[10] As shown in Figure 5.9, in the absence of carcinoembryonic antigen (CEA), no immunocomplex is formed, and both antibodies remain free, with red or green spots visible in the image in Figure 5.9A. Upon formation of the immunocomplex, the red fluorescence of QD 655 is less than 200 nm away from the green fluorescence of QD 565. This distance falls within the diffraction limit, meaning that the red and green fluorescence cannot be spatially resolved by optical microscopy and appear as merged yellow spots (as shown in Figures 5.9B, C, and D). As the antigen concentration

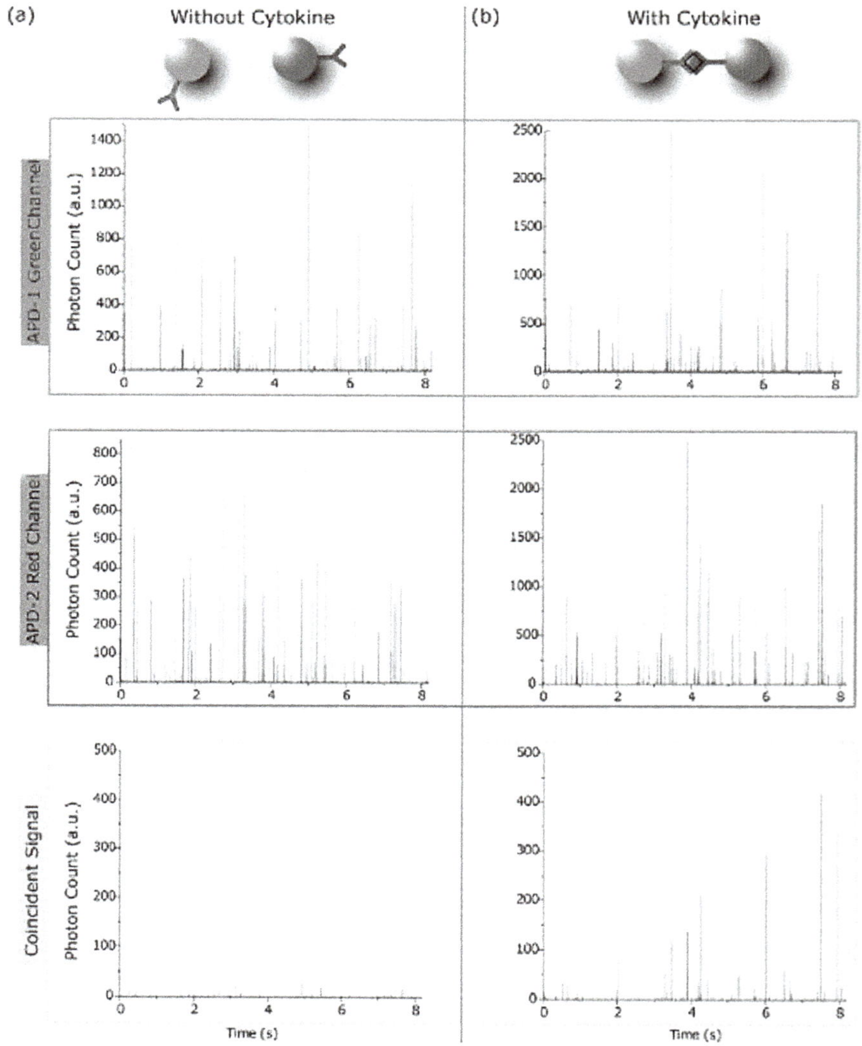

FIGURE 5.8 Real-time photon burst data from antibody-conjugated nanoparticles in a microfluidic channel: Signals were simultaneously recorded from APD-1 (green channel), APD-2 (red channel), and the coincidence channel (yellow). (A) Data from a mixture of green and red probes in the absence of target protein, and (B) Data in the presence of the target protein.

Source: Reproduced from Agrawal, A.; Zhang, C.Y.; Byassee, T.; Tripp, R.A.; Nie, S.M. Counting single native biomolecules and intact viruses with color-coded nanoparticles, *Analytical Chemistry*, 2006, 78, 1061–1070 (ref 9). Copyright 2021 American Chemical Society.

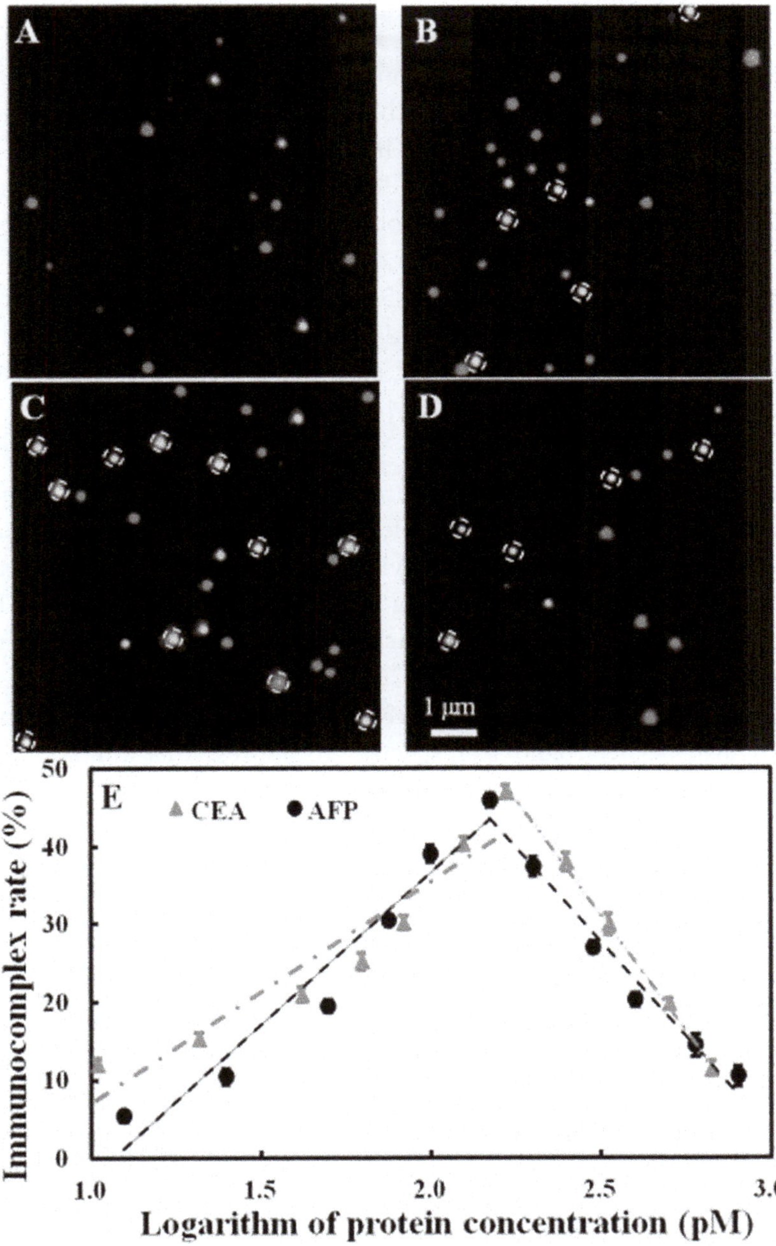

FIGURE 5.9 Fluorescence images of CEA-induced dual-color QD-labeled immunocomplexes at different CEA concentrations: (A) 0 pM, (B) 41.67 pM, (C) 166.67 pM, (D) 333.33 pM, and (E) calibration curve for CEA detection.

Source: Reproduced from Liu, X.J.; Huang, C.H.; Zong, C.H.; Liang, A.Y.; Wu, Z.J.; Zhang, Y.S.; Zhang, Q.Q.; Zhao, W.F.; Gai, H.W. A single-molecule homogeneous immunoassay by counting spatially overlapping two-color quantum dots with wide-field fluorescence microscopy, *ACS Sensors*, 2018, 3, 2644–2650 (ref 10). Copyright 2021 American Chemical Society.

increases, the number of yellow spots increases until the ratio of CEA to CA-QD655 reaches 1:1. After this point, further increases in CEA concentration result in a decrease in the number of yellow spots. The ratio of yellow spots to red spots is defined as the immunocomplex ratio. Quantification is performed by plotting the immunocomplex ratio on the vertical axis and the logarithm of the protein concentration on the horizontal axis, as shown in Figure 5.9E. At a signal-to-noise ratio of 3σ, the LODs for CEA and AFP were 6.1 pM and 9.5 pM, respectively. These LODs were comparable to those of most single-particle counting-based heterogeneous immunoassays.

Labeling Using Tags with Similar Spectra The spectral overlap of distinctly spectral labels is relatively simple and straightforward to achieve. However, due to the limited sensitivity of color CCDs, the LOD remains higher, preventing it from reaching the typical sensitivity levels of DIs. In single-molecule and single-particle imaging, a more commonly used camera is the electron-multiplying CCD (EMCCD), which offers superior sensitivity. However, because the EMCCD operates in grayscale, it cannot distinguish between colors or spectra. To address this limitation, our research group has developed a single-molecule, single-particle spectral imaging technique using a transmission grating.[11–13] As shown in Figure 5.10A, a transmission grating is placed in front of the EMCCD. Fluorescence from single molecules is divided into two beams by the grating: a zero-order spectral beam and a first-order spectral beam. The zero-order beam appears as a dot (the zero-order spot), and the first-order beam as a streak on the EMCCD chip (the first-order streak). Notably, after passing through the grating, the zero-order spectral spot retains its original propagation direction. The distance (L) between the zero-order spectral spot and the first-order spectral streak is related to the fluorescence wavelength (λ) and the properties of the transmission grating, as described by the equation 5.2:

$$L = \lambda S/d \tag{5.2}$$

where S is the distance between the grating and the EMCCD chip, and d is the grating constant (1/70 mm). Figure 5.10B presents single-particle spectral imaging of QDs with varying fluorescence emission wavelengths. To conserve space, the region between the zero-order spot and the first-order streak, where no relevant information is present, has been omitted.

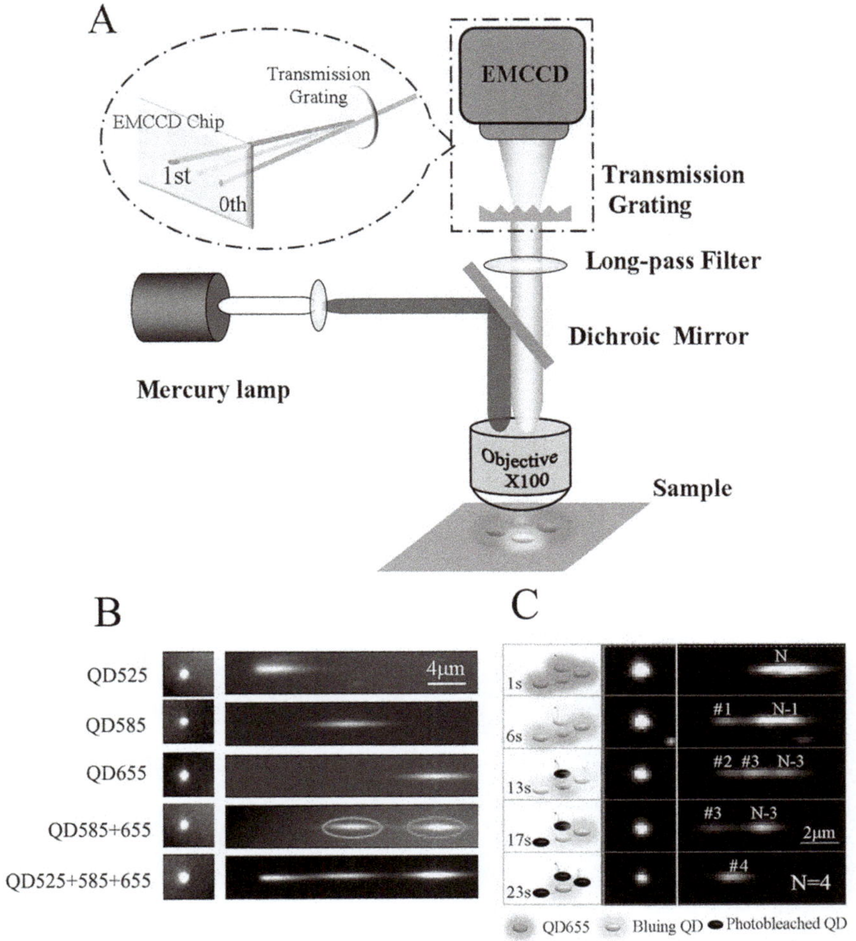

FIGURE 5.10 Spectral imaging of single QDs using a transmission grating-based microscopy system: (A) Equipment setup scheme, (B) Representative spectral images of single QDs, and (C) Separation of first-order streaks from aggregated QDs.

The figures clearly show that the distance between the first-order spot and the zero-order streak varies with the fluorescence wavelength of the QDs. Specifically, as the fluorescence wavelength increases, the separation between the first-order spot and the zero-order streak becomes greater. In cases where two or more QDs with different fluorescence wavelengths spatially overlap, the zero-order spots of the individual QDs cannot be distinguished; however, their first-order streaks may still be resolved. We observed three key optical properties of individual QDs under continuous illumination, as depicted in Figure 5.11. First, the fluorescence

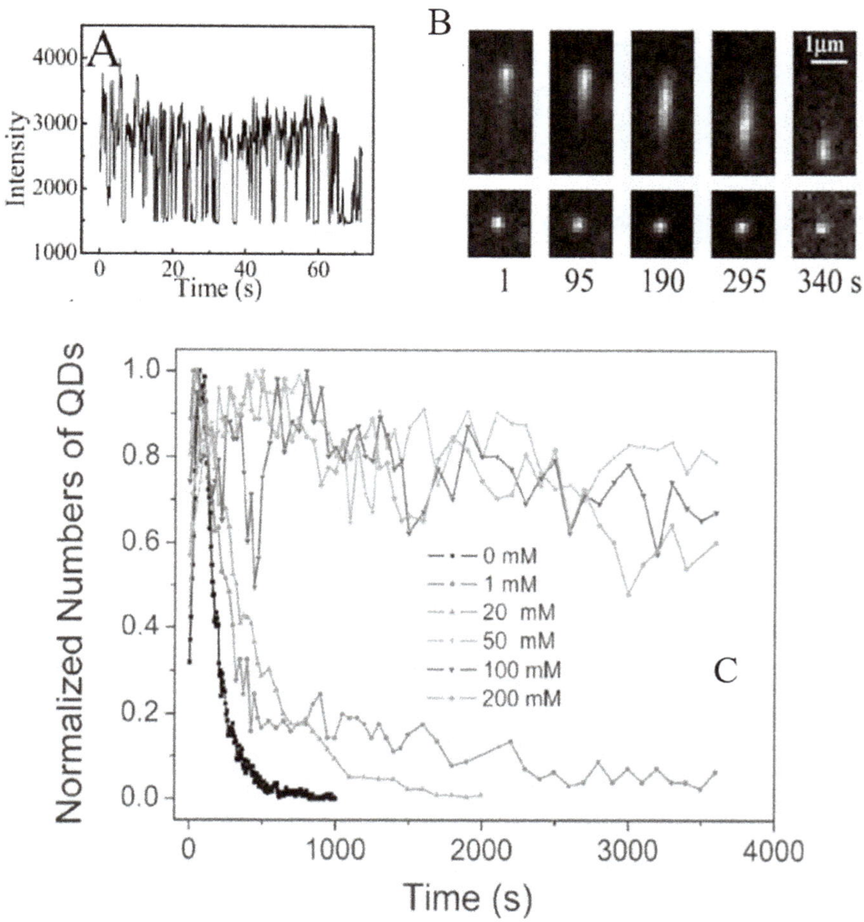

FIGURE 5.11 Optical properties of individual QDs under continuous illumination: (A) blinking, (B) spectral blue shift, and (C) the effect of the spiked monoethanolamine on individual QD photobleaching.

Source: Reproduced from Chen, H.P.; Gai, H.W.; Yeung, E.S. *Chemical Communications*, 2009, 1676–1678 (ref 11), with permission of The Royal Society of Chemistry.

intensity of the QD exhibits fluctuations between bright and dark states, a phenomenon known as blinking (Figure 5.11A). Second, the first-order streak shifts toward the zero-order spot, a process referred to as spectral blue shift (Figure 5.11B). Third, the number of QDs in the field of view gradually decreases over time, a process known as photobleaching (Figure 5.11C). QD blinking occurs when the electron, after optical excitation, becomes temporarily trapped by surface defects, thereby hindering its return to the valence band and subsequent fluorescence emission. The spectral blue shift is attributed to photoxidation, which leads to a reduction in the core size of the QD. The magnitude of the blue shift is dependent on both the excitation wavelength and the type of QD. For QD 525, QD 585, and QD 655, the blue shift typically ranges around 15 nm, 30 nm, and 55 nm, respectively. Although QDs are generally photostable, they are not entirely resistant to photobleaching. The use of mercaptoethylamine has been shown to mitigate both photobleaching and spectral blue shift effects in QDs.

During the investigation of the spectral properties of individual QDs, we observed variations in the rate of spectral blue shift, even among QDs with the same fluorescence wavelength. These differences manifested as a splitting of the first-order streak. Figure 5.10C illustrates the spectral blue shift process of an aggregate consisting of four QDs. Notably, the first-order streak split sequentially, occurring a total of three times. This variation in the blue shift rate is attributed to slight differences in the oxidation rates of the individual QDs. We developed a homogeneous DI based on the labeling of QDs of the same color.[14] In the absence of antigens, both the detection and capture antibodies remained unbound, each associated with a single QD. In this case, the spectra of QDs only exhibited a blue shift without any splitting, as shown in the left column of Figure 5.12. The first-order streak shifted toward the zero-order spot before eventually photobleaching. However, in the presence of antigens, a sandwich structure forms within the immunocomplex, which contained two QDs. Despite the dimer having an overall higher fluorescence intensity, it was not sufficient to distinguish it clearly. Under continuous illumination, the first-order streak of the dimer gradually split into two distinct fringes before ultimately bleaching (right column of Figure 5.12). This splitting occurred because each QD within the immunocomplex underwent blue shift at different rates. The splitting of the first-order streak allowed for the identification of the immunocomplex. When using CEA and alpha-fetoprotein (AFP) as target analytes, we observed that as the antigen concentration increased, the formation of dimers also increased, reaching a maximum when the ratio of capture antibody to antigen was approximately 1:1, and then decreased. The concentration of CEA exhibited a logarithmic linear relationship with the dimer ratio, with linearity observed in the concentration ranges of 0.014–43.892 pM and 43.892–1365.521 pM, corresponding to a total detectable range exceeding five orders of magnitude. Similarly, AFP demonstrated a linear relationship within the ranges of 0.018–51.376 pM and

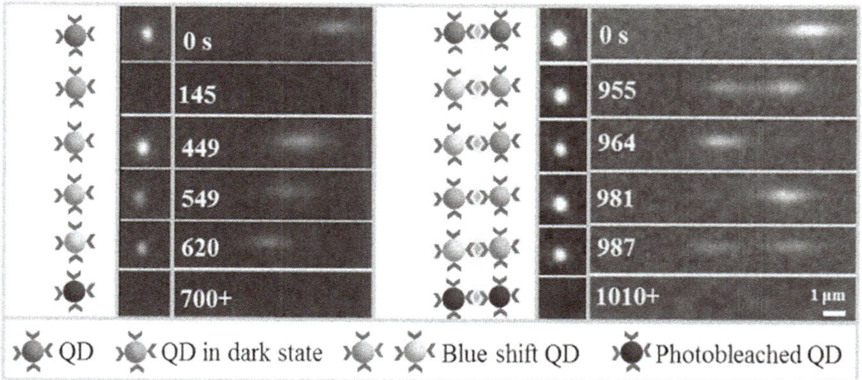

FIGURE 5.12 Typical spectral blue shift patterns of an antibody-modified QD (A) and a QD nanoaggregate (capture antibody-modified QD-CEA detection antibody-modified QD) (B).

Source: Reproduced from Liu, X.J.; Huang, C.H.; Dong, X.L.; Liang, A.Y.; Zhang, Y.S.; Zhang, Q.Q.; Wang, Q.; Gai, H.W. *Chemical Communications*, 2018, 54, 13103–13106 (ref 14), with permission of The Royal Society of Chemistry.

51.376–1644.047 pM. Under a signal-to-noise ratio of 3, the LODs for CEA and AFP were found to be 6.7 fM and 3.4 fM, respectively.

Multicomponent Assays Building upon previous studies, we proposed a homogeneous DI capable of conducting two-component tests using both single-color and dual-color QD labeling simultaneously.[15] In this assay, both the capture and detection antibodies against AFP were labeled with QD 585, and the AFP immunocomplex, formed by the binding of AFP with the antibodies, contained two QD 585s. The AFP immunocomplex was identified through the splitting of the first-order streak, as shown in Figure 5.13. For CEA, its capture antibody was labeled with QD 655, while its detection antibody was labeled with QD 585. The resulting CEA sandwich immunocomplex contained one QD 655 and one QD 585. These QDs could be easily distinguished due to their distinct spectral differences. Free antibodies, which did not form complexes, label only one QD, resulting in a single first-order streak. By analyzing and quantifying the dimer ratios of these two types of QDs, both CEA and AFP were simultaneously quantified in a single experiment. At a signal-to-noise ratio of 3, the LODs for CEA and AFP were 0.10 pM and 0.02 pM, respectively. While the sensitivity was lower compared to single-component assays, it

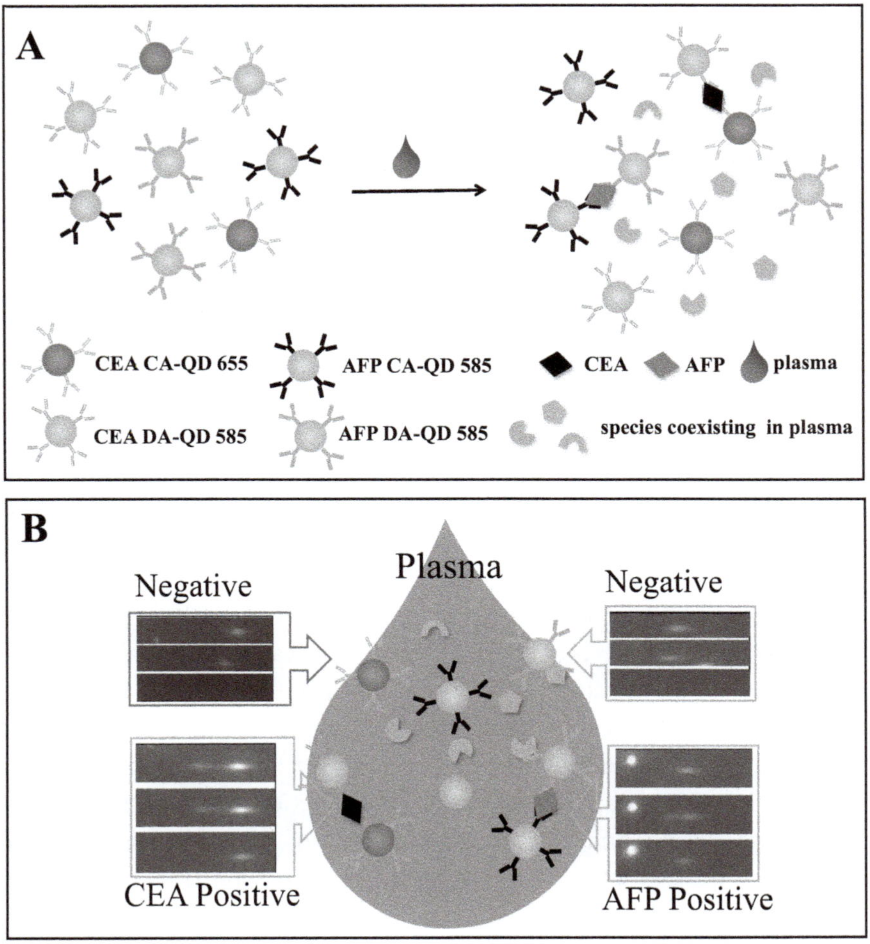

FIGURE 5.13 Dual immunocomplex formation and identification: (A) formation of two distinct immunocomplexes induced by CEA and AFP and (B) identification of immunocomplexes based on their unique spectral signatures.

Source: Reproduced from Liu, X.J.; Sun, Y.Y.; Lin, X.Y.; Pan, X.Y.; Wu, Z.J.; Gai, H.W. Digital duplex homogeneous immunoassay by counting immunocomplex labeled with quantum dots, *Analytical Chemistry*, 2021, 93, 3089–3095 (ref 15). Copyright 2021 American Chemical Society.

remained one order of magnitude higher than that of color CCD-based immunoassays. Theoretically, this encoding method can detect N(N+1)/2 tumor biomarkers, where N represents the number of QD types. For example, using two QD types allows for the detection of three components. However, in practice, the fluorescence intensity of QDs varies by type, and it can be challenging to obtain high-quality images for all QDs simultaneously under identical conditions. As the number of QD types increases, mutual interference between them becomes more pronounced, leading to missed detection of immunocomplexes and reduced quantification accuracy. Furthermore, in order to observe the splitting of the first-order streak, it is necessary for the QDs within the field of view to undergo photobleaching. This process extends the imaging time, which is not ideal for rapid detection.

To enhance the multiplexed capability and address the challenges associated with long imaging times using single-color QDs and fluorescence intensity mismatches with different color QDs, we leveraged GNPs in combination with QDs, exploiting their "invisibility" to one another.[16] This approach enabled the development of a 3-plex homogeneous DI. In this assay, 70 nm GNPs functionalized with detection antibodies specific to CEA, AFP, and PSA acted as scattering probes, while QD 525, QD 585, and QD 655, each functionalized with a capture antibody specific to one target, served as fluorescence probes, as shown in Figure 5.14A. After the immunoreaction, 5 μL of the reaction mixture was deposited onto a glass slide and covered with a cover slip. The device was then sealed with nail polish and observed using a microscope equipped with both fluorescence and dark-field modes. The same spot was imaged sequentially in both modes, and the resulting images were merged for analysis. Although multiple immunocomplexes existed in the reaction mixture, the single-particle scattering and fluorescence spectra of each immunocomplex overlapped in distinct patterns. These overlapping spectra were utilized to differentiate between the various immunocomplexes. Figure 5.14B illustrates the types of species that are likely present in the reaction mixture. Panel C provides a schematic diagram of a single QD fluorescence spectral image, where red is used to represent the fluorescence mode, distinguishing it from the green scattering mode (Panel d), which does not represent the actual color of the QD. Notably, the scattering spectrum of GNPs was broader (550±50 nm) compared to the fluorescence spectrum of QDs, resulting in a first-order streak for GNPs that was longer than that of the QDs. Panel d shows the merged image combining both fluorescence and scattering modes. Panel f represents a spectrum derived from the two-dimensional spectral image. Rows 1, 2, and 3 in Figure 5.14F show the schematic representation of free QDs, while row 4 illustrates the schematic of free GNPs. In these representations, the corresponding fluorescence and scattering spectra are depicted with a single zero-order spot and a first-order streak for each type of probe. Row 5 illustrates an immunocomplex

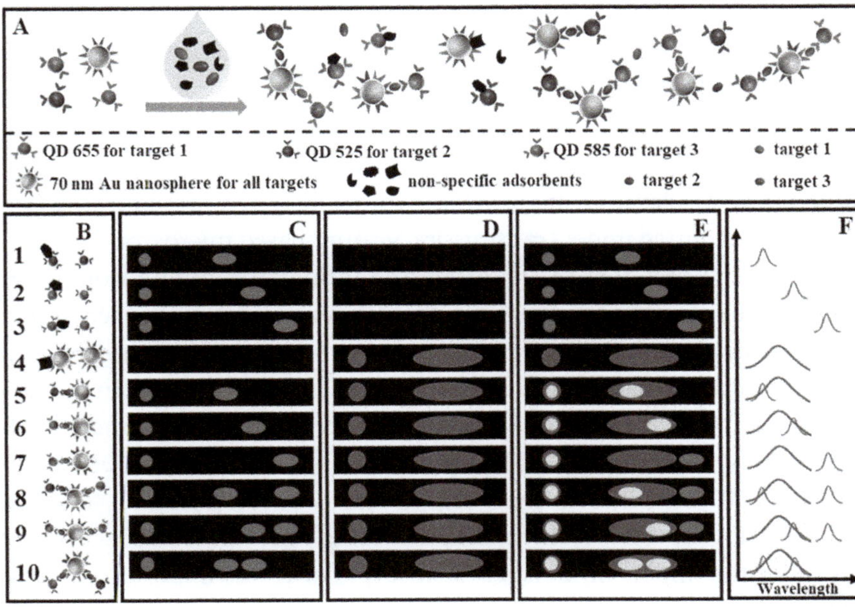

FIGURE 5.14 Principle of multiplexed immunoassay: (A) three-plex homogeneous immunoreaction, (B) possible probes present in the immunoreaction, (C) fluorescence spectral patterns shown in red, (D) scattering spectral patterns shown in green, (E) merged images of panels (C) and (D), and (F) corresponding converted spectra from panel (E).

Reproduced from Liu, X.J.; Lin, X.Y.; Pan, X.Y.; Gai, H.W. Multiplexed homogeneous immunoassay based on counting single immunocomplexes together with dark-field and fluorescence microscopy. *Analytical Chemistry* 2022, 94, 5830–5837 (ref 16). Copyright 2022 American Chemical Society.

formed by one GNP and one QD 525, which exhibits both fluorescence and scattering spectra. The fluorescence spectrum has a maximum emission wavelength of 525 nm, and upon overlaying, a yellow hue appears on the left side of the scattering spectrum. Row 6 shows an immunocomplex formed by one GNP and one QD585, where after overlapping, yellow appears on the right side of the scattering spectrum. Row 7 depicts an immunocomplex consisting of one GNP and one QD 655, with the fluorescence spectrum appearing outside the scattering spectrum after overlap. Row 8 illustrates an immunocomplex formed by one GNP, one QD 525, and one QD 655. After overlapping, yellow appears both on the left side and outside of the scattering spectrum. Rows 9 and 10 show two additional scenarios. Notably, in this experiment, an immunocomplex consisting of one GNP and three types of QDs was not observed. In summary, immunocomplexes formed by tumor biomarkers consisted of one GNP and one or more QDs. Consequently, these immunocomplexes exhibited two distinct signals—fluorescence and scattering—allowing them to be differentiated from free tags or nonspecifically bound tags, which only displayed a single signal. To quantify the three targets, two steps were required under transmission grating-based microscope spectral imaging: The first step in the assay was to identify immunocomplexes based on the overlap of scattering and fluorescence signals. The second step involved determining the type of target molecule in the complex and quantifying them based on the position of the first-order streak. In the PBS buffer, the LODs for CEA, AFP, and PSA were 30 fM, 10 fM, and 50 fM, respectively. In serum, the LODs for CEA, AFP, and PSA were 100 fM, 20 fM, and 60 fM, respectively. These values were very similar to those obtained in the PBS system, demonstrating that the assay effectively mitigates the impact of nonspecific adsorption. It was important to note that after conjugating QD 525 or QD 585 to GNPs, the fluorescence intensity of the QDs slightly decreased. However, this reduction did not interfere with the ability to identify their single-particle fluorescence signals.

DIFFUSION DIFFERENCE-BASED RECOGNITION

The assay, in which immunocomplexes are recognized based on the differences in diffusion between immunocomplexes and free antibodies, was first reported by Noji's research group in 2019[17] and subsequently developed into a dual-component tumor biomarker immunoassay in 2020.[18]

Basic Principle

Figure 5.15 illustrates the process of a diffusion-based homogeneous immunoassay.[17] Magnetic beads, each with a diameter of 550 nm, were conjugated with capture antibodies. Following incubation, a mixture of the capture antibody-modified magnetic beads and the antigen was injected into the reaction channel of the microfluidic chip. The reaction channel contains 5.5 million micro-reactor chambers, each with a diameter of 2.5 micrometers and a depth of 0.9 micrometers (Figure 5.15A). The bottom of each microchamber is coated with detection antibodies via flexible connecting arms. The chip is then placed in a magnetic field. Driven by both magnetic force and fluid shear force, the magnetic beads are drawn into the detection antibody-coated microchambers. The ratio of microchambers

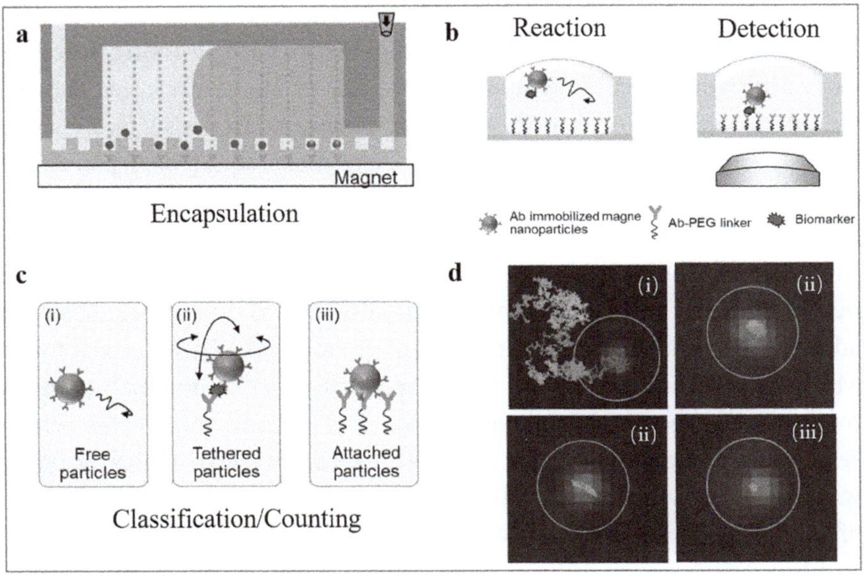

FIGURE 5.15 Principle of a digital homogeneous nonenzymatic immunosorbent assay: (A) design of the flow cell featuring a femtoliter reactor array, (B) immunoreaction and detection process, (C) typical nanoparticle motion patterns: (i) free diffusion, (ii) tethered motion, and (iii) nonspecific attachment, and (D) representative nanoparticle trajectories during antigen-antibody interactions, including (i) free diffusion, (ii) tethered diffusion, and (iii–iv) nonspecific attachment.

Source: Reproduced from Akama, K.; Iwanaga. N.; Yamawaki, K.; Okuda, M.; Jain, K.; Ueno, H.; Soga, N.; Minagawa, Y.; Noji, H. Wash- and amplification-free digital immunoassay based on single-particle motion analysis, *ACS Nano*, 2019, 13, 13116–13126 (ref 17). Copyright 2019 American Chemical Society.

containing one magnetic bead to the total number of one million magnetic beads was approximately 25.6%. After a five-minute immunoreaction, the movement trajectory of the bead centers was recorded using a dark-field microscope equipped with a 60× objective lens and a high-speed camera operating at 5,000 frames per second (Figure 5.15B). The magnetic beads exhibited different movement behaviors, which corresponded to distinct diffusion patterns. Free magnetic beads that had not bound to the microchambers displayed free diffusion (Figure 5.15Ci). Magnetic beads that formed sandwich-structured immunocomplexes exhibited restricted Brownian motion (Figure 5.15Cii). Nonspecifically adsorbed magnetic beads were firmly attached to the chamber bottom and remained stationary without displacement (Figure 5.15Ciii). The trajectories corresponding to these three motion patterns were shown in Figure 5.15D. Among the beads exhibiting restricted Brownian motion, 2% of the trajectories showed biased movement due to the beads being connected to the surface via two binding sites. For quantitative analysis, the mean squared displacement (MSD) of each magnetic bead was measured. In the absence of antigen, the MSD divided into two groups: one showed a linear increase with time, corresponding to beads in free diffusion, and the other remained nearly constant, corresponding to beads adsorbed to the surface. After the addition of the antigen, a saturation group emerged, characterized by an MSD less than 0.03 μm^2 after 0.02 seconds. This limitation in movement was due to the 80 nm length of the flexible connecting arms, which restricted the bead's movement range. The diffusion coefficient of the beads was calculated within the 0–0.02s timeframe using MSD. The distribution of the diffusion coefficients showed two peaks in the absence of antigen and three peaks upon antigen introduction. However, the use of the diffusion coefficient for classification required storing large amounts of data, which was impractical for real-world applications. Instead, the square root of MSD (RMSD) distribution could achieve the same differentiation effect as the diffusion coefficient distribution, offering a more efficient and suitable method for classification.

Figure 5.16 illustrates the distribution of RMSD and the corresponding quantitative results at various PSA concentrations. Peak 1, Peak 2, and Peak 3 represent nonspecifically adsorbed stationary magnetic beads, free-diffusing magnetic beads, and magnetic beads within immunocomplexes, respectively. The proportion of Peak 2 in the total number of detected magnetic beads is defined as the "active fraction," which serves as the quantitative parameter. Even in the absence of the antigen, Peak 2 remains observable, indicating the presence of nonspecific adsorption. The impact of nonspecific adsorption becomes more pronounced as the incubation time in the microchamber increases. However, a short reaction time—such as 5 minutes—can reduce some of this interference. Without the use of microwell-based partitioning, the signal-to-noise ratio would decrease by a factor of 80. Figure 5.16D demonstrates that this assay exhibits a linear range spanning over three orders of magnitude, with an LOD of 0.093 pg/mL. However, the broadening of the RMSD peaks, and the overlap between these peaks can affect the accuracy of quantification. Peak broadening was primarily caused by the variation in particle size, with the magnetic bead size CV being 29%. Although using polystyrene

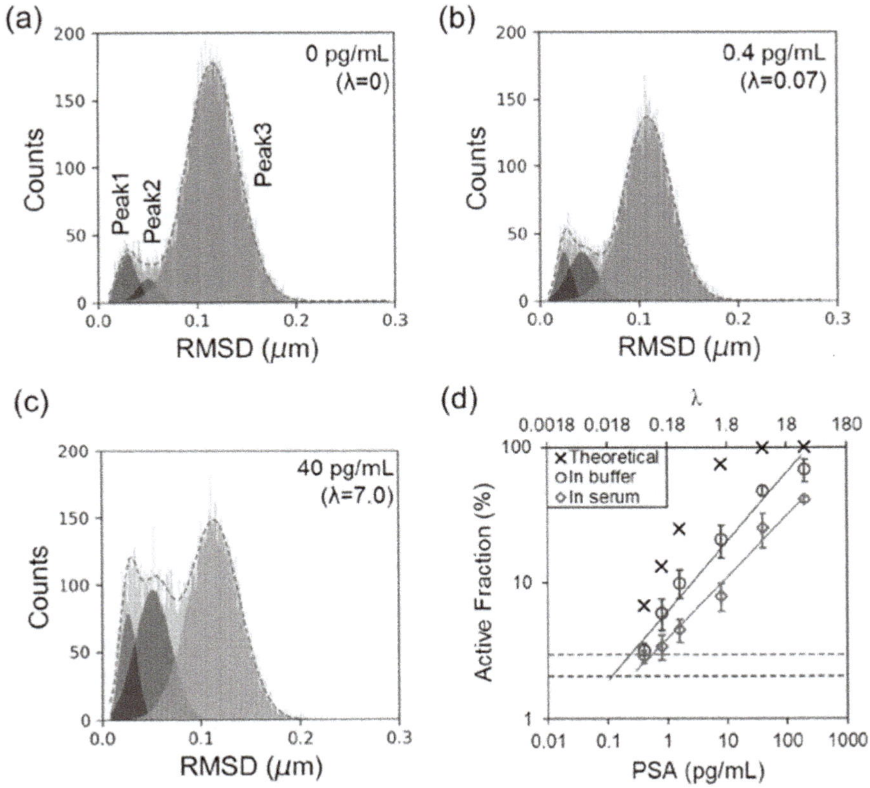

FIGURE 5.16 PSA quantification using a digital homogeneous nonenzymatic immunosorbent assay: (A–C) RMSD distributions at different PSA concentrations, and (D) Corresponding calibration curve.

Source: Reproduced from Akama, K.; Iwanaga. N.; Yamawaki, K.; Okuda, M.; Jain, K.; Ueno, H.; Soga, N.; Minagawa, Y.; Noji, H. Wash- and amplification-free digital immunoassay based on single-particle motion analysis, *ACS Nano*, 2019, 13, 13116–13126 (ref 17). Copyright 2019 American Chemical Society.

microspheres with a smaller size CV of 8% could reduce peak broadening, it resulted in a decreased microsphere loading efficiency in the microchamber. In the absence of magnetic force assistance, the loading efficiency decreases by a factor of 570.

To account for the heterogeneity of magnetic beads, Noji's research group proposed a correction method that combines fluorescence intensity and the ratio of the long-to-short axes of the confidence ellipse.[19] Alexa488 dye was conjugated to the magnetic beads, allowing for the recording of bead movement along with changes in fluorescence intensity. A corrected

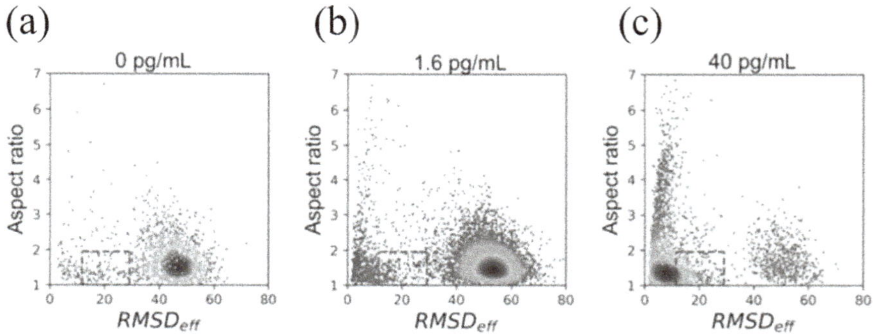

FIGURE 5.17 PSA concentration correction using multiparameter analysis at different target levels: (a) 0 pg/mL, (b) 1.6 pg/mL, and (c) 40 pg/mL.

Source: Reproduced from Akama, K.; Noji, H. *Analyst*, 2021, 146, 1303–1310 (ref 19), with permission of The Royal Society of Chemistry.

parameter, termed $RSMD_{eff}$, was derived by multiplying the RMSD by the square root of the fluorescence intensity. Additionally, the long-to-short axis ratio (AR) of the 95% confidence ellipse, based on the distribution of the magnetic bead movement trajectories, was used as another corrected parameter. These two parameters were employed to assess the two-dimensional distribution in the cases of no antigen and excess antigen, respectively. A positive signal region was manually defined as having an AR between 1 and 2, and $RSMD_{eff}$ between +10 times the variance of the multibonding $RSMD_{eff}$ mean and −3 times the variance of the free diffusion $RSMD_{eff}$, as shown by the red box in Figure 5.17. Although this region was not physically distinct, it helped reduce the LOD and the variation coefficient of the quantification. By applying this correction, the LOD of PSA decreased from 0.13 pg/mL (using RMSD alone) to 0.034 pg/mL, improving sensitivity by a factor of 3.9. For a PSA sample at 0.064 pg/mL, the coefficient of variation of the percentage of positive particles decreased from 35% to 10%.

Two-Component Assays

Building on the single-component test, a homogeneous two-component immunoassay was developed by labeling magnetic beads with different fluorescent dyes.[18] PSA and IL-6 were selected as model tumor markers. The magnetic beads were conjugated with Alexa488 and Alexa647, respectively, serving as two separate probes. The probes (each 1 million beads) were incubated with PSA and IL-6 samples for 1 hour, after which

10 μL of the reaction solution was injected into the detection channel. Fluorocarbon oil and magnetic force were applied to encapsulate the magnetic beads into the microchambers. The surfaces of the microchambers were coated with detection antibodies specific to both markers. After a 5-minute reaction in the microchambers, fluorescence microscopy was used to capture 50 frames per chamber, each with an exposure time of 10 ms and a 100 ms interval. ImageJ software was employed to analyze the movement trajectories and RMSD distributions to determine the number of magnetic beads within immunocomplexes. The LODs for PSA and IL-6 were 0.06 pg/mL (18 fM) and 0.042 pg/mL (19 fM), respectively, which were comparable to the results obtained from the single-component test.

When comparing this diffusion-difference multicomponent immunoassay with the ones based on the spectral difference, we found that both assays exhibited similar performance, with LODs in the tens of femtomolar range and dynamic ranges spanning approximately three orders of magnitude. Moreover, both assays required complex image process, and machine learning was necessary to accelerate data interpretation. Strictly speaking, the diffusion-difference, two-component immunoassay cannot yet be classified as a true multicomponent assay, as it does not simultaneously measure both analytes in a single test. Instead, the Alexa488 and Alexa647 channels must be imaged separately. Furthermore, due to limitations imposed by antibody density on the microchamber surface and the constraints of fluorescent dye detection, the number of detectable components in diffusion-difference homogeneous DI is unlikely to increase substantially. In contrast, the spectral-difference assay has successfully detected up to three components, although further efforts are required to expand the number of detectable components in this system.

BINDING DIFFERENCE-BASED RECOGNITION

The Single Molecule Analysis Group at the University of Michigan developed a single-molecule recognition strategy known as equilibrium Poisson sampling (SiMREPS), which monitors dynamic binding events between fluorescent probes and target molecules immobilized on a surface.[20] Binding is a transient and reversible process. When a fluorescent probe is used, fluorescence is observed during binding and disappears during dissociation. The fluorescence intensity over time generates a characteristic "fingerprint" that can distinguish specific from nonspecific binding

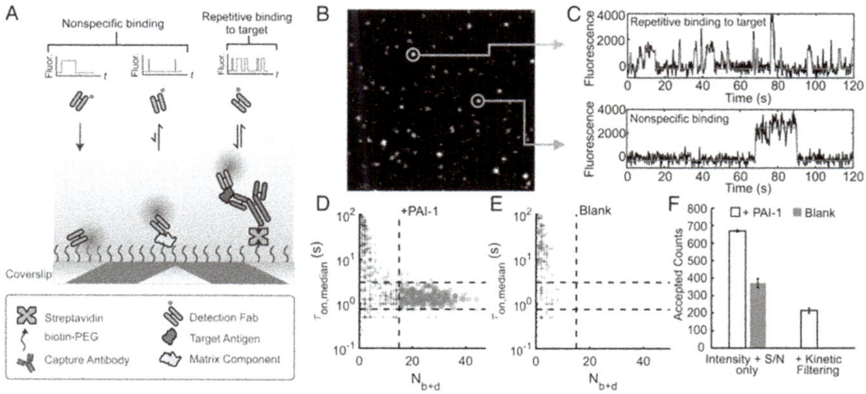

FIGURE 5.18 Principle of SiMREPS-based DI: (A) binding of a detection Fab fragment to a surface-immobilized target antigen, (B) single-particle image, (C) representative kinetic fingerprints showing repetitive binding to the same antigen molecule (top) and nonspecific binding events, (D, E) plots of N_{b+d} versus τ_{on} in the presence (D) and absence (E) of the target antigen PAI-1, and (F) effect of kinetic filtering on the number of accepted PAI-1 counts.

Source: Reproduced with permission from Proceedings of the National Academy of Sciences USA Chatterjee, T.; Knappik, A.; Sandford, E.; Tewari, M.; Choi, S.W.; Strong, W.B.; Thrush, E.P.; Oh, K.J.; Liu, N.; Walter, N.G.; Johnson-Buck, A. *Proceedings of the National Academy of Sciences*, 2020, 117, 22815–22822 (ref 21).

events. SiMREPS is applicable to any interaction where one of the reactants can be immobilized on the surface, including DNA hybridization, immunoreactions, enzyme catalysis, and other molecular interactions. Figure 5.18A illustrates the basic principles of SiMREPS as applied in DIs.[21] Captured antibodies were immobilized on the substrate surface, and low-binding constant detection antibodies were selected to specifically bind to the antigen captured by the capture antibody, thereby forming a sandwich structure. To minimize interference from free fluorescent probes in the solution, total internal reflection microscopy was employed to observe individual molecular fluorescence. The fluorescence intensity of each spot was monitored over time, and changes in fluorescence intensity reflected the binding process. Specific and nonspecific binding events show distinct fluorescence changes over time (Figure 5.18B and C). In specific binding, the frequency of binding and dissociation events was much higher than in nonspecific binding. Binding events (N_b) were characterized by transitions from bright to dark, while dissociation events (N_d) were marked by transitions from dark to bright. Using N_{b+d} and the average fluorescence lifetime (τ) as judgment parameters, specific and nonspecific binding could be differentiated. As shown in Figure 5.18D and E, nonspecific binding exhibited a smaller N_{b+d} and a longer average fluorescence lifetime compared to specific binding, consistent with the transient, reversible nature of specific binding. As the data collection time increased, more binding and dissociation events occurred, enhancing the distinction between specific and nonspecific binding. Typically, a collection time of two to five minutes was required for reliable results. In immunoassay measurements, the majority of the single-molecule signals raised from nonspecific binding (Figure 5.18F). If nonspecific binding was not effectively filtered out, it could significantly compromise the accuracy of the measurements. For four markers, the LODs ranged from 680 aM to 6.5 fM, which was 55 to 383 times lower than the LODs of ELISA. The LOD for IL-34, after filtering out false positives, was 6.5 fM, whereas without filtering, it was 340 fM. Clinical detection required only 2 μL of serum, with a dynamic range spanning 3.5 orders of magnitude.

Filtering out false-positive results caused by nonspecific binding significantly enhances detection accuracy. However, the primary factor limiting further improvements in detection sensitivity is the capture efficiency (detection efficiency). The overall antigen capture efficiency ranges from 0.5% to 1.5%. Given that each imaging field (100 microns square) represents only 0.1% of the total capture area, the detection efficiency of a test with nine imaging fields is only 0.01% of the total antigen in the solution. Increasing the size of the imaging field and improving the detection coverage can further lower the LOD.

The need to screen for effective detection antibodies is another limitation of this assay. Since the dynamic dissociation process must be monitored, it is preferable for the binding to be weak ($k_{off} \sim 0.1$–$0.01\ s^{-1}$). Therefore, weak-binding antibodies for the antigen must be carefully selected. Although this involves additional screening steps, the success rate for identifying weak-binding antibodies is much higher than for selecting strong-binding ones.

Tao et al. had developed a series of wash-free quantification assays based on dynamic changes in the number of molecules.[22,23] The basic process began with the immobilization of capture antibodies on a substrate, followed by the addition of the antigen to be tested. Detection antibodies, labeled with GNPs, were then introduced. From the onset of the immunoreaction, the change in the number of sandwich-structured immunocomplexes formed on the substrate surface was monitored. Interference from free-diffusing detection antibodies in the solution was minimized using a unique differential imaging algorithm, ensuring that only those detection antibodies bound to the surface are recorded. The imaging techniques employed included surface plasmon resonance imaging (SPRI) and bright-field imaging.

Figure 5.19 illustrates the SPRI DI process.[22] A gold-coated glass slide was modified with capture antibodies to serve as the sensing chip.

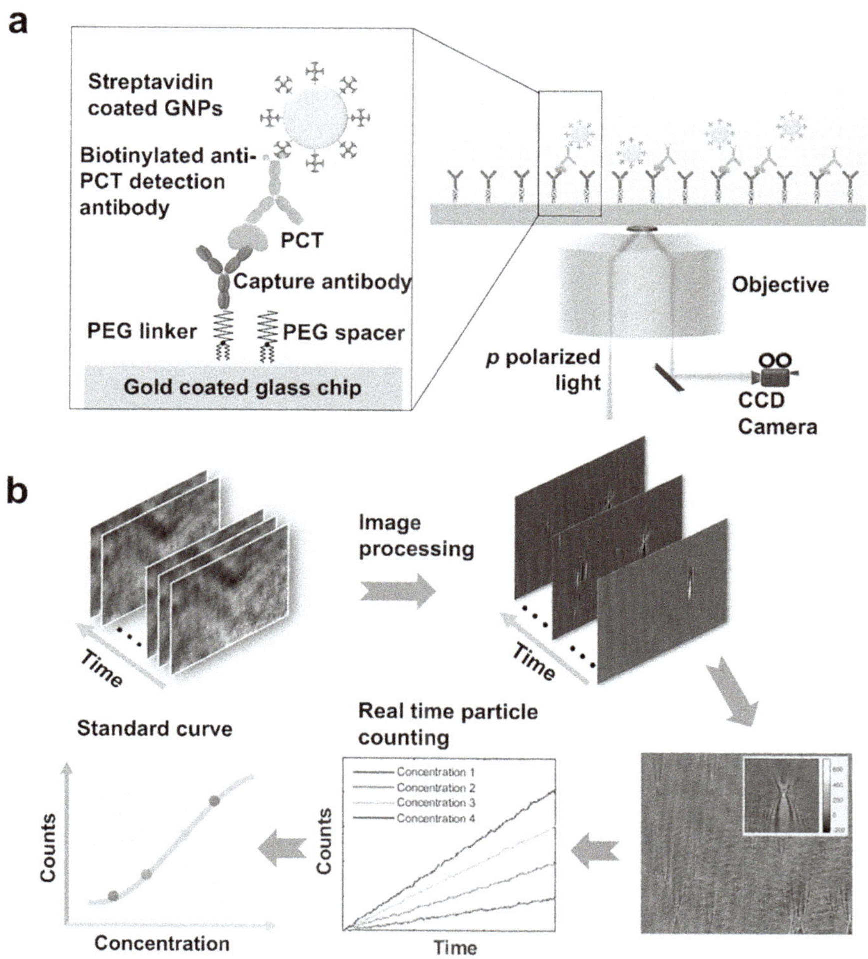

FIGURE 5.19 Principle of time-resolved DI: (A) experimental setup and (B) tracking of individual binding and unbinding events of single gold nanoparticles (GNPs) using an image processing algorithm.

Reproduced from Jing, W.W.; Wang, Y.; Yang, Y.Z.; Wang, Y.; Ma, G.Z.; Wang, S.P.; Tao, N.J. Time-Resolved Digital Immunoassay for Rapid and Sensitive Quantitation of Procalcitonin with Plasmonic Imaging. ACS Nano 2019, 13, 8609–8617 (ref 22). Copyright 2019 American Chemical Society.

Monochromatic polarized light was introduced into a 60× objective lens, and its incident angle was adjusted to a specific position that allowed total internal reflection to occur on the gold film surface of the chip, generating surface plasmon resonance (SPR) and producing plasmonic waves. Since SPR was highly sensitive to changes in the dielectric constant of the gold film surface, when GNPs bound to the surface, the resonance condition changed. This caused the plasmonic waves to excite the GNPs, producing scattering signals. However, the intensity of this scattering signal was relatively low, only a few times higher than the background noise. Therefore, a sequence of images must be collected and processed through denoising and averaging to obtain a single-particle signal with a clear diffraction pattern. Before the immunoreaction reached equilibrium, the number of GNPs bound to the gold film surface increased over time. The higher the sample concentration, the greater the slope of this increase. At a given time, a standard curve was constructed by plotting the number of particles against the concentration. This assay did not require washing, as the sequence image averaging and differential processing effectively filtered out non-binding GNPs. Additionally, images were captured in a sub-nanometer depth evanescent field, ensuring that free GNPs not entering the evanescent field could not be imaged. After 10 minutes of incubation (image collection) within a total detection time of 25 minutes, the LOD for procalcitonin was found to be 2.8 pg/mL, with a dynamic range spanning from 4.2 to 12,500 pg/mL.

Wang et al. achieved bright-field imaging of single 60 nm GNPs by processing sequential images using differential and averaging methods.[23] The quantification strategy employed was similar to that used in the SPRI DI, where the number of GNPs bound to the surface was monitored in real-time. The sequential image processing effectively removed free GNPs, eliminating the need for washing steps. However, nonspecifically adsorbed GNPs on the substrate had not been excluded, which could still affect the accuracy of the quantification.

The research group led by Yu Hui at Shanghai Jiao Tong University, based on SPRI imaging, distinguished between specific and nonspecific binding by monitoring the binding time.[24] Capture antibodies were immobilized on a solid-phase substrate, while detection antibodies were labeled with magnetic beads or GNPs. SPRI was used to record the binding time of individual magnetic beads or GNPs on the substrate. At the same time, the binding speed of these particles to the surface was controlled using magnetic or electric fields. In blank experiments, magnetic beads or

GNPs either quickly collided and detached or adsorbed onto the surface for an extended period, with an average binding time of 4.11 ± 0.26 seconds. Consequently, a binding time threshold of 5 seconds was set to differentiate between specific and nonspecific binding. Magnetic beads or GNPs with a binding time of less than 5 seconds were considered nonspecifically adsorbed, while those with a binding time between 5 and 150 seconds were regarded as specifically bound. Nonspecific binding was then filtered out. Under the influence of an electric field, the binding time of Aβ1–42 decreased from 45.05 seconds to 29.63 seconds, reducing the LOD to 1.08 fM. When manipulated by a magnetic field, the binding time decreased to 17.59 seconds, and the LOD was further reduced to 0.05 fM.

OUTLOOK

The impact of the separation and elution steps in heterogeneous immunoassays on sensitivity must be considered from two perspectives. On one hand, the removal of free labeled probes and nonspecifically adsorbed substances reduces background noise, thereby enhancing detection sensitivity. On the other hand, these steps can cause the binding equilibrium to shift toward dissociation, which impedes the formation of immunocomplexes and, consequently, reduces sensitivity. Homogeneous DIs, which do not require elution or separation steps, enable quantification through a single mixing operation. Compared to heterogeneous immunoassays, they offer several advantages: (1) Shorter measurement time: The elimination of washing steps saves time, and the faster reaction rate in solution results in an overall reduction in measurement time. (2) Higher measurement accuracy: Fewer procedural steps result in less cumulative error. The solution-phase reaction minimizes errors associated with protein conformational changes caused by surface interactions, thereby enhancing measurement accuracy. (3) Wider application range: In the absence of elution steps, analytes with weak to moderate interactions with the antibody can still form stable immunocomplexes. (4) Easier instrument miniaturization: This approach facilitates the development of portable, bedside diagnostic instruments for personalized healthcare.

A comprehensive consideration of the three main categories of homogeneous DI technologies—optical signal differences, diffusion differences, and binding differences—reveals that both diffusion- and binding-difference-based assays use solid-phase surfaces as immunoreaction sites. As such, they cannot be fully classified as homogeneous immunoassays, as described earlier. Although these assays do not require elution or separation steps, it is more accurate to refer to them as wash-free or one-step immunoassays. Furthermore, both types of assays involve the collection of a series of images, either long or short, with quantification achieved through tracking and processing the sequence of images. Additionally, the impact of relatively large labeled tags, such as GNPs or magnetic beads, on the accuracy of quantitative results remains unclear.

In comparison to other homogeneous DIs, we believe that spectral difference–based homogeneous immunoassays possess the following characteristics: (1) The identification of immunocomplexes relies on the spatial distribution of spectra rather than signal intensity, significantly enhancing identification accuracy and eliminating the need for background intensity correction. (2) Immunoreactions occur in solution rather than at interfaces, emphasizing the inherent advantages of homogeneous reactions. (3) The use of a naturally partitioning method removes the need for complex physical partitioning, thereby reducing costs. (4) This approach is at the forefront of multicomponent testing within homogeneous DIs. (5) The QD labels used in this method are comparable in size to protein molecules and are significantly smaller than magnetic beads, microspheres, UCNPs, and noble metal nanoparticles, resulting in superior immunoreaction efficiency. (6) Recognition can be completed with a single or, at most, two imaging acquisitions of the target region, eliminating the need for prolonged imaging times. (7) Dual-labeling strategies reduce the likelihood of nonspecific adsorption. In summary, spectral difference–based homogeneous immunoassays are highly suited for development into specialized instruments due to these advantages.

The future development of homogeneous DIs should focus on enabling immunoreactions to occur in a pure liquid phase, utilizing small-sized tag or tag-free detection, and facilitating rapid identification of nonspecific adsorption. Additionally, efforts should be directed toward enhancing miniaturization, portability, and automation, while also enabling multicomponent detection capabilities.

REFERENCES

1. Fredriksson S, Gullberg M, Jarvius J, Olsson C, Pietras K, Gustafsdottir SM, Ostman A, Landegren U. Protein detection using proximity-dependent DNA ligation assays, *Nature Biotechnology*, 2002, 20, 473–477.
2. Byrnes SA, Huynh T, Chang TC, Anderson CE, McDermott JJ, Oncina CI,

Bernhard H. Weigl BHHH, Nichols KP. Wash-free, digital immunoassay in polydisperse droplets, *Analytical Chemistry*, 2020, 92, 3535–3543.

3. Kim D, Garner OB, Ozcan A, Carlo DD. Homogeneous entropy-driven amplified detection of biomolecular interactions, *ACS Nano*, 2016, 10, 7467–7475.
4. Lukianova-Hleb E, Hu Y, Latterini L, Tarpani L, Lee S, Drezek RA, Hafner JH, Lapotko DO. Plasmonic nanobubbles as transient vapor nanobubbles generated around plasmonic nanoparticles, *ACS Nano.* 2010 4, 2109–2123.
5. Liu Y, Ye HH, Huynh HD, Xie C, Kang PY, Kahn JS, Qin ZP. Digital plasmonic nanobubble detection for rapid and ultrasensitive virus diagnostics, *Nature Communications*, 2022, 13, 1687.
6. Li X, Wei L, Pan LL, Yi ZY, Wang X, Ye ZJ, Xiao LH, Li H-W, Wang JF. Homogeneous immunosorbent assay based on single-particle enumeration using upconversion nanoparticles for the sensitive detection of cancer biomarkers, *Analytical Chemistry*, 2018, 90, 4807–4814.
7. Liu GL, Long YT, Choi Y, Kang T, Lee LP. Quantized plasmon quenching dips nanospectroscopy via plasmon resonance energy transfer, *Nature Methods*, 2007, 4, 1015–1017.
8. Liu XZJ, Zhang YS, Liang AY, Ding HW, Gai HW. Plasmonic resonance energy transfer from a Au nanosphere to quantum dots at a single particle level and its homogenous immunoassay, *Chemical Communications*, 2019, 55, 11442–11445.
9. Agrawal A, Zhang CY, Byassee T, Tripp R A, Nie SM. Counting single native biomolecules and intact viruses with color-coded nanoparticles, *Analytical Chemistry*, 2006, 78, 1061–1070.
10. Liu XJ, Huang CH, Zong CH, Liang AY, Wu ZJ, Zhang YS, Zhang QQ, Zhao WF, Gai HW. A single-molecule homogeneous immunoassay by counting spatially overlapping two-color quantum dots with wide-field fluorescence microscopy, *ACS Sensors*, 2018, 3, 2644–2650.
11. Chen HP, Gai HW, Yeung ES. Inhibition of photobleaching and blue shift in quantum dots, *Chemical Communications*, 2009, 1676–1678.
12. Shi XB, Xie ZQ, Song YH, Tan YJ, Yeung ES, Gai HW. Super-localization spectral imaging microscopy of multi-color quantum dot complex, *Analytical Chemistry*, 2012, 84, 1504–1509.
13. Shi XB, Dong SL, Li MM, Liu XJ, Zhang QQ, Zhao WW, Zong CH, Zhang YW, Gai HW. Counting quantum dot aggregates for the detection of biotinylated proteins, *Chemical Communications*, 2015, 51, 2353–2356.
14. Liu XJ, Huang CH, Dong XL, Liang AY, Zhang YS, Zhang QQ, Wang Q, Gai HW. Asynchrony of spectral blue-shifts of quantum dot based digital homogeneous immunoassay, *Chemical Communications*, 2018, 54, 13103–13106.
15. Liu XJ, Sun YY, Lin XY, Pan XY, Wu ZJ, Gai HW. Digital duplex homogeneous immunoassay by counting immuno-complex labeled with quantum dots, *Analytical Chemistry*, 2021, 93, 3089–3095.

16. Liu XJ, Lin XY, Pan XY, Gai HW. Multiplexed homogeneous immunoassay based on counting single immunocomplexes together with dark-field and fluorescence microscopy, *Analytical Chemistry*, 2022, 94, 5830–5837.
17. Akama K, Iwanaga N, Yamawaki K, Okuda M, Jain K, Ueno H, Soga N, Minagawa Y, Noji H. Wash- and amplification-free digital immunoassay based on single-particle motion analysis, *ACS Nano*, 2019, 13, 13116–13126.
18. Akama K, Noji H. Multiplexed homogeneous digital immunoassay based on single-particle motion analysis, *Lab Chip*, 2020, 20, 2113–2121.
19. Akama K, Noji H. Multiparameter single-particle motion analysis for homogeneous digital immunoassay, *Analyst*, 2021, 146,1303–1310.
20. Johnson-Buck A, Su X, Giraldez MD, Zhao MP, Tewari M, Walter NG. kinetic fingerprinting to identify and count single nucleic acids, *Nature Biotechnology*, 2015, 33, 730–732
21. Chatterjee T, Knappik A, Sandford E, Tewari M, Strong WB, Thrush EP, Oh KJ, Liu N, Walter NG, Johnson-Buck A. Direct kinetic fingerprinting and digital counting of single protein molecules, *Proceedings of the National Academy of Sciences of the United States of America*, 2020, 117, 22815–22822.
22. Jing WW, Wang Y, Yang YZ, Wang Y, Ma GZ, Wang SP, Tao NJ. Time-resolved digital immunoassay for rapid and sensitive quantitation of procalcitonin with plasmonic imaging, *ACS Nano*, 2019, 13, 8609–8617.
23. Yi Wang, Yunze Yang, Chao Chen, Shaopeng Wang, Hui Wang, Wenwen Jing, and Nongjian Tao, One-step digital immunoassay for rapid and sensitive detection of cardiac troponin I, *ACS Sensors*, 2020, 5, 1126–1131
24. Zeng Q, Zhou XY, Yang YT, Sun Y, Wang JA, Zhai CH, Li JH, Yu H, Dynamic single-molecule sensing by actively tuning binding kinetics for ultrasensitive biomarker detection, *Proceedings of the National Academy of Sciences of the United States of America*, 2022, 119, 2120379119

CHAPTER 6

Applications of Digital Immunoassays

TUMOR BIOMARKER DETECTION

Malignant tumors are a leading cause of death worldwide and often remain asymptomatic in their early stages. By the time noticeable symptoms appear, the disease is usually at an advanced stage, at which point curative treatments are often no longer effective. In these cases, palliative care becomes the main focus, aiming to alleviate symptoms, manage pain, and improve the patient's quality of life. Malignant tumors are highly heterogeneous, and their propensity for relapse, metastasis, and drug resistance presents significant challenges during treatment. The World Health Organization's "Early Cancer Diagnosis Guidelines" emphasize the importance of early screening and diagnosis for improving treatment outcomes. Monitoring tumor markers is one strategy for achieving early detection and diagnosis of malignancies. The development of new detection techniques and tumor biomarkers represents a critical step forward in enhancing early screening and diagnosis. Among these innovations, digital immunoassay (DI) is currently considered one of the most promising techniques for the detection of tumor markers.

Tumor biomarkers are biochemical substances that reflect the presence and progression of tumors. They can be produced in two ways: either synthesized and released directly by tumor cells or abnormally produced by the body in response to tumor presence. These biomarkers play a crucial

DOI: 10.1201/9781003729211-6

role in disease risk assessment, tumor screening, differential diagnosis, prognosis evaluation, predicting treatment response, and monitoring disease progression. Clinically, imaging techniques, pathological diagnosis, and tumor biomarker detection are primarily used for diagnosing cancer and monitoring its progression. Imaging techniques, such as X-ray computed tomography (CT), magnetic resonance imaging (MRI), and positron emission tomography (PET), have limitations in detecting tumors smaller than 2 cm and require expensive, sophisticated equipment. Pathological diagnosis, which involves examining tissue obtained through biopsy, surgery, or excision, is considered the gold standard for diagnosing malignant tumors, including determining their type, stage, grade, and molecular phenotype. However, pathological diagnosis provides only localized tumor information, and due to the high heterogeneity of malignant tumors, there is a risk of missing certain tumor types by sampling only a portion of the tumor. Additionally, repeated sampling to monitor tumor progression is challenging. Given these limitations, imaging and pathological techniques are not ideal for screening high-risk populations or for continuous monitoring of cancer progression. Therefore, the development of highly specific, sensitive, and high-throughput tumor detection methods is essential to improve screening efficiency, monitor cancer progression in a timely manner, and enhance treatment outcomes. Many disease-related proteins exist at levels below the limitations of detection (LODs) of conventional methods, presenting a challenge in measuring them in biological samples. Among early diagnostic methods, immunoassays are particularly popular due to their excellent specificity for corresponding targets.

Protein tumor markers primarily include carcinoembryonic antigen (CEA), alpha-fetoprotein (AFP), and prostate-specific antigen (PSA). These markers are commonly used as model targets for the development of DIs. Additionally, exosomes have emerged as an important research target in DIs. This section provides an overview of these protein biomarkers and exosomes.

CEA

CEA was first discovered in 1965 in the patient of colorectal cancer. It is a single-chain glycoprotein consisting of 641 amino acids, with a molecular weight ranging from 150 to 200 kDa. Subsequent studies revealed that CEA was also abnormally expressed in other cancers, including lung and pancreatic cancer. In healthy individuals, the concentration of serum

CEA is typically very low, around 2.5 ng/mL (12.5 pM). However, the average CEA concentration in smokers is approximately twice that in nonsmokers. Abnormal upregulation of CEA levels is a common feature of many malignant tumors. As a broad-spectrum tumor marker, elevated CEA levels in serum can indicate the presence of various cancer types.

In 2009, Chen et al. developed an electrochemical counting-based enzyme immunoassay for the detection of CEA.[1] The anti-CEA capture antibody was immobilized onto a substrate surface, while streptavidin and anti-CEA detection antibodies were simultaneously conjugated to magnetic beads. These magnetic beads formed an immunocomplex with CEA and the immobilized anti-CEA antibody. Following a washing step to remove excess beads, the magnetic beads were eluted from the substrate and subsequently bound to biotinylated alkaline phosphatase (ALP) via streptavidin-biotin affinity. After removing any excess ALP, the enzyme substrate, diphosphorylated phenyl (DPP), was added. The mixture was then injected into a capillary tube maintained at 40°C using an ice-water bath syringe. The ALP catalyzed the conversion of DPP to phenol around each magnetic bead. As the magnetic beads passed through an electrode, each bead generated an electrochemical peak and the number of peaks corresponds to the quantity of CEA molecules present. The assay exhibited an LOD of 5.0×10^{-17} mol/L and a linear detection range from 5.0×10^{-17} to 1.0×10^{-14} mol/L.

In 2016, Poon et al. developed an immunoassay that utilized capture antibody-modified Au nanoparticles (GNPs) and detection antibody-coated Ag nanoparticles (AgNPs) to form sandwich-structured immunocomplexes.[2] The scattering intensity of the aggregate formed by GNPs and AgNPs was significantly stronger than that of the individual nanoparticles. The average of scattering intensity, recorded using dark-field microscopy, was directly related to the amount of antigen present, enabling the establishment of a standard curve for the quantitative detection of tumor markers. As the concentration of the antigen increased, more immunocomplexes were formed, resulting in higher scattering intensity. The assay demonstrated linear detection ranges for CEA, AFP, and PSA of 0–300 pM, with LODs of 1.7 pM, 3.3 pM, and 5.9 pM, respectively. Unlike typical DIs that count individual immunocomplexes, this immunoassay relied on the measurement of scattering intensity, categorizing it as a nonhomogeneous soft discretization technique.

In 2017, Huang et al. developed a dual-component immunoassay for the simultaneous detection of CEA and AFP.[3] In this method, anti-CEA and anti-AFP capture antibodies were immobilized onto a flat substrate surface. Samples were then applied along with anti-CEA detection antibody-labeled AgNPs with a diameter of 54 nm and anti-AFP detection antibody-modified GNPs with a diameter of 63 nm. Following the reaction, unbound antibodies were washed away, and the resulting immunocomplexes were imaged using dark-field microscopy. The number of luminescent particles was counted, with CEA-induced complexes appearing blue and AFP-induced complexes appearing green. The LOD for both antigens was 0.5 ng/mL, with a quantitative range from 0.5 to 10 ng/mL. Similarly, Li et al. developed an immunoassay based on magnetic beads.[4] Biotinylated capture antibodies and detection antibodies formed an immunocomplex with the presence of antigen. The streptavidin-labeled magnetic beads were added to the immunoreaction solution, generating the compound of magnetic bead-capture, antibody-antigen-detection antibody. After magnetic separation and enrichment, the secondary antibody-modified GNPs were added to the magnetic bead suspension, forming magnetic bead-capture, antibody-antigen-detection antibody-secondary antibody-GNP. Magnetic separation and enrichment again, GNPs were dissociated from the immunocomplex using urea and observed using dark-field microscopy, where each GNP represented a single antigen molecule. The LODs for CEA, AFP, and PSA were 5 ng/mL, 1 ng/mL, and 1 ng/mL, respectively, with dynamic ranges of 5–30 ng/mL, 1–25 ng/mL, and 1–20 ng/mL, respectively.

Nanoparticles are frequently employed for dynamic tracking in addition to their role in static imaging. Chen et al. proposed a method for quantifying disease biomarkers by analyzing the diffusion coefficients of GNPs.[5] In the absence of the antigen, the GNPs conjugated with capture antibodies remained monodispersed. However, upon the introduction of CEA, the GNPs aggregated, leading to an increase in their size and a corresponding decrease in their diffusion coefficient. By analyzing the histogram of these diffusion coefficients, it was feasible to distinguish between aggregated and individual particles. The ratio of these diffusion coefficient differences was then used to quantify CEA, with detectable concentrations ranging from 50 to 750 pM.

In 2018, we developed a homogeneous DI using same colored quantum dots (QDs) as tags and electron-multiplying CCD (EMCCD) as detector.[6] The assay demonstrated LODs for CEA and AFP of 6.7 fM and 3.4 fM, respectively. The dynamic range of the assay spanned from 13.6 fM to 1399.3 pM.

In 2020, we proposed an ultrasensitive protein counting method based on nonrandom counting, where all target molecules were individually counted using microspheres as signal labels.[7] By combining multiple rounds of evaporation-induced particle sedimentation with grid-assisted multiframe imaging, we were able to accurately count the number of microspheres released from all immunocomplexes. This absolute quantification assay achieved an LOD for CEA of 4.9 aM, with a dynamic range spanning from 5 aM to 500 aM.

Tang et al. employed solid-state nanopores for the detection of CEA.[8] Aptamer-modified magnetic nanoparticles (MNPs) bound to CEA, and the resulting complexes were driven through the nanopores by an electric field. The distinct size difference between the CEA-Apt-MNPs and the Apt-MNPs led to varying degrees of current blockage. The frequency of current blockages induced by CEA-Apt-MNPs was found to be directly proportional to the CEA concentration. The LOD was 0.6 ng/mL, with a linear detection range extending from 0.01 to 1 nM (2–200 ng/mL).

AFP

Hepatocellular carcinoma (HCC), the third leading cause of cancer-related deaths worldwide, is the fifth most common cancer globally. Treatment options for HCC are limited, and the overall survival rate for patients remains low. Early diagnosis and treatment of liver cancer are critical to improving patient outcomes. AFP was first discovered in fetal serum and, in the 1960s, was found to be associated with HCC. AFP is one of the most widely used biomarkers for liver cancer. It is a glycoprotein consisting of 590 amino acids, with a molecular weight of 69–70 kDa and a half-life of 5 to 6 days. In healthy adults, AFP concentrations are typically below 20 ng/mL, whereas in liver cancer patients, levels can exceed 400 ng/mL. Several studies have demonstrated that abnormal AFP levels often correlate with the presence of liver cancer. The combination of serum AFP detection with imaging techniques is commonly employed for the clinical screening of HCC. AFP also serves as a model antigen and is frequently utilized in the development of DIs.

In 2016, Ahn et al. fabricated 100 nm diameter nanoislands that were coated with a 20 nm gold-chromium alloy. The Au-coated nanoislands were modified with anti-AFP capture antibodies. After the nanoislands reacted with the antigen and excess antigen was washed away, 20 nm AgNPs labeled with detection antibodies were introduced. Total internal reflection (TIR) microscopy with super-resolution was employed to differentiate the nanoislands containing immunocomplexes based on their significantly higher scattering intensity compared to those without immunocomplexes. The intensity difference increased with higher antigen concentrations, and this difference served as the quantitative basis for detection. The LOD for this highly sensitive assay was 7.04 zM, which is $100 \backsim 5 \times 10^9$ times lower than previous immunoassays for AFP.

Tian et al. encapsulated β-galactosidase (β-GaL) and its substrate into droplets generated by microfluidic chips.[10] By counting the number of luminescent droplets under a microscope, they were able to quantify the enzyme concentration. AFP immunocomplexes on magnetic beads were labeled with β-GaL. After magnetic enrichment and separation, the unbound β-GaL was encapsulated into droplets, and therefore AFP was indirectly quantified. The LOD for this assay was determined to be 20 fM.

PSA

Prostate cancer is an epithelial malignancy that originates in the prostate gland. It is the second most common cancer in men worldwide, following lung cancer. In China, however, the majority of prostate cancer cases are diagnosed at advanced stages, leading to missed opportunities for optimal treatment. This delay contributes to a higher mortality rate compared to other countries. Early diagnosis is crucial, as patients diagnosed at an early stage have a significantly better chance of receiving effective treatment, thereby improving their five-year survival rates. Consequently, early screening of high-risk populations is essential for timely diagnosis and enhanced patient outcomes. PSA is the most commonly used biomarker for diagnosing prostate cancer. PSA is a serine protease composed of 237 amino acid residues, with a molecular weight of 26,079 Da. In peripheral blood, PSA exists in two forms: free PSA (fPSA) and complexed PSA (cPSA). The total PSA (tPSA) level in serum is the sum of these two forms, with cPSA accounting for approximately 80% of the total. Under normal conditions, only a small amount of PSA enters the peripheral blood, as its passage is restricted by the basement membrane. However, when this barrier undergoes pathological changes, more PSA is released into the bloodstream, leading to an abnormal rise in PSA levels. However, PSA lacks tumor specificity, and elevated PSA levels can also be seen in other urological conditions. When the tPSA level is between 4 and 10 ng/mL, only about 22% of patients are diagnosed with prostate cancer. The low specificity of PSA and the potential for false positives lead to a significant number of unnecessary biopsy procedures. The use of the fPSA to tPSA (fPSA/tPSA) ratio has been shown to significantly improve both the sensitivity and specificity of prostate cancer diagnosis. When the fPSA ratio is <0.1, the likelihood of prostate cancer increases to approximately 56%. Generally, a fPSA ratio >0.16 is considered within the normal range. Clinically, radical prostatectomy is a common treatment for localized prostate cancer. Measuring serum PSA levels is a crucial tool for post-surgery monitoring, as PSA levels typically rise in the event of recurrence.

As early as 2010, Rissin et al. employed microarray-discretizationed digital enzyme immunoassay to detect PSA in 25% bovine serum, achieving an LOD of 50 aM.[11] To monitor prostate cancer recurrence, they measured serum PSA levels in patients after prostatectomy and found the average level to be around 0.4 fM (14 fg/mL). At low concentrations, the number of luminescent microspheres was used as the quantitative basis, classifying the assay as a DI. At higher concentrations, the luminescence intensity of the microspheres served as the quantitative basis, placing the assay within the category of analog signaling. By combining both assay modes, the detection range was extended to six orders of magnitude, from 8 fg/mL to 100 pg/mL (250 aM to 3.3 pM).[12] Noji et al. fabricated a polydimethylsiloxane (PDMS) microarray on the scale of millions and subsequently encapsulated microsphere-immunocomplex-enzyme compounds within the microarray.[13] This approach achieved an LOD for PSA of 2 aM (60 ag/mL). Chen et al. visually detected proteins without the need for the excitation and emission typically required in traditional fluorescence-based sensing.[14] Platinum nanoparticles (PtNPs) were used as signal tags for the immunocomplex that was encapsulated in microarrays to catalyze the decomposition of H_2O_2 to produce bubbles. These bubbles were observed and recorded using a smartphone, with each microbubble representing a single immunocomplex molecule. The LOD for PSA in prostatectomy patients was found to be 2.1 fM (0.06 pg/mL). Walter et al. employed the Simoa platform to measure PSA expression differences at the single-cell level.[15] Briefly, a cell lysis buffer and test reagents were added to the selected living cells, and the resulting mixture was loaded into a microarray. The fluorescence intensity of each well was then measured. The LOD was 0.0043 ± 0.0022 pg/mL, and this value corresponded to approximately 12,000 PSA molecules in 140 μL. Using $LNCaP_A$ and $LNCaP_B$ cells as targets, the results obtained through short tandem repeat sequences revealed a 12% genetic drift between the two cell types. They found that the PSA level in each cell line spanned two orders of magnitude, with an average of 2.15×10^6 molecules per cell for the $LNCaP_A$ cells, and 7.04 × 10^4 molecules per cell for the $LNCaP_B$ cells. Assuming a cell volume of 2 pL, the corresponding concentrations were calculated to be 1.79 μM and 0.0586 μM, respectively.

Shim et al. designed a multilayer microfluidic device capable of manipulating highly monodispersed fL droplets at generation frequencies of up to 1.3 MHz.[16] The droplet signals, catalyzed by single enzyme molecules, were observed within a few minutes. By applying and releasing external pressure, the fL droplets were reloaded. Due to the high frequency of droplet generation, the entire generation process took only about 10 seconds, indicating that the rate of repeated tests was not constrained. This device achieved an LOD for PSA as low as 46 fM. In 2014, Poon et al. coated the bottom of a flow cell with poly-L-lysine.[17] Subsequently, the negatively charged PSA absorbed onto this positively charged surface. Anti-PSA detection antibodies conjugated with 65 nm AgNPs were then introduced into the flow cell. These AgNPs bound to PSA were detected using dark-field microscopy, and the particle was counted. The assay demonstrated an LOD of 9 pM and a dynamic range of 0–100 pM, making it suitable for serum PSA detection. In 2017, Tan et al. employed gold nanorods (GNRs) and magnetic beads to label anti-PSA detection antibodies and capture antibodies, respectively.[18] After the formed sandwich-structured immunocomplex was magnetically separated from the excess detection antibodies, the GNRs were then released and deposited onto an amino-modified glass slide. The GNRs adsorbed to the slide were visualized using dark-field microscopy, and their count was used for quantification. The assay demonstrated a linear range of 0.01–10 pg/mL in buffer, with an LOD of 8 fg/mL (~226 aM).

The Gorris et al. developed a DI utilizing upconversion nanoparticles (UCNPs) as tags.[19,20] In 2017, they encapsulated $NaYF_4$ UCNPs with silicon for the detection of PSA.[19] Quantification of PSA was achieved through both digital counting and intensity measurements. In 25% serum, the LOD using digital counting was 1.2 ng/mL (42 fM), with a dynamic range of 10 pg/mL to 1 ng/mL. When intensity measurements were employed, the LOD was 20.3 pg/mL (700 fM), with a dynamic range of 100 pg/mL to 10 ng/mL. In 2019, they replaced silicon with streptavidin-PEG to enhance the immunoassay performance.[20] The UCNPs were linked to the sandwich-structured immunocomplex via biotin binding. The hydrophobic interactions and spatial steric hindrance provided by PEG inhibited both the aggregation and nonspecific adsorption of the UCNPs. This modification significantly enhanced detection performance, making it suitable for testing serum samples. The DI demonstrated an LOD of 0.023 pg/mL (0.8 fM), while the intensity-based assay achieved an LOD of 0.41 pg/mL (14 fM), with a combined dynamic range spanning from 0.1 to 1,000 pg/mL. Li et al. developed a simple and flexible single-particle counting immunoassay for quantifying PSA.[21] They employed luminescence resonance energy transfer (LRET) between UCNPs and GNPs. Carboxyl-functionalized UCNPs were conjugated with anti-PSA detection antibodies and served as the LRET donor, while 50 nm GNPs, modified with anti-PSA capture antibodies, acted as the LRET acceptor. Due to the strong overlap between the fluorescence spectrum of UCNPs and the absorption spectrum of GNPs, LRET occurred when the UCNPs and GNPs were brought into close proximity after immunocomplex formation, resulting in significant quenching of UCNP fluorescence. The immunoreaction mixture was then applied to a glass slide, and the fluorescent particles were counted. The number of the illuminating particles, which inversely correlated with the target concentration, allowed for the quantification of PSA. In buffer solution, the assay achieved an LOD of 1.0 pM, which was well below the critical value of PSA in patient serum samples. In serum, a similar LOD of 2.3 pM was obtained. The assay demonstrated a dynamic range from 0 to 500 pM.

Table 6.1 summarizes the performance of several DIs used for detecting protein tumor biomarkers. The immunoassays within the same category are listed in chronological order of their publication. Generally, the sensitivity of heterogeneous DIs exceeds that of homogeneous DIs, with LODs approximately 100 times lower.

Exosome

Exosomes are small extracellular vesicles ranging from 40 to 100 nm in size and characterized by a lipid bilayer membrane structure. Secreted by various cells, exosomes are found in a wide array of biological fluids, including blood, urine, and saliva. They serve as carriers of molecular information from their parent cells, transporting a variety of bioactive molecules such as proteins, lipids, cytokines, and genetic material. Consequently, exosomes play a significant role in intercellular communication. Tumor cells, in particular, release exosomes in greater quantities compared to normal cells, making exosomes derived from tumor tissues valuable for cancer diagnosis, classification, and prognosis. However, during the early stages of cancer, the abundance of exosome subpopulations in liquid biopsy samples, particularly in peripheral blood, is relatively low, which poses a challenge for detection. Existing ensemble methods for detecting exosome

TABLE 6.1 Detection of Tumor Biomarkers Using Digital Immunoassays

Target	Immunoassay Mode	Discretization Mode	Separation Technique	Detection Probe	Detection Technique	LOD	Dynamic Range	Matrix	Ref.
CEA	Heterogeneous	Soft discretization: Time	Surface wash plus magnet bead	Alkaline phosphatase	Electrochemistry	50 aM	50 aM–10 fM	Buffer	[1]
	Heterogeneous	Soft discretization: Space	Surface wash	Silver nanoparticle	Dark-field imaging	0.5 ng/mL	0.5–10 ng/mL	Buffer	[3]
	Heterogeneous	Soft discretization: Space	Magnet bead	God nanoparticle	Dark-field imaging	5 ng/mL	5–30 ng/mL	Serum	[4]
	Heterogeneous	Soft discretization: Time	Magnetic nanoparticle	Magnetic nanoparticle	Current	0.6 ng/mL	0.01–1nM (2–200 ng/mL)	Serum	[8]
	Heterogeneous	Soft discretization: Space	Microsphere centrifugation	Microsphere	Bright-field imaging	4.9 aM	5–500 aM	Plasma	[7]
	Homogeneous	Soft discretization: Space		Gold nanoparticle plus silver nanoparticle	Dark-field imaging	1.7 pM	0–300 pM	Serum	[2]
	Homogeneous	Soft discretization:Space		Same colored quantum dot	Fluorescence imaging	6.7 fM	13.6 fM –1399.3 pM	Serum	[6]
	Homogeneous	Molecule diffusion		Gold nanoparticle	Dark-field imaging		50~750 pM	Buffer	[5]
	Homogeneous	Soft discretization: Space		Quantum dot	Fluorescence imaging	20 fM	30 fM–3pM	Plasma	[22]
	Homogeneous	Soft discretization: Space		Gold nanoparticle plus quantum dot	Dark-field plus fluorescence imaging	10 fM	0.1–2.2 pM	Plasma	[23]

(*continued*)

TABLE 6.1 Detection of Tumor Biomarkers Using Digital Immunoassays (*Continued*)

Target	Immunoassay Mode	Discretization Mode	Separation Technique	Detection Probe	Detection Technique	LOD	Dynamic Range	Matrix	Ref.
AFP	Heterogeneous	Soft discretization: Space	Magnet bead	Gold nanoparticle	Dark-field imaging	1 ng/mL	1–25 ng/mL	Serum	[4]
	Heterogeneous	Soft discretization: Space	Surface wash	Gold nanoparticle	Dark-field imaging	0.5 ng/mL	0.5–10 ng/mL	Buffer	[3]
	Heterogeneous	Hard discretization: Droplet	Magnet bead	B-galactosidase	Fluorescence imaging	~20 fM		Buffer	[10]
	Homogeneous	Soft discretization: Space		Gold nanoparticle plus silver nanoparticle	Dark-field imaging	3.3 pM	0–300 pM	Serum	[2]
	Homogeneous	Soft discretization: Space		Same colored quantum dot	Fluorescence imaging	3.4 fM	13.6 fM–1399.3 pM	Serum	[6]
	Homogeneous	Soft discretization: Space		Quantum dot	Fluorescence imaging	10 fM	30 fM–3 pM	Plasma	[22]
	Homogeneous	Soft discretization: Space		Gold nanoparticle plus quantum dot	Dark-field plus fluorescence imaging	20 fM	0.1–2.2 pM	Plasma	[23]
PSA	Heterogeneous	Hard discretization: Microwell	Magnet bead	B-galactosidase	Fluorescence imaging	1.5 fg/ml (50aM)	200 aM–2.5 fM	Serum	[11]
	Heterogeneous	Hard discretization: Microwell	Magnet bead	B-galactosidase	Fluorescence imaging	8 fg/mL	8 fg/mL–100 pg/mL	Serum	[12]
	Heterogeneous	Hard discretization: Microwell	Magnet bead	B-galactosidase	Fluorescence imaging	2 aM (60 ag/mL)	~5 aM–1 fM	Buffer	[13]
	Heterogeneous	Hard discretization: Droplet	Microsphere centrifugation	B-galactosidase	Fluorescence imaging	46 fM (1.2 pg/mL)	0.046-4.6 pM	Buffer	[16]

Heterogeneous	Soft discretization:Space	Surface wash	Silver nanoparticle	Dark-field imaging	9 pM	0–100 pM	Serum	[17]
Heterogeneous	Soft discretization:Space	Magnet bead	Gold nanoparticle	Dark-field imaging	1 ng/mL	1-20 ng/mL	Serum	[4]
Heterogeneous	Soft discretization:Space	Magnet bead	Gold nanorods	Dark-field imaging	8 fg/mL (~226 aM)	0.01-10 pg/mL	Buffer	[18]
Heterogeneous	Soft discretization:Space	Surface wash	Upconversion nanoparticle	Fluorescence imaging	1.2 ng/mL (42 fM)	10 pg/mL–1 ng/mL	Serum	[19]
Heterogeneous	Soft discretization:Space	Surface wash	Upconversion nanoparticle	Fluorescence imaging	23 fg/mL (800 aM)	0.1–1000 pg/mL	Serum	[20]
Heterogeneous	Hard discretization: Microwell	Magnet bead	Platinum nanoparticle	Bright-field imaging	2.1 fM (0.06 pg/mL)	0.06–1pg/mL	Buffer	[14]
Homogeneous	Soft discretization:Space		Gold nanoparticle plus silver nanoparticle	Dark-field imaging	5.9 pM	0–300 pM	Serum	[2]
Homogeneous	Soft discretization:Space		Upconversion nanoparticle plus gold nanoparticle	Fluorescence imaging	1.0 pM	0–500 pM	Buffer	[21]
Homogeneous	Soft discretization:Space		Gold anoparticle plus quantum dot	Dark-field plus fluorescence imaging	60 fM	0.1–2.2 pM	Plasma	[23]

often lack sufficient sensitivity to identify these low-abundance populations. To address this limitation, counting-based DIs have been adapted for detecting exosome.

In 2018, Liu et al. developed a single-exosome-counting immunoassay using droplet discretization.[24] The detection process and typical images are shown in Figure 6.1. Using MDA-MB-231 breast cancer exosomes as

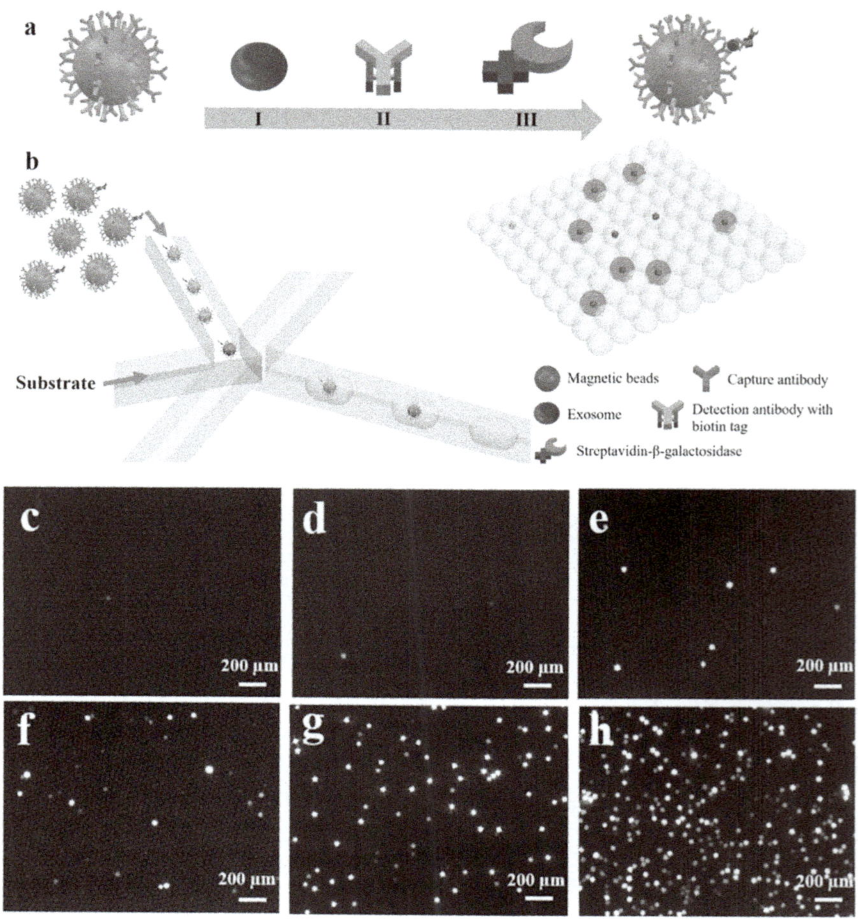

FIGURE 6.1 Droplet-based single-exosome-counting enzyme-linked immunoassay (droplet digital ExoELISA) for exosome quantification: (A) immunoreaction on a magnetic bead, (B) generation of microdroplets, (C) image of the negative control without target exosomes, and (D–H) images of target exosomes at varying concentrations.

Source: Reproduced from Liu, C.C.; Xu, X.N.; Li, B. et al. Single-exosome-counting immunoassays for cancer diagnostics, *Nano Letters*, 2018, 18, 4226–4232 (ref 24). Copyright 2018 American Chemical Society.

the target, antibodies against the exosome marker CD63 were immobilized onto magnetic beads for exosome capture. The captured exosomes on the magnetic beads then reacted with anti-GPC-1 antibodies, forming a complex of magnetic bead-anti-CD63 antibody-exosome-anti-GPC-1 antibody. GPC-1, a protein significantly expressed on cancer exosome membranes, has considerable potential for early cancer diagnosis. The anti-GPC-1 antibody was conjugated to β-GaL via a biotin-streptavidin reaction. The resulting complex (magnetic bead-anti-CD63 antibody-exosome-anti-GPC-1 antibody-β-GaL) was introduced into droplet generation chips, where it was encapsulated in droplets along with the enzyme substrate fluorescein di-β-D-GaL (FDG). Under fluorescence microscopy, droplets containing target exosomes emitted fluorescence, while droplets containing only magnetic beads or those without magnetic beads showed no fluorescence. The number of fluorescent droplets was then quantified to determine the exosome count. This immunoassay was applied to detect GPC-1-positive exosomes in clinical serum samples. Figure 6.2 compares the exosome levels in healthy individuals, patients with benign breast disease, breast cancer patients, and the exosome levels before and after surgery. While certain amounts of GPC-1-positive exosomes were detected in both healthy individuals and noncancer patients, the levels were significantly elevated in breast cancer patients. Although exosome levels decreased markedly after surgery, they remained slightly elevated compared to healthy individuals.

The exosome concentrations in clinical samples typically range around $10^8/\mu L$, while conventional methods are capable of detecting only up to 10^4 exosomes. This disparity presents a significant challenge in rapidly screening all exosomes and identifying rare exosome populations. To address this, Yang et al. developed a high-throughput exosome counting technique capable of producing up to 20 million droplets per minute.[25] Fluorescent paramagnetic microspheres (5 μm in diameter), modified with capture antibodies, were used to capture target exosomes. The captured exosomes then bound to the compound of detection antibody and horseradish peroxidase (HRP). Both the magnetic beads and the HRP substrate were encapsulated into droplets. After a five-minute reaction, the droplets passed through a detection zone where two lasers were co-focused. Signals emitted from the green fluorescence of the microspheres and the red fluorescence from the HRP enzyme substrate were interpreted as signals corresponding to the presence of exosomes. The exosomes isolated from human neuronal cells were used as the standard sample and

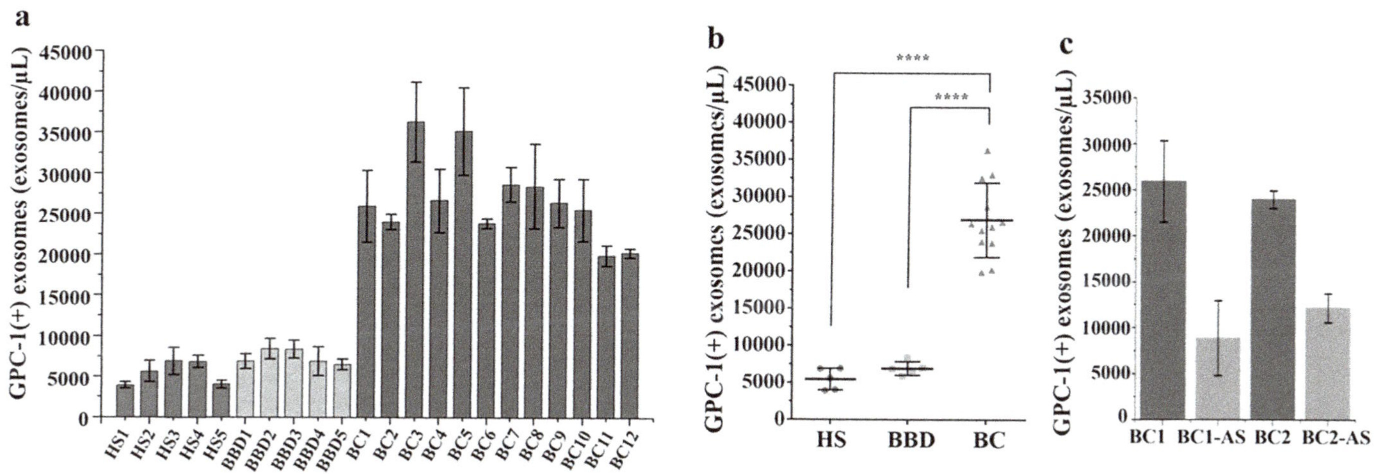

FIGURE 6.2 Clinical analysis of GPC-1(+) exosomes using droplet digital ExoELISA: (A) quantification of GPC-1(+) exosomes from various serum samples, (B) scatter plot of GPC-1(+) exosome levels in the breast cancer (BC) group, and (C) comparison of GPC-1(+) exosome levels in two breast cancer patients before and after surgery.

Source: Reproduced from Liu, C.C.; Xu, X.N.; Li, B. et al. Single-exosome-counting immunoassays for cancer diagnostics, *Nano Letters*, 2018, 18, 4226–4232 (ref 24). Copyright 2018 American Chemical Society.

were serially diluted in PBS buffer, ranging from 0 to 10^5 exosomes/μL. The LOD and dynamic range of the assay were determined to be 9 exosomes/μL and 9–5 × 10^5 exosomes/μL, respectively. To assess the assay's specificity, exosomes from human neurons were diluted in bovine serum. The presence of bovine exosomes (2 × 10^7 exosomes/μL) did not interfere with the detection of human exosomes, and the LOD in this matrix was found to be 11 exosomes/μL.

Morasso et al. employed single-molecule array (Simoa) to directly identify CD9-CD63-positive exosomes in plasma, achieving a quantification limit (LOQ) of 2–3 ng/μL.[26] They measured exosome levels in the plasma of 181 subjects (95 BC patients and 86 healthy controls (HCs)). The median exosome level in the healthy individuals was 613.0 ng/μL (range: 30.0–6862.0 ng/μL), whereas the exosome level in the BC patients was significantly elevated, with a median of 1779.1 ng/μL (range: 72.1–22805.9 ng/μL). Given that the total plasma exosome level was used to generate the standard curve and assuming that only 30% of the exosomes originated from the target, the expected LOQ would be approximately 10^4 exosomes/μL.

Wei et al. Combined Simoa with microarray discretization to detect CD9- and EpCAM-positive exosomes.[27] Microbeads were utilized to capture CD9-positive and EpCAM-positive EVs, which were then detected using anti-CD63 antibodies. The LODs for the two types of EVs were 34 particles/μL for CD9-positive exosomes and 25 particles/μL for EpCAM-positive exosomes. The authors compared the plasma exosome levels in 163 colorectal cancer (CRC) patients, 46 HCs, and 51 patients with adenomas. The levels of both CD9-positive and EpCAM-positive exosomes were significantly higher in CRC patients compared to HCs. The diagnostic performance, as evaluated by receiver operating characteristic (ROC) curve analysis, revealed that the use of CD9-CD63 (area under the curve (AUC) = 0.96) and EpCAM-CD63 (AUC = 0.90) as biomarkers surpassed the performance of traditional tumor markers, such as CEA and CA125.

Huang et al. imaged individual extracellular vesicle using lanthanide-doped UCNPs and TIRF microscopy.[28] They immobilized anti-CD81, anti-CD63, and anti-CD9 antibodies on the surface of glass slides to capture the corresponding EVs. Subsequently, anti-EpCAM antibody-modified UCNPs were used to detect the captured EVs. The number of UCNPs was directly counted to determine the concentration of exosomes. The LODs for CD9+EpCAM+, CD63+EpCAM+, and CD81+EpCAM+ exosomes were 1.8 × 10^6/mL, 1.1 × 10^7/mL, and 8.9 × 10^6/mL, respectively.

NEURODEGENERATIVE DISEASE BIOMARKER DETECTION

Neurodegenerative diseases are chronically progressive, irreversible disorders characterized by cognitive and motor impairments resulting from the gradual loss of neuronal structure or function. These diseases include Alzheimer's disease (AD), Parkinson's disease (PD), amyotrophic lateral sclerosis (ALS), and others. The pathogenesis of neurodegenerative diseases is highly complex and not yet fully understood. Furthermore, effective treatments remain unavailable. However, early diagnosis has the potential to significantly delay disease progression and enhance treatment efficacy, underscoring the critical importance of early detection in managing neurodegenerative diseases. When neuronal membrane structures are damaged, specific biomarkers such as neurofilament light chain (NfL), amyloid-β (Aβ), tau protein, and phosphorylated tau protein (ptau) are released into the cerebrospinal fluid (CSF) and can subsequently cross the blood–brain barrier into peripheral blood. As the disease progresses to more advanced stages, biomarker levels in body fluids change to varying extents. Monitoring these biomarker level changes can significantly aid in clinical diagnosis. While these biomarkers are most abundant in CSF, sampling CSF requires a lumbar puncture, a procedure that carries certain risks and is less acceptable to patients. In contrast, blood sampling is less invasive, easier to perform, and can be repeated more readily. However, due to the blood–brain barrier, the concentration of these biomarkers in peripheral blood is typically low, making detection challenging for traditional immunoassays. Single-molecule-level immunoassays, with their high sensitivity, are particularly advantageous for the early diagnosis of neurodegenerative diseases. As a result, neurodegenerative diseases represent one of the most promising applications for DIs.

Aβ

AD is one of the most prevalent neurodegenerative diseases worldwide. It leads to personality changes and a progressive decline in cognitive function, with severe memory loss being the most prominent symptom. In China, the prevalence of AD among individuals aged 65 and older is 3.21%, meaning that over 8 million people are diagnosed with the disease. Furthermore, as the aging population increases, the incidence of AD continues to rise each year. AD not only significantly affects the quality of life for patients but also imposes a substantial socioeconomic burden. According to AD diagnostic guidelines, relevant pathological changes associated with the disease appear in the brain at least ten years before the onset of cognitive symptoms. This has driven researchers to actively seek out biomarkers associated with AD pathology. The two primary pathological markers of AD are Aβ plaques and neurofibrillary tangles resulting from the excessive phosphorylation of tau protein. Consequently, Aβ and tau proteins have long been recognized as specific pathological biomarkers for AD. Both Aβ and tau are detectable using PET imaging or immunoassays in CSF up to 15–20 years before clinical symptoms manifest.

The extracellular accumulation of Aβ into oligomers and fibrillar deposits in brain tissue is a hallmark pathological feature of AD. Aβ is a protein typically composed of 39–42 amino acids, with a molecular weight of approximately 40 kDa. As a naturally disordered protein, Aβ does not possess a specific quaternary structure in solution and exists as a mixture of various forms. Among these forms, Aβ40 and Aβ42 are the most frequently used biomarkers in AD research. Traditional immunoassays have not revealed significant differences in plasma Aβ levels between AD patients and healthy controls (HCs). However, in 2016, Janelidze et al. achieved ultrahigh sensitivity detection of plasma Aβ42 and Aβ40 using the Simoa. Both Aβ42 and Aβ40 were captured with the same antibody targeting the N-terminus but detected with distinct antibodies targeting different C-termini. The LODs for Aβ42 and Aβ40 were 0.019 pg/mL and 0.16 pg/mL, respectively, with LOQs of 0.167 pg/mL and 1.939 pg/mL, which were far lower than those of traditional immunoassays. A cohort study involving 719 subjects demonstrated a positive correlation between plasma Aβ42 and CSF Aβ42, as well as plasma Aβ40 and CSF Aβ40. AD patients exhibited significantly lower plasma levels of Aβ42, Aβ40, and the Aβ42/Aβ40 ratio compared to controls, individuals with subjective cognitive decline (SCD), and those with mild cognitive impairment (MCI), as shown in Table 6.2. These findings challenge the previous notion that plasma Aβ levels are not useful for AD diagnosis. Vergallo et al. investigated the accuracy of the Aβ40/Aβ42 ratio in predicting brain Aβ-PET status in a cohort of elderly individuals with subjective memory complaints (n = 276).[30] Using the Simoa platform, they determined that the optimal threshold for the Aβ40/Aβ42 ratio was 17.82. At this threshold, the AUC, sensitivity, and specificity were 0.794, 78.1%, and 74.9%, respectively. Chatterjee et al. examined the accuracy of the Aβ42/Aβ40 ratio in distinguishing

TABLE 6.2 Detection of Plasma and CSF Aβ Using DI

	Control n = 274	SCD n = 174	MCI n = 214	AD n = 57
Gender, %women	61	55	44[a]	60
Age	73(5)	70(6)[b]	71(5)	76(5)[b,cd]
10-word delayed recall (errors)*	2.0(1.9)	3.4(2.2)[b]	6.5(2.3)[b,c]	8.5(1.5)[b,cd]
MMSE	29.1(0.9)	28.4(1.4)[b]	27.1(1.8)[b,c]	22.1(3.5)[bcd]
SUVR of [^{18}F] flutemetamol	1.3(0.3)	1.4(0.4)	1.7(0.5)[b,e]	N/A
Plasma Aβ42,pg/ml	19.6(5.2)	18.8(5.4)	18.8(6.1)	13.2(7.3)[bc,d]
Plasma Aβ40,pg/ml	276.7(66.1)	276.9(69.1)	287.6(77.0)	244.3(105.8)[a,c,d]
Plasma Aβ42/Aβ40	0.073(0.023)	0.070(0.025)	0.066(0.015)[f]	0.057(0.022)[b,c,g]
CSF Aβ42, pg/ml	554.0(195.1)	588.8(253.4)	470.1(232.3)[b,c]	289.5(103.8)[b,c,d]
CSF Aβ40, pg/ml	4688.5(1650.0)	4966.5(1750.5)	4765.3(1884.8)	4387.2(1761.6)[h]
CSF Aβ42/ Aβ40	0.123(0.036)	0.123(0.045)	0.104(0.044)[b,c]	0.070(0.022)[b,c,d]

Source: Reprinted by permission from Springer Nature Limited, Janelidze, S.; Stomrud, E.; Palmqvist, S.; et al. *Scientific Reports*, 2016, 6, 26801 (ref 29). Copyright 2016.

Note: Data are shown as mean (SD) unless otherwise specified. AD, Alzheimer's disease; ADAS-cog, Alzheimer's Disease Assessment ScaleCognition; MCI, mild cognitive impairment; CSF, cerebrospinal fluid; MMSE, Mini-Mental State Examination; SCD, subjective cognitive decline; SUVR, standardized uptake value ratios. Plasma and CSF biomarkers were analyzed with univariate general linear models controlling for age and gender; statistical significance was set to $p<0.008$ to account for Bonferroni correction

[a] Compared with control, $p<0.001$
[b] Compared with control, $p<0.0001$
[c] Compared with SCD, $p<0.0001$
[d] Compared with MCI, $p<0.0001$
[e] Compared with SCD, $p<0.001$
[f] Compared with control, $p = 0.002$
[g] Compared with MCI, $p = 0.003$
[h] Compared with SCD, $p = 0.003$
* From the ADAS-cog, subtest 3.

Aβ-positive (Aβ+) and Aβ-negative (Aβ-) participants.[31] Using the PET scan standard uptake ratio as a reference to classify Aβ+ and Aβ- participants, they compared the plasma levels of Aβ42, Aβ40, and the Aβ42/Aβ40 ratio between these two groups, all measured using the Simoa platform. The study found no significant differences in the plasma levels of Aβ42 and Aβ40 between Aβ+ and Aβ- participants. However, the plasma Aβ42/Aβ40 ratio was significantly lower in Aβ+ participants. Notably, the decrease in the plasma Aβ42/Aβ40 ratio for Aβ+ individuals was modest, ranging from 10% to 20%, compared to a more substantial decrease of 40%–60% in the CSF Aβ42/Aβ40 ratio. Additionally, the plasma Aβ42/Aβ40 testing was susceptible to procedural interference. These findings suggest that while the plasma Aβ42/Aβ40 ratio shows some potential in differentiating Aβ+ from Aβ- participants, its diagnostic value in distinguishing AD patients from non-AD patients requires further validation. The combination of Aβ and ptau may offer enhanced diagnostic precision for early detection of AD.

Tau Protein

Tau is a microtubule-binding protein with a molecular weight ranging from 48 to 67 kDa. It promotes the formation of microtubule bundles, thereby stabilizing them. Tau is primarily expressed in neurons of the central nervous system, with smaller amounts found in astrocytes and oligodendrocytes. In patients with neurodegenerative diseases or severe brain injuries, tau levels are elevated in the CSF, suggesting that tau may serve as a specific biomarker for brain damage. This has led to the hypothesis that detecting tau in peripheral blood, after crossing the blood–brain barrier, could provide a convenient method for assessing brain injury. However, serum tau levels are typically below the pg/mL range, which has limited its use as a reliable biomarker for brain injury. The advent of digital immunoassays has created new opportunities for detecting tau in blood, potentially overcoming this limitation.

In 2013, Randall et al. developed a Simoa-based assay for the detection of serum tau.[32] Tau and pTau were captured using Tau5 as the capture antibody, with HT7 and BT2 serving as detection antibodies. This assay achieved an exceptionally low LOD of 0.02 pg/mL in serum, over 1,000 times more sensitive than traditional immunoassays, which typically detected in the range of 30–60 pg/mL. They monitored the changes in total tau levels in the serum of 25 patients who had experienced cardiac arrest and were subsequently resuscitated, tracking these levels over a period of 108 hours. These values were used to predict clinical outcomes at 6 months. Tau levels in the serum of the patients increased to varying degrees, ranging from <10 pg/mL to several thousand pg/mL. The pattern of increase differed among individuals: some patients exhibited an early rise within 24 hours, others showed a delayed increase between 24 and 48 hours, and a subset displayed a biphasic pattern. The relationship between tau levels and clinical outcomes was assessed using the AUC for the time-course of tau levels. For patients whose tau levels increased within 24 hours, no significant association with clinical outcomes was observed. However, in patients exhibiting delayed increases in tau, a strong correlation with 6-month outcomes was found, with a sensitivity of 91% and a specificity of 100%. In the same year, Zetterberg et al. quantified plasma tau levels in patients with AD using DIs.[33] This pioneering work initiated the exploration of the relationship between blood tau and AD. In 2021, a review by a research team from Sichuan University conducted a statistical analysis of plasma tau and pTau181 levels measured using Simoa, synthesizing results from 41 studies.[34] The review primarily addressed three key questions: (1) What are the normal levels of plasma tau and plasma ptau? (2) How much higher are the levels of tau and ptau in AD patients compared to healthy individuals? (3) What is the diagnostic accuracy of using tau and pTau as biomarkers for AD? For the first question, the average plasma tau level in healthy individuals was found to be 3.07 pg/mL. To further examine the impact of age and sex on tau levels, the studies were grouped by age: individuals under 40 years (young), between 40 and 60 years (adults), and over 60 years (older), as well as by sex ratio: a low male group (with the male-to-total ratio <40%), an equal male group (with the male-to-total ratio between 40% and 60%), and a high male group (with the male-to-total ratio >60%). It was found that tau levels were not significantly associated with age but showed a notable correlation with sex. Specifically, the average tau levels were as follows: low male group: 4.15 pg/mL, equal male group: 2.64 pg/mL, and high male group: 3.24 pg/mL. Additionally, the average plasma pTau181 level across 7 cohorts (comprising 1,424 healthy individuals) was found to be 11.18 pg/mL. For the second question, plasma tau levels in 1,189 AD patients from 14 cohorts were found to be higher than those in 1,611 HCs, with a weighted average difference of 0.61 pg/mL. When analyzed by sex, differences in tau levels were observed as follows: 0.48 pg/mL in the low male group, 0.99 pg/mL in the equal male group, and 1.30 pg/mL in the high male group. For the third question, the diagnostic accuracy of plasma tau in AD was assessed across five cohorts from four studies. The pooled sensitivity and specificity for tau in diagnosing AD were 0.75 and 0.69, respectively, with an AUC of 0.77 for the summary receiver operating characteristic (SROC) curve. In four other studies, five additional cohorts evaluated the diagnostic accuracy of plasma pTau181. The pooled sensitivity and specificity for pTau181 in diagnosing AD were 0.89 and 0.86, respectively, with an AUC of 0.93. These findings clearly indicate that the diagnostic accuracy of plasma pTau181 for AD is superior to that of tau and is highly comparable to that of PET and CSF pTau181. However, two important questions remain unanswered, including when plasma pTau181 levels exceed physiological thresholds in AD and how plasma pTau181 correlates with the spatiotemporal progression of characteristic AD pathology.

In 2021, the Scholl group from the University of Gothenburg, Sweden, conducted a longitudinal cohort study that tracked plasma pTau181 levels over a period of at least six years using the Simoa platform.[35] The study aimed to compare changes in plasma pTau181 levels with amyloid pathology, CSF pTau181 levels, and other existing diagnostic methods to better understand the natural progression of plasma pTau181 in the spectrum of sporadic AD. The prospective research cohort comprised 1,067 individuals, including cases of cognitively normal (CN), MCI, and AD. They found: (1) Dynamic changes in plasma pTau181 levels were strongly associated with β-amyloid pathology. The correlation between baseline plasma pTau levels and baseline PET, with the strongest correlation observed in individuals with MCI, followed by those with AD. The correlation between baseline plasma pTau and changes in PET, as well as the relationship between changes in plasma pTau and the changes in PET. In both cases, MCI showed the strongest correlation, followed by CN individuals, while no significant correlation was found in AD patients. These findings suggest that the relationship between plasma pTau levels and Aβ deposition is stronger during the earlier stages of disease progression. Plasma pTau181 levels began to rise and surpass the threshold before CSF Aβ levels reach their cutoff point. (2) Elevated plasma pTau181 levels were associated with increased tau-PET signals after six years. The distribution of these signals closely overlapped with the regions most affected by neurofibrillary tangles (NFTs) in AD. (3) Changes in plasma pTau181 levels reflected the dynamic process of CSF pTau181, with plasma pTau181 levels rising before Aβ biomarker levels become abnormal. These findings strongly support the potential of plasma pTau181 as a biomarker for diagnosing and screening AD. Some researchers have proposed that other pTau might serve as biomarkers for AD, potentially with superior performance compared to pTau181. For instance, Palmqvist et al. suggested that plasma pTau217 was more effective in distinguishing AD from other neurodegenerative diseases and showed a stronger correlation with PET imaging results than pTau181.[36a] In 2025, the superiority of plasma pTau217 (AUC = 0.942) over pTau181 (AUC = 0.909) in differentiating AD from other disease groups was demonstrated using DIs.[36b] In 2021, Ashton et al. evaluated the accuracy and feasibility of plasma pTau231 as a biomarker for AD using the Simoa assay.[37] The quantitative range for pTau231 was 0.25–64 pg/mL, with a minimum limit of quantification around 2 pg/mL. However, measurement errors exceeded 20% when sample concentrations were below 2 pg/mL. Their findings demonstrated that plasma pTau231 could effectively distinguish AD patients from Aβ-negative cognitively intact elderly individuals, with AUC ranging from 0.92 to 0.94. It also differentiated AD from non-AD neurodegenerative disorder patients with an AUC of 0.93, and from Aβ-negative MCI individuals with an AUC of 0.89. The authors concluded that pTau231 is as effective as pTau181 in identifying AD, both in clinical settings and neuropathological assessments, with the added benefit that pTau231 levels rise earlier

Neurofilament Light Chain

NfL is dispersed in the cytoplasm of neurons and, as a subunit of neurofilaments, plays a crucial role in maintaining neuronal structural stability. Under normal conditions, the levels of NfL released by axons are relatively low. However, its release gradually increases with age, doubling between the ages of 20 and 50, and doubling again by age 70. In pathological conditions, such as axonal damage, the release of NfL increases significantly. The released NfL can cross the blood–brain barrier and enter peripheral blood from the CSF. Elevated levels of blood NfL have been observed in a variety of neurodegenerative diseases. Compared to other biomarkers, blood NfL levels show a strong correlation with those in CSF, indicating that blood NfL levels reflect central nervous system pathology with minimal interference from peripheral sources.

To determine which of the three immunoassays—enzyme-linked immunosorbent assay (ELISA), electrochemiluminescence (ECL) immunoassay, and Simoa—was most sensitive and reliable in measuring NfL in both serum and CSF, Kuhle et al. compared their performance.[38] The lower LOQ for these assays were 78.0 pg/mL, 15.6 pg/mL, and 0.62 pg/mL, respectively. Over 50% of serum samples could not be reliably quantified using either ELISA or ECL. Simoa demonstrated the strongest correlation between serum and CSF NfL levels, followed by ECL and ELISA. In CSF, NfL measurements showed strong correlations across all three assays, whereas in serum, Simoa and ECL exhibited a relatively strong correlation, with poor agreement between Simoa and ELISA, as well as between ECL and ELISA. Subsequently, Gisslén et al. explored the feasibility of using NfL as a biomarker for central nervous system (CNS) injury in individuals infected with Human Immunodeficiency Virus (HIV).[39] They demonstrated that while plasma NfL levels were 50 times lower than CSF NfL levels, they could still be reliably measured using the Simoa platform. Notably, NfL levels were significantly elevated in both plasma and CSF in patients with HIV-associated dementia and those with low CD4+ T-cell counts. Rojas et al. investigated the potential of using plasma NfL as a diagnostic marker for progressive supranuclear palsy (PSP).[40] The mean plasma NfL level in 12 healthy individuals was 17.5 pg/mL, while in 15 PSP patients, it was 31 pg/mL. All measurements were conducted using the Simoa platform. Using a threshold of 20 pg/mL, the sensitivity for diagnosing PSP was 0.80, and the specificity was 0.83. Rohrer et al. compared serum NfL levels between patients with frontotemporal dementia (FTD) and HCs to evaluate NfL as a potential diagnostic marker for FTD.[41] The mean serum NfL level in FTD patients was 79.9 pg/mL ± 51.3 pg/mL, while in HCs, it was 19.6 pg/mL ± 8.2 pg/mL. These data indicated that NfL levels were significantly elevated in FTD patients; however, NfL levels varied considerably across different FTD subtypes. Mattsson et al. measured plasma NfL levels in patients with AD using the Simoa platform, with a quantitative range spanning from 2.2 ng/L to 1,620 ng/L.[42] They compared plasma and CSF NfL levels, along with cognitive function assessments and neuroimaging results, across three groups: HCs (n = 193), MCI patients (n = 197), and AD patients (n = 180). Plasma NfL levels were significantly elevated in both MCI (42.8 ng/L) and AD patients (51.0 ng/L) compared to controls (34.7 ng/L). Additionally, plasma NfL levels showed positive correlations with several other disease markers. Plasma NfL demonstrated diagnostic accuracy for AD dementia, with an AUC of 0.87, which was comparable to that of CSF NfL. Byrne et al. investigated the potential of plasma NfL as a prognostic marker for Huntington's disease in a three-year follow-up study.[43] They compared plasma NfL levels, cognitive function, motor function, and brain volume between healthy individuals and CAG expansion mutation carriers. Baseline NfL levels in gene mutation carriers (3.63 ± 0.54 pg/mL) were significantly higher than those in HCs (2.68 ± 0.52 pg/mL). Higher baseline NfL levels were significantly correlated with subsequent cognitive decline, a decrease in overall functional ability, and brain atrophy. In 2017, Pileh et al. examined the correlation between the NfL levels of both CSF and serum and multiple sclerosis (MS) using the Simoa.[44] In the neurodegenerative disease control group (n = 27) and the MS patient group (n = 39), the average CSF NfL levels were 341 (standard deviations (SD): 267) pg/mL and 1475 (SD: 2358) pg/mL, respectively, while the average serum NfL levels were 8.2 (SD: 3.58) pg/mL and 17.0 (SD: 16.94) pg/mL, respectively. In the fingolimod treatment group (n = 243), the average serum NfL level decreased significantly from 20.4 (SD: 10.7) pg/mL to 13.5 (SD: 7.3) pg/mL over 12 months and remained stable over the following 24 months. Siller et al. tracked serum NfL levels and MRI findings in patients diagnosed with MS over a period of 6 to 37 months.[45] They found that baseline serum NfL levels were strongly associated with T2 lesion volume on MRI. Patients with higher baseline NfL levels experienced a more rapid increase in T2 lesion volume and a faster decline in brain parenchymal volume. Oskar et al. used the Simoa platform to assess NfL levels in blood samples from three cohorts: patients with atypical Parkinsonian disorders (APD), PD, and HCs.[46] They found a strong correlation between NfL levels in blood and CSF ($\rho = 0.73–0.84$, $p \leq 0.001$). Blood NfL levels effectively distinguished PD from APD, with AUC values of 0.81, 0.85, and 0.91 across the three cohorts. However, blood NfL levels could not differentiate PD from HCs, highlighting the limitations and prerequisites of using blood-based biomarkers for diagnosis. Ashton et al. recruited 2,269 participants from two independent multicenter cohorts (Ackl and Lund) in Figure 6.3.[47] The subjects were categorized into five groups: a cognitive impairment group, a continuous AD group, other neurodegenerative disease groups, a Down syndrome group, and a depression group. Plasma NfL levels were measured for each group, with the results presented in Figure 6.3. In both multicenter cohorts, plasma NfL levels were significantly elevated in the cognitive impairment, Down syndrome with AD, and ALS groups compared to the cognitively unimpaired (CU) Aβ-negative group. However, no significant increase was observed in the PD, Down syndrome, or depression groups. The authors further explored the diagnostic value of NfL in distinguishing between different neurodegenerative diseases and cognitive impairments. The accuracy of using NfL to differentiate between CU Aβ− and CU Aβ+, SCD, and MCI was relatively poor (AUC = 52%–65%). However, the accuracy for identifying AD dementia was higher (KCL cohort, AUC =

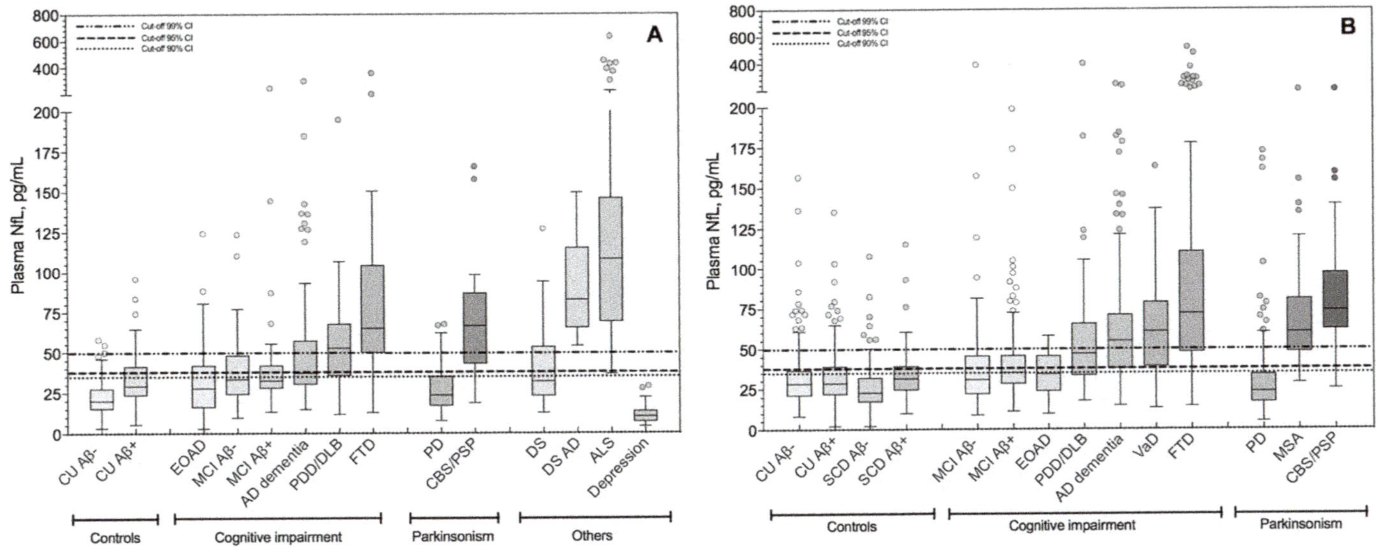

FIGURE 6.3 Plasma NfL concentrations in diagnostic and control groups: (A) KCL cohort (n = 805) and (B) Lund cohort (n = 1464).

Source: Reprinted by permission from Springer Nature Limited, Ashton, J.N.; Janelidze, S.; Ahmad, A.L.K. et al. *Nature Communications*, 2021, 12, 3400–3412 (ref 47). Copyright 2021.

79%; Lund cohort, AUC = 80%). In the KCL cohort, distinguishing atypical PD, Down syndrome with AD, FTD, and ALS from CU Aβ– showed strong performance (AUCs > 80%). Finally, the authors proposed age-related plasma NfL thresholds to predict various neurodegenerative disorders, including Down syndrome, depression, and cognitive impairment.

Gleerup et al. measured saliva NfL levels in 152 patients and 17 HCs using the Simoa.[48] Of the 162 samples, NfL was quantifiable, with levels ranging from 1.8 to 2.3 pg/mL. No significant differences were observed between healthy individuals, MCI, AD, and non-AD groups, suggesting no correlation between saliva NfL levels and plasma NfL or CSF Aβ42, pTau, or tau levels. The authors concluded that NfL in saliva does not reflect neurodegeneration in the brain and is unsuitable as a diagnostic biomarker for neurodegenerative diseases.

Glial Fibrillary Acidic Protein

Glial fibrillary acidic protein (GFAP) is an intermediate filament protein and a key structural component of astrocytes, the star-shaped glial cells in the brain. GFAP is rich in glutamic acid and aspartic acid, with a molecular weight of 50–52 kDa. The expression of GFAP in astrocytes is dynamic, changing in response to cellular stimulation. Astrocytes make up 50%–60% of glial cells in the brain, a number more than five times greater than that of neurons. Serum GFAP levels reflect the extent and severity of neuronal damage. Historically, it was believed that GFAP levels in CSF were not capable of significantly distinguishing AD patients from HCs. However, recent findings from DIs have demonstrated that GFAP levels in the blood of AD patients, particularly those with continuous spectrum stages, are elevated.

In 2019, Oeckl et al. measured serum GFAP levels in 28 AD patients, 35 behavioral variant frontotemporal dementia (bvFTD) patients, 11 PD patients, 19 dementia with Lewy bodies (DLB) patients, and 34 HCs using the Simoa.[49] The GFAP levels in these groups were 376, 211, 186, 343, and 157 pg/mL, respectively. GFAP levels were significantly elevated in AD and DLB patients compared to HCs. Additionally, CSF GFAP levels were also increased in various neurodegenerative diseases. The AUC for serum GFAP in distinguishing AD from controls and AD from bvFTD was 0.91 and 0.85, respectively, indicating strong diagnostic potential. A study conducted by Elahi et al. also confirmed that plasma GFAP levels were higher in both early-onset and late-onset AD patients compared to HCs, while NfL levels were particularly elevated in early-onset AD patients.[50] In

2020, Verberk et al. compared plasma levels of Aβ, GFAP, and NfL with PET imaging to identify clinical AD spectrum pathologies.[51a] Their research included 70 individuals with SCD, 50 with MCI, and 132 with AD dementia. The AUC values for identifying Aβ-positive individuals were as follows: 73% for the Aβ42/Aβ40 ratio, 81% for GFAP, 71% for NfL, and 88% for a combined AUC for Aβand GFAP. For the SCD + MCI group, the AUC values for identifying Aβ-positive individuals were 67% for the Aβ42/Aβ40 ratio, 76% for GFAP, 63% for NfL, and 84% for the combined AUC for Aβand GFAP. These findings suggest that a combination of plasma Aβ and GFAP levels can be used to screen for Aβ positivity. Additionally, they found that GFAP and NfL could be useful for monitoring the severity of AD. Asken et al. confirmed that plasma GFAP levels were positively correlated with Aβ-PET centiloid measurements, whereas NfL did not show such a correlation.[52] A significant difference in serum GFAP levels was found between patients with frontotemporal lobar degeneration (FTLD) and HCs, with GFAP levels correlating with disease severity.[53] Chatterjee et al. evaluated the risk of CN elderly individuals developing AD based on plasma GFAP levels.[54] Using PET image as a reference for Aβ positivity, plasma GFAP levels were significantly higher in the Aβ-positive group (240.12 ± 124.88 pg/mL) compared to the Aβ-negative group (151.42 ± 58.49 pg/mL). When age, sex, and APOE ε4 risk factors were incorporated into a basic model, the AUC for distinguishing Aβ-positive (Aβ+) from Aβ-negative (Aβ-) individuals was 0.78. Adding plasma GFAP levels to this model increased the AUC to 0.91, and further including the Aβ42/Aβ40 ratio raised the AUC to 0.92. Verberk et al. recruited a cohort of 300 CN elderly individuals and monitored changes in serum GFAP and NfL levels over an average period of three years.[51b] Of these, 27 individuals developed dementia, and their baseline serum GFAP (327 ± 141 pg/mL) and NfL (14.8 ± 6.5 pg/mL) levels were significantly higher compared to those of the remaining participants (GFAP: 195 ± 123 pg/mL; NfL: 10.7 ± 5.9 pg/mL). Furthermore, the individuals who developed dementia exhibited a steeper rate of change over time. Cicognola et al. investigated the association between plasma GFAP levels and the conversion of MCI to AD dementia by tracking changes in GFAP levels over an average period of 4.7 years in 160 MCI patients.[55] Baseline GFAP levels were found to reflect abnormalities in CSF Aβ42/Aβ40 and Aβ42/T-tau ratios, with AUC values of 0.79 and 0.80, respectively. The AUC for GFAP in predicting MCI conversion to AD was 0.84. Additionally, the rate of change in GFAP levels was steeper in Aβ-positive individuals and MCI converters compared to Aβ-negative and stable MCI individuals.

Alpha-Synuclein

Alpha-Synuclein (α-syn) is a neuronal protein composed of 140 amino acids, with a molecular weight of 14.4 kDa. It is predominantly found in presynaptic nerve terminals. In healthy neurons, α-syn exists primarily as a monomer, but in diseased states, it becomes abnormally expressed and aggregates into soluble oligomers and insoluble fibrils. The precise relationship between α-syn aggregation and neuronal degeneration, as well as its toxic characteristics, remains unclear. However, it is generally accepted that soluble oligomers are the most cytotoxic form of α-syn. These oligomers are more readily accessible to different parts of neurons compared to the fibrillar amyloid aggregates that accumulate inside the cell. A hallmark of PD pathogenesis is the deposition of misfolded α-syn into Lewy bodies within the brain. Consequently, α-syn has been proposed as a potential biomarker for the early diagnosis, disease monitoring, and prognosis of PD. PD is the second most common neurodegenerative disorder after AD, affecting over 1% of individuals aged 65 and older. Studies have shown that, due to Lewy body formation, α-syn levels in CSF are lower in patients with PD or PD dementia. However, blood levels of α-syn show variability depending on the immunoassay used, and there is no consensus regarding these measurements.

Ng et al. assessed α-Syn levels in the plasma of 221 individuals, including 51 HCs and 170 patients with PD using the Simoa platform.[56] Their data indicated that plasma α-Syn levels did not show a significant correlation with age, sex, or disease duration. The average α-Syn level in PD patients (15.5 ± 8.48 μg/L) was 16% higher than in HCs (13.1 ± 7.77 μg/L). α-Syn levels were not associated with the Hoehn and Yahr (H&Y) stages but tended to decrease in stages 3 and 4. No difference in α-Syn levels was observed between controls and patients with a motor score below 23, while patients with a motor score above 24 exhibited higher α-Syn levels compared to controls. Furthermore, patients with lower Mini-Mental State Examination (MMSE) scores had elevated α-Syn levels. The AUC for distinguishing PD from HCs using α-Syn was 0.599, but when MMSE ≤ 25 was included, the AUC increased to 0.63.

Large deposits and misfolding of phosphorylated α-syn, particularly phosphorylation at serine 129 (pS129 α-syn) which constitutes a major component of these aggregates, have been identified in postmortem brain samples. Consequently, pS129 α-syn has been proposed as a potential biomarker for the diagnosis of PD. Cariulo et al. employed the Sigulex single-molecule counting immunoassay to measure the levels of α-syn and pS129 α-syn in CSF and plasma.[57] The LODs, lower LOQs, and upper LOQs for pS129 α-syn were 0.15, 4.19, and 2560 pg/mL, respectively, and for total α-syn, they were 12.4, 65.5, and 16,000 pg/mL. Their findings revealed that matrix effects made it challenging to detect pS129 α-syn in CSF, with only 70% recovery observed even when pS129 α-syn was added to the sample. However, detection of pS129 α-syn in plasma was more straightforward. Moreover, plasma pS129 α-syn levels were found to be influenced by phosphatase activity, with a ten-fold increase in pS129 α-syn levels when phosphatase inhibitors were added during sample collection. In a small cohort of five PD patients, plasma pS129 α-syn levels were significantly higher than those observed in five HCs.

Youssef et al. compared the performance of three immunoassays—BioLegend chemiluminescence, MesoScale Discovery electrochemiluminescence, and Quanterix DI—for measuring α-Syn levels in plasma.[58] The results from all three assays indicated that α-Syn levels were higher in patients with PD compared to HCs, although the extent of the increase varied depending on the assay used. Specifically, the BioLegend chemiluminescence, MesoScale Discovery electrochemiluminescence, and Quanterix DI showed increases in α-Syn levels of 10%, 13%, and 30%, respectively. The p-values from the Quanterix assay were significantly higher than those obtained from the other two methods. The data from all three assays did not show any correlation with patient age, sex, disease duration, or H&Y stage. However, the release of α-Syn due to red blood cell lysis had a more pronounced effect on the performance of the BioLegend and MesoScale methods compared to the Quanterix assay.

A Panel of Biomarkers

Using a panel of multiple biomarkers for diagnosis is an effective approach to improving diagnostic accuracy. Various researchers have explored the potential of combining different biomarkers. Sugarman et al. evaluated the accuracy of combining plasma NfL and total tau levels to distinguish between normal cognition (NC), MCI, and AD dementia.[59] The baseline NfL levels in the AD group were higher than those in the MCI and NC groups, with log-transformed differences of 0.55 and 0.68, respectively. Baseline total tau levels were also slightly elevated in the AD group compared to the MCI and NC groups, although no significant difference was found between the MCI and NC groups. In terms of diagnostic accuracy, the AUC for NfL and total tau alone in distinguishing MCI from NC, AD from NC, and AD from MCI were nearly identical. Moreover, the combination of the two biomarkers did not improve the AUC. Wu et al. evaluated the accuracy of plasma pTau181, NfL, and Aβ levels in diagnosing AD in the Chinese population.[60] When used individually, pTau181 demonstrated the highest diagnostic accuracy, with an AUC of 0.885. A diagnostic model combining total tau, Aβ40, Aβ42, pTau181, NfL, and clinical data (model 9) achieved an AUC of 0.951. A simplified version of this model, model 12, which combines pTau181, Aβ42, and clinical features, achieved an AUC of 0.933. The sensitivity for distinguishing AD from normal controls was 0.786, while the specificity was 0.942.

Huang et al. explored the relationship between Aβ burden, objective-defined subtle cognitive decline (Obj-SCD), and plasma levels of pTau181, NfL, and Aβ.[61] Using the Simoa, they measured biomarker levels in plasma samples from 65 individuals diagnosed with Aβ-negative NC (Aβ-NC), 58 with Aβ-positive NC (Aβ+NC), 65 with Aβ-negative Obj-SCD (Aβ-Obj-SCD), and 48 with Aβ-positive Obj-SCD (Aβ+Obj-SCD). The study found that the Aβ+Obj-SCD group exhibited significantly higher levels of pTau181 and NfL compared to the Aβ-NC group. The AUC for pTau181 and NfL alone were 0.763 and 0.748, respectively. When combined, the AUC increased to 0.814, suggesting improved diagnostic accuracy.

Wolf et al. conducted a study involving 4,444 nondemented participants from the Rotterdam Study.[62] Plasma samples and clinical data spanning from 2002 to 2005 were obtained, and levels of total tau, NfL, Aβ40, and Aβ42 were measured using the Simoa HD-1 analyzer platform. The participants were followed up until January 2016 to assess the relationship between blood biomarker levels and the risk of developing all-cause dementia and AD dementia. Over a follow-up period of up to 14 years, 549 participants developed dementia, including 374 with AD dementia. The study found that higher baseline levels of Aβ42 were associated with a reduced risk of both all-cause dementia (hazard ratio [HR] 0.6) and AD dementia (HR 0.59), while higher baseline NfL levels were linked to an increased risk of all-cause dementia (HR 1.59) and AD dementia (HR 1.50). A combination of low Aβ42 and high NfL levels was associated with an even higher risk for both all-cause dementia (HR 9.5) and AD dementia (HR 15.7). No significant associations were observed between total tau or Aβ40 levels and dementia risk. Longitudinal analysis revealed that individuals who developed AD exhibited a 3.4-fold faster increase in NfL levels compared to those who remained nondemented. In fact, AD diagnosis occurred, on average, 9.6 years after plasma biomarker values deviated from controls. These findings suggest that plasma NfL and Aβ42 levels can be used as biomarkers to assess dementia risk, with NfL serving as a potential marker for monitoring the progression of AD dementia, despite its lack of specificity.

Simren et al. assessed the diagnostic and monitoring utility of plasma biomarkers, including pTau181, NfL, Aβ, and GFAP, in individuals with MCI, AD dementia, and CU controls.[63] While Aβ and total tau showed limited diagnostic value for distinguishing MCI and AD, pTau181, NfL, and GFAP exhibited significantly higher diagnostic value for both MCI and AD. The ROC analysis demonstrated varying diagnostic accuracies for these biomarkers. Specifically, pTau181 displayed the highest AUC for differentiating AD dementia from CU (AUC = 0.91), followed by its ability to distinguish AD from MCI (AUC = 0.75) and MCI from CU (AUC = 0.71).

Li et al. aimed to develop a biomarker panel for diagnosing PD syndrome (PDS) using a cohort that included a PD group (n = 45), a multiple system atrophy (MSA) group, a PSP group, and an age-matched HC.[64] They employed the Simoa platform to measure plasma levels of NfL, α-Syn, total tau, Aβ42, and Aβ40. The average values for each biomarker are summarized in Table 6.3. The resulting biomarker panel, which consisted of α-syn, Aβ42, Aβ40, the Aβ42/40 ratio, and NfL, showed strong diagnostic performance, with AUC of 0.983 for distinguishing PDS from HC and 0.977 for distinguishing PD from HC. Overall, the panel's diagnostic

TABLE 6.3 Detection of Plasma Biomarkers for Neurodegenerative Diseases Using Simoa

Biomarker (pg/mL)	**PD (n = 45)**	**MSA (n = 13)**	**PSP (n = 8)**	**HC (n = 33)**
NfL	20.43 ± 14.09	86.53 ± 33.74	53.9 ± 22.97	16 ± 5.18
Total tau	1.08 ± 0.66	0.51 ± 0.18	0.55 ± 0.17	0.93 ± 0.63
Aβ42	10.1 ± 2.99	5.3 ± 1.17	5.83 ± 2.28	5.26 ± 2.01
Aβ40	167.35 ± 49.41	119.5 ± 23.78	136.39 ± 30.76	103.9 ± 31
Aβ42/40	0.04 ± 0.01	0.04 ± 0.01	0.04 ± 0.01	0.05 ± 0.01
α-synuclein	43.5 ± 33.63	44.01 ± 18.51	53.75 ± 52.18	10.36 ± 5.51

Source: Reprinted with permission from Li, Q.; Li, Z.; Han, X.X. et al., *Frontier in Neuroscience*, 2022, https://doi.org/10.3389/fnins.2022.805953 (ref 64).

accuracy outperformed that of individual biomarkers in differentiating PD and APD.

CARDIOVASCULAR DISEASE BIOMARKER DETECTION

Cardiovascular disease (CVD) is an umbrella term that encompasses a range of heart and vascular conditions, including coronary artery disease, cerebrovascular disease, rheumatic heart disease, and others. According to the China Cardiovascular Health and Disease Report 2021, CVD is the leading cause of mortality in China, responsible for 46.74% of deaths in rural areas and 44.26% in urban areas. The mortality rate due to CVD continues to rise. Currently, an estimated 330 million people in China are living with CVD, including 245 million with hypertension, 13 million with stroke, 11.39 million with coronary artery disease, and 8.9 million with heart failure. CVD places a significant and growing economic burden on both individuals and society. The primary risk factors for CVD include hypertension, dyslipidemia, diabetes, chronic kidney disease, metabolic syndrome, and air pollution.

CVD are heterogeneous and multifaceted, and their biomarkers provide valuable insights into the presence and severity of these conditions, serving as sensitive and specific indicators of CVD. Key biomarkers include those associated with myocardial injury, such as cardiac troponin (cTn), markers of myocardial ischemia like D-dimer, cardiac function markers such as B-type natriuretic peptide (BNP and NT-proBNP), and inflammatory biomarkers like C-reactive protein (CRP) and tumor necrosis factor-alpha (TNF-α). However, it is important to note that inflammatory biomarkers are not specific to cardiovascular disease, and their associated DIs are typically discussed in the context of inflammatory biomarkers in separate chapters. This section primarily focuses on the detection and application of biomarkers for myocardial infarction, specifically cTn. Given the acute and complex nature of CVD, as well as the wide array of cardiovascular biomarkers, clinical tests must not only provide rapid diagnoses to enable timely treatment but also facilitate the simultaneous detection of multiple biomarkers to ensure accurate diagnosis. Single-molecule counting immunoassays offer significant advantages over traditional immunoassays, enhancing the sensitivity, specificity, and multiplexing capabilities of cardiovascular biomarker testing. cTn is a structural protein found in cardiac muscle cells. It binds with myosin and actin to form the thin filaments of myofibrils, playing a crucial role in the contraction and relaxation of the heart muscle. cTn consists of three subunits, each with distinct molecular weights: the calcium-binding subunit cTnC, the inhibitory subunit cTnI (22.5 kDa), and the myosin-binding subunit cTnT (37 kDa). Under normal physiological conditions, cTn is entirely stored within cardiac muscle cells and does not cross the cell membrane into the bloodstream. However, when myocardial injury occurs and myofibrils are damaged, cTn is released into the bloodstream. It is important to note that cTnC is not suitable as a heart-specific biomarker, as it is also expressed in skeletal muscle, with an amino acid sequence identical to that of the cTnC found in cardiac muscle.

cTnI is widely regarded as the gold standard biomarker for diagnosing acute myocardial infarction (AMI). Within three to four hours following myocardial injury, cTnI concentrations in the blood increase rapidly, ranging from several ng/L to thousands of ng/L. In 2000, the European Society of Cardiology and the American College of Cardiology redefined the diagnostic criteria for AMI, recommending that the 99th percentile of cTn concentrations in reference populations be used as a diagnostic cutoff, with intra-assay variability of less than 10%. This standard emphasizes the need for tests to be highly sensitive, rapid, and suitable for bedside or emergency care settings. In addition to its role in diagnosing acute myocardial injury, cTnI can also serve as a biomarker for nonacute myocardial damage, such as monitoring drug-induced cardiotoxicity or detecting cardiac rejection in heart transplant recipients.

In 2006, Singulex introduced the flow single-molecule-counting immunoassay and assessed its performance in analyzing cTnI levels.[65] Neumann et al. evaluated the sensitivity of three immunoassays for predicting CVD risk by monitoring cTnI levels in 7,899 participants without a history of major adverse cardiovascular events (MACE) over a 14-year period.[66] The sensitivity of the three assays varied: the LODs for the STAT cTnI immunoassay, the STAT highly sensitive immunoassay, and Singulex's single-molecule ultrasensitive immunoassay were 10 pg/mL, 1.9 pg/mL, and 1.0 pg/mL, respectively. The corresponding measurement values for a 10% CV were 32 pg/mL, 5.2 pg/mL, and 0.78–1.6 pg/mL. The ultrasensitive single-molecule immunoassay detected cTnI levels in 93.9% of the population, and higher cTnI levels were associated with an increased risk of future myocardial disease. Subsequently, Neumann et al. used the ultrasensitive cTnI immunoassay to diagnose 1,534 suspected cases of myocardial infarction (MI) and established lower cutoff values.[67] Patients with plasma cTnI levels below 1 ng/L, or below 2 ng/L with a rise of less than 1 ng/L after one hour, were excluded from the MI diagnosis. A single cTnI measurement upon admission excluded over 25% of patients, while two measurements excluded 50%, with a negative predictive value of 99.7%. Among the excluded patients, the rate of MI occurrence one year later was approximately 1%. Conversely, patients with plasma cTnI levels above 25 ng/L upon admission or those with an increase of more than 6 ng/L after one hour were triaged into the non-ST-elevation myocardial infarction (NSTEMI) group. Nearly 20% of patients fell into this category, with a positive predictive value of 79.5%. In this group, the rate of MI occurrence after one year was 11.6%. The single-molecule ultrasensitive immunoassay provides valuable technical support for the rapid triage of suspected MI patients, facilitating timely diagnosis and management.

Jing et al. developed two homogeneous DIs based on time and spatial differences using bright-field imaging to record the number of GNP-labeled sandwich immunocomplexes. In the time-difference-based immunoassay, the detection antibody, labeled with GNPs, reacted with the sample and was subsequently added to a sensor coated with capture antibodies.[68] Individual GNPs binding to the sensor surface were counted, allowing for the detection of the antibody-cTnI-antibody complex. When applied to undiluted human blood, this assay provided results within 10 minutes, with an LOD of 5.7 ng/L and a LOQ of 6.6 ng/L. In the spatial-difference-based immunoassay, the researchers designed a microfluidic chip with four distinct regions.[69] After injecting 1 μL of plasma sample, the formed immunocomplexes created a concentration gradient of cTnT along the channel. The detection antibodies in the various regions were imaged and counted, and the differences in counts between regions were used to quantify cTnT. This assay was completed within 30 minutes, with an LOD of 1.8 ng/L and a LOQ of 6.2 ng/L. These immunoassays demonstrate innovative approaches to DIs for the rapid and sensitive detection of cardiovascular biomarkers.

Jarolim et al. developed a DI for cTnI using the Simoa platform.[70] The assay provided results within 45 minutes, with an LOD of 0.01 ng/L (SD = 0.0053 ng/L) and a LOQ of 0.079 ng/L (SE = 0.034–0.039). The measurement for 10% CV was 2.0 ng/L. In a cohort of 97 HCs, the median cTnI concentration was 0.65 ng/L, with a minimum value of 0.072 ng/L. In contrast, among 362 heart failure patients, the median cTnI concentration was significantly higher, at 15.3 ng/L. Empana et al. utilized the Simoa platform to measure plasma cTnI levels in 9,503 volunteers without CVD.[71] The platform demonstrated an LOD of 0.013 pg/mL, with 99.6% of participants having detectable cTnI levels. The median cTnI concentration in this cohort was 0.63 pg/mL. After an average follow-up of 8.34 years, 516 participants had 612 CVD-related visits. cTnI levels were strongly associated with an increased number of CVD-related visits, demonstrating the strongest predictive ability for CVD-related visits. These findings suggest that cTnI can serve as an independent biomarker for the primary prevention of CVD, with higher cTnI levels correlating with a greater likelihood of future CVD-related healthcare visits.

INFLAMMATORY BIOMARKER DETECTION

Inflammatory biomarkers are often represented by cytokines. They are polypeptide or protein molecules secreted by both immune and nonimmune cells in response to various stimuli. Cytokines function as key mediators of intercellular communication. By binding to their specific receptors on cell surfaces, cytokines play a crucial role in regulating immune responses, modulating anti-inflammatory reactions, promoting wound healing, influencing tumor growth, and governing processes such as cell proliferation and differentiation. Based on their functions, cytokines are classified into several categories, including TNF, interleukins (IL), interferons (IFN), growth factors (GF), colony-stimulating factors (CSF), and others. When the body experiences infection, inflammation, or other immune responses, cytokine concentrations often change, making them useful biomarkers for a variety of diseases. For instance, TNF-α, IL-6, and IL-8 levels are significantly elevated in the plasma of HIV patients, while IFN-γ concentrations are notably increased in the serum of individuals with active tuberculosis. Cytokines also play a critical role in disease prognosis and treatment. The concentration profiles of specific cytokines can serve as distinct "fingerprints" for particular diseases. For example, it has been reported that the concentrations of 19 different cytokines vary between benign and malignant thyroid diseases.[72] In healthy individuals, the concentrations of most cytokines are typically below the LODs of traditional immunoassays and can only be detected in patients with acute diseases. As a result, highly sensitive immunoassays are necessary for early diagnosis, before symptoms appear. DIs are particularly suited to meet these requirements, offering enhanced sensitivity and enabling earlier detection of cytokines at low concentrations.

Tumor Necrosis Factor-α

TNF-α is a pleiotropic pro-inflammatory cytokine primarily produced by activated mononuclear macrophages. It consists of 157 amino acids and has a molecular weight of 17 kDa. As a critical multifunctional cytokine, TNF-α plays a key role in coordinating inflammation, immune responses, and neurohormonal regulation. It stimulates the production of other pro-inflammatory factors, such as IL-1, IL-6, and IL-8. Under normal conditions, TNF-α is secreted by the thymus, but it can also be released in response to inflammatory stimuli from various tissues, including the heart and kidneys. Physiologically, TNF-α concentrations are typically very low, around 10^{-10} mol/L. However, when its concentration increases to 10^{-8} mol/L, it can lead to pathological conditions, for example, microvascular coagulation. Due to its role in these processes, TNF-α is considered a valuable protein marker for the early detection of inflammatory diseases and tumors.

Wang et al. developed a novel and highly sensitive heterogeneous immunoassay for the fluorescent detection of TNF-α.[73] In their study, they successfully combined the signal amplification of hybrid chain reaction (HCR) with single-molecule detection. As illustrated in Figure 6.4, TNF-α antibodies were immobilized on a glass substrate. TNF-α and biotinylated TNF-α antibodies were sequentially added to form a sandwich-structured immunocomplex on the substrate. Streptavidin and biotinylated DNA primers were then introduced to the immunocomplex. Streptavidin acted as a bridge, connecting a biotinylated immunocomplex to three primers. These primers initiated the HCR, generating three long double-helix DNAs. The fluorescent dye SYBR Green I was inserted into the grooves of the double-stranded DNAs, emitting significantly stronger fluorescence upon binding. This enhanced fluorescence allowed for the imaging and counting of the immunocomplexes using an inverted fluorescence microscope. The detection range for TNF-α in actual serum samples was found to be between 50 fM and 1 pM. The immunoassay demonstrated excellent precision, reproducibility, high specificity, and minimal matrix effects. Moreover, by adjusting the corresponding antibodies, the assay can be easily adapted to detect other low-abundance biomarkers.

Wu et al. optimized antibody pairs for the detection of 10 cytokines using the Simoa platform.[74] The LODs for these cytokines ranged from 0.09 to 5.92 fM. They measured the levels of the 10 cytokines in healthy human serum to establish baseline values. In a subsequent study, Wu et al. addressed the challenge of antibody instability, which can affect

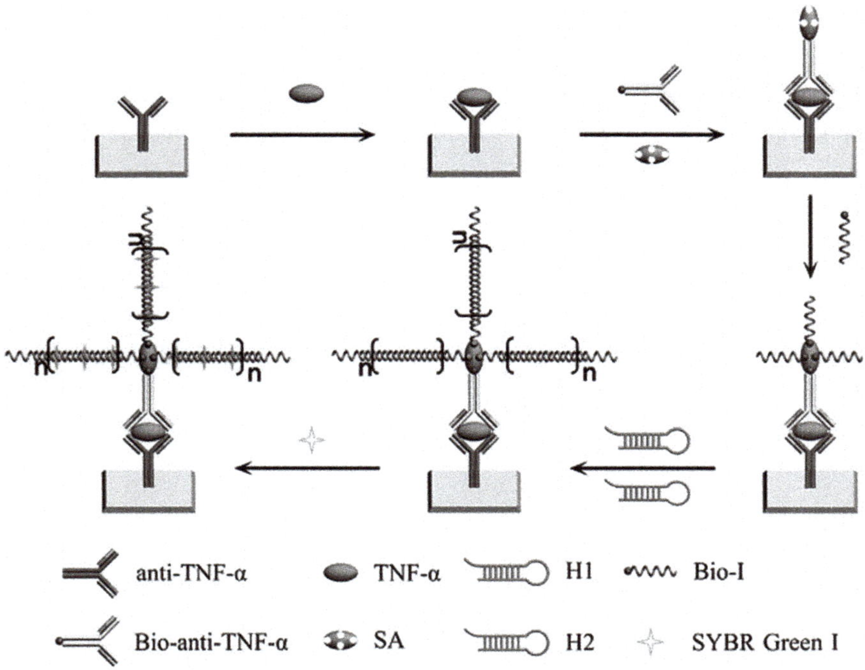

FIGURE 6.4 Schematic illustration of the single-molecule detection (SMD) method based on hybridization chain reaction.

Source: Reprinted from *Biosens. Bioelectron.*, Vol. 60, Dai S.; Feng C.; Li W. et al.; Quantitative detection of tumor necrosis factor-alpha by single molecule counting based on a hybridization chain reaction, pp. 180–184 (ref 73). Copyright 2014, with permission from Elsevier.

measurement reproducibility, by replacing the detection antibodies with slow off-rate modified aptamers (SOMAmers).[75] SOMAmers offer several advantages, including high binding affinity and chemical stability. In this modified assay, the capture antibody remained unchanged, while SOMAmers replaced the detection antibodies to form a magnetic bead-capture, antibody-antigen-SOMAmers-galactosidase immunocomplex within the microplate of the Simoa platform. Upon addition of a fluorogenic substrate to the microchamber, the magnetic beads with the bound antigen emitted light. The immunoassay using SOMAmers demonstrated significantly improved sensitivity, achieving a lower LOD of 0.67 fM (0.012 pg/mL) for TNF-α. In 2017, Wu et al. conducted a study to monitor the levels of 15 cytokines in healthy human serum over a 14-week period, establishing reference baselines using both single and multiplex Simoa platforms.[76] The cytokine levels for each individual remained stable throughout the 14-week period, with minimal fluctuation within individuals. Interindividual variability was classified into two categories: a low-variability group and a high-variability group. The low-variability group included cytokines such as IL-15, TNF-α, IL-12 p70, IL-17A, GM-CSF, IL-12 p40, IL-10, IL-7, IL-1α, and IL-5. The high-variability group included IL-8, IFN-γ, IL-2, IL-6, and IL-1β. Cytokines in the low-variability group may serve as markers for establishing standard threshold ranges, while those in the high-variability group are more appropriate for personalized treatment monitoring. Additionally, the study revealed challenges in detecting IL-1α and GM-CSF, with detection rates of 24% and 20%, respectively.

IL

ILs are a group of cytokines primarily produced by lymphocytes and mononuclear phagocytes. Initially discovered in white blood cells, they were named for their role in mediating communication between these cells. ILs play crucial roles in regulating immune responses, including hematopoietic cell development and differentiation. They are also involved in tissue repair, apoptosis, antitumor responses, and inflammation-related pathological damage. ILs are numbered according to the order of their discovery, with IL-1 being the first identified. To date, over 40 ILs have been identified, with their molecular characteristics and functions varying significantly. For further details, relevant literature provides comprehensive information.[77]

Crohn's disease (CD) is a chronic inflammatory bowel disease characterized by inflammation that can affect any part of the gastrointestinal tract. During periods of acute inflammation, the concentrations of cytokines undergo significant changes, and these proteins are believed to play a key role in the progression of chronic inflammatory diseases. In a study by Song et al. sensitive digital ELISA was employed to measure the plasma concentrations of TNF-α and IL-6 in CD patients both before and after anti-TNF-α treatment.[78] Plasma samples were collected from 17 CD patients prior to treatment, and their Crohn's Disease Activity Index (CDAI) was assessed. Follow-up measurements were conducted at 12 weeks post-treatment, and plasma samples from age- and gender-matched controls were also analyzed. The results indicated that the plasma concentrations of TNF-α and IL-6 in CD patients were 3.6 ± 0.9 pg/mL and 10.9 ± 11.2 pg/mL, respectively. Compared to the control group, no significant difference in TNF-α levels was observed between patients and controls, while IL-6 levels were significantly elevated in the patient group. Following treatment, the average reductions in TNF-α and IL-6 concentrations were 46% and 58%, respectively. This study is the first to use digital ELISA to quantitatively measure TNF-α and IL-6 levels in the plasma of CD patients, with LODs of 0.008 pg/mL and 0.006 pg/mL, respectively. The high sensitivity of digital ELISA enabled precise quantification of cytokine levels post-treatment, highlighting its potential as a valuable tool for monitoring therapeutic effects in CD. In 2015, they developed a fully automated 6-plex Simoa platform for the simultaneous detection of six cytokines (IL-6, TNF-α, GM-CSF, IL-10, IL-1β, and IL-1α) in plasma.[79] The LODs for each cytokine were as follows: 10 fg/mL for IL-6, 24 fg/mL for TNF-α, 16 fg/mL for GM-CSF, 26 fg/mL for IL-10, 13 fg/mL for IL-1β, and 26 fg/mL for IL-1α. The system was used to compare plasma cytokine levels in CD patients before and after TNF-α antibody therapy, as well as in serum samples from patients with type 1 diabetes. It was found that only IL-6 exhibited a significant difference between CD patients and HCs. Following treatment, the concentrations of IL-6, TNF-α, and GM-CSF decreased by 52%, 34%, and 25%, respectively. The 6-plex assay for IL-6 showed a high correlation with results from the single-plex assay, whereas the correlation for TNF-α was relatively weak. Furthermore, in patients with type 1 diabetes, serum concentrations of TNF-α, GM-CSF, IL-10, and IL-1β were found to be elevated compared to healthy individuals.

IL-12p70 expression levels are commonly used to assess the pharmacodynamic effects of checkpoint inhibitor therapy. However, IL-12p70 is present at very low concentrations, typically ranging from 60 to 1,000 fg/mL. Gupta et al. optimized the experimental conditions for detecting IL-12p70 using the Simoa platform, achieving a LOQ of 0.08 pg/mL.[80] This LOQ demonstrated the platform's suitability for plasma IL-12p70 detection. IL-12p70 was successfully detected in 27 out of 29 samples, with a CV of less than 20%. Further monitoring of plasma IL-12p70 levels in melanoma patients 15 days after receiving atezolizumab revealed increased levels in most patients, suggesting that IL-12p70 could serve as a prognostic biomarker following cancer immunotherapy. These findings support the use of DIs for effectively monitoring changes in IL-12p70 expression.

Song et al. introduced a pre-equilibrium DI (PE-dELISA) in which the immunoreactions are terminated at an early stage (15–300 seconds) to enable rapid detection.[81] For IL-6, the LOD was 25.9 pg/mL with a

15-second reaction time, and the dynamic range extended up to 10 ng/mL. In multiplex assays, only 15 μL of sample was required, with an LOD of less than 0.4 pg/mL. The entire process, from sample collection to result output, was completed within 4 hours, making it ideal for time-course monitoring in intensive care units (ICUs). Using this assay, the researchers continuously monitored changes in IL-6, TNF-α, IL-1β, and IL-10 levels in the serum of two ICU patients with COVID-19 over two weeks. Both patients experienced severe cytokine storms and acute respiratory distress syndrome but showed significant therapeutic improvement after treatment with a selective cytopheretic device. IL-6 and IL-6/IL-10 ratios exhibited a decreasing trend. Regarding the effect of tocilizumab on COVID-19 patients, significant individual variability was observed, with CVs of various parameters around 100%. Notably, IL-6 levels were significantly elevated in the treatment group. The assay was further refined by combining machine learning with dual-color labeling and spatial encoding to enhance throughput and accuracy.[82] This improved assay allowed the simultaneous detection of 12 cytokines in serum within 40 minutes (with 5 minutes for incubation), requiring only 15 μL of sample and achieving an LOD of less than 5 pg/mL. This enhanced assay can be applied to monitor cytokine level changes during chimeric antigen receptor T-cell therapy.

Gao et al. fabricated a microfluidic platform in which each channel was functionalized with specific capture antibodies and labeled detection antibodies with 100 nm AgNPs.[83] Immunoreactions occurred within these channels, with each channel corresponding to a particular antigen. The presence of an antigen was indicated by the silver nanoparticles, which were imaged using dark-field microscopy. The amount of antigen was quantified using machine learning algorithms. Only 3 μL of serum was required to detect six cytokines, with detection limits of 0.91 pg/mL for IL-1β, 0.47 pg/mL for IL-2, 0.46 pg/mL for IL-6, 1.36 pg/mL for IL-10, 0.71 pg/mL for TNF-α, and 1.08 pg/mL for IFN-γ. A total of 1,440 tests were completed within 5 hours using five chips, which included six repeated runs of 40 serum samples from critically ill COVID-19 patients. The results, shown in Figure 6.5, revealed that patients 1–25, who received no treatment, had elevated IL-6 levels compared to healthy individuals, with 52%, 60%, 64%, 52%, and 80% of patients exhibiting elevated levels of IL-1β, IL-2, TNF-α, and IFN-γ, respectively. These elevated levels suggest the presence of a potential cytokine storm. In contrast, patients 26–35 and 36–40, who received treatment with tocilizumab and selective cytopheretic device, respectively, demonstrated significantly higher cytokine levels.

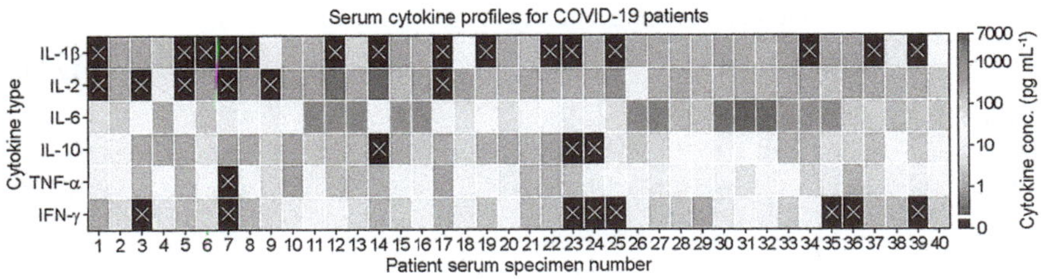

FIGURE 6.5 Heatmap of serum cytokine levels in COVID-19 patients detected using a machine learning-assisted microfluidic nanoplasmonic DI.

Note: The cytokine levels were based on the average value of six replicate tests. Scales indicate value of cytokine concentration. Symbol "×" as "not applicable" represents that the cytokine concentration is lower than the LOD of our immunoassay.

Source: Reproduced from Gao, Z.; Song, Y.; Hsiao, T.Y. et al. Machine-learning-assisted microfluidic nanoplasmonic digital immunoassay for cytokine storm profiling in covid-19 patients, *ACS Nano*, 2021, 15, 18023–18036 (ref 83). Copyright 2022 American Chemical Society.

Walt et al. developed a flow-based DI, termed MOSAIC, by integrating magnetic beads with DNA rolling circle amplification.[84] In contrast to the Simoa assay, MOSAIC eliminates the need for a physical microplate array for discretization, achieving higher detection efficiency, enhanced sensitivity, and superior multiplex detection capabilities through its flow-based detection technique. The single-plex and multiplex detection capabilities for cytokines are summarized in Tables 6.4A, 6.4B and 6.5. In a study measuring IFN-γ levels in saliva, MOSAIC successfully detected IFN-γ in 65% (17/26) of the samples, compared to 42% (11/26) detection by the Simoa assay.

TABLE 6.4A Limitation of Detection of Single-Component MOSAIC (aM)

Number of Magnet Beads	IL-10	IFN-γ	IL-6	IL-1β	IL-8	IL-12p70
100,000	255.5	99.9	421.3	152	109.9	5.8
50,000	43.3	20.7	263.7	124	98.8	16.7
20,000	15.9	43.1	227.9	517.1	105.2	6.9
10,000	24.8	18.9	877.1	425.5	37.9	9.6
5,000	22.7	34.8	607.6			
2,000	30.1					

Source: Reproduced from Wu, C.; Dougan, T.J.; Walt, D.R. High-throughput, high-multiplex digital protein detection with attomolar sensitivity, *ACS Nano*, 2022, 16, 1025–1035 (ref 84). Copyright 2022American Chemical Society.

TABLE 6.4B Limitation of Quantification of Single-Component MOSAIC (aM)

	IL-10	IFN-γ	IL-6	IL-1β	IL-8	IL-12p70
100,000	809.1	304.5	900.1	467.9	359.8	19.9
50,000	124.7	67.4	809.9	401.5	262.2	52.9
20,000	45.9	107.5	699.0	1,280.40	322.4	23.5
10,000	72.1	47	2,447.0	1,236.40	120.2	30.5
5,000	67.6	78	1,578.20			
2,000	77.3					

Source: Reproduced from Wu, C.; Dougan, T.J.; Walt, D.R. High-throughput, high-multiplex digital protein detection with attomolar sensitivity, *ACS Nano*, 2022, 16, 1025–1035 (ref 84). Copyright 2022American Chemical Society.

TABLE 6.5 Comparison between Eight-Component and Four-Component DIs

	LOD (aM)		LLOQ (aM)	
	MOSAIC	Simoa (Bead Number: 125000)	MOSAIC	Simoa
IFN-γ	166.0 [136.6–194.4]	308.7 [240.3–717.6]	524.9 [434.3–614.3]	1053.2 [714.4–2,192.7]
IL-1β	445.5 [230.4–520.9]	322.1 [161.1–364.7]	1357.8 [748.9–1380.6]	976.3 [555.9–1,175.1]
IL-5	942.0 [539.4–1373.3]	532.0 [352.1–1139.5]	2995.2 [1523.4–3470.6]	1683.2 [1175.6–3920.6]
IL-6	707.2 [87.5–1570.6]	426.7 [344.8–670.2]	1919.7 [297.7–4687.8]	1350.7 [1043.1–1856.8]
IL-10	59.1 [33.2–107.2]	377.4 [216.4–381.0]	157.2 [106.3–287.1]	1106.7 [682.4–1257.3]
IL-12 p70	54.6 [23.0–271.4]	83.6 [80.8–125.3]	163.1 [72.7–807.8]	312.0 [287.8–375.5]
IL-18	11141.6 [4922.8–997.9]	1497.2 [1319.0–2460.8]	29356.8 [16539.1–114204.2]	4265.7 [3958.0–7678.9]
VEGF	3056.7 [1965.9–3673.4]	986.4 [766.5–1163.7]	10325.7 [6963.4–10831.0]	2,864.1 [2583.8–3002.8]

Source: Reproduced from Wu, C.; Dougan, T.J.; Walt, D.R. High-throughput, high-multiplex digital protein detection with attomolar sensitivity. *ACS Nano*, 2022, 16, 1025–1035 (ref 84). Copyright 2022American Chemical Society.

In 2023, Su et al. integrated a DI with tissue chips to monitor cytokine secretion by brain endothelial cells in situ.[85] As illustrated in Figure 6.6, mouse brain microvascular endothelial cells are cultured in the upper layer of the chip, while a movable three-component DI microplate array is positioned in the lower layer. The nanomembrane in the middle, along with the endothelial cells, simulates the barrier between brain microvessels and the brain. Upon endotoxin stimulation in the upper layer, the endothelial cells secrete cytokines, some of which traverse the barrier and bind to the capture proteins immobilized on the microplate. Notably, the DI eliminates the need for magnetic beads. Instead, specific cytokine antibodies are bound to distinct regions of the chip. After cytokine capture, enzyme-labeled antibodies and substrates are directly added to the chip, addressing the low dispersion efficiency typically associated with magnetic beads. The LODs for MCP1, IL-6, and KC were 0.576, 0.101, and 0.433 pg/mL, respectively. This assay enables cytokine measurement at multiple time points on both sides of the barrier within ten hours. The secretion and barrier-crossing rates varied across cytokines, with cytokines on the vascular side reaching steady state after six hours, while those on the brain side continued to increase over the full ten-hour period. The assay provides valuable technical support for continuous monitoring of cytokine distribution in organ-on-chip models.

Interferon

IFNs are glycoproteins with the properties of antiviral, antiproliferative, immunoregulatory, and antitumor. They are primarily classified into three types based on their structure and origin: leukocyte IFN (IFN-α), fibroblast IFN (IFN-β), and immune cell IFN (IFN-γ). Additionally, IFNs are categorized into type I and type II based on their stability under acidic and heat conditions. Type I IFNs, including IFN-α and IFN-β, are resistant to both acid and heat, while type II IFN, primarily IFN-γ, loses its activity under acidic and heat conditions.

Rodero et al. were the first to conduct ultrasensitive detection of interferons using the Simoa platform, achieving an LOD of 0.23 fg/mL for Type I IFN-α.[86] They compared the serum levels of IFN-α in 20 healthy volunteers, 27 patients with type I interferonopathies, 43 patients with juvenile dermatomyositis (JDM), 72 patients with systemic lupus erythematosus (SLE), and 30 patients with retinal vasculopathy and leukoencephalopathy (RVCL) due to specific chromosomal mutations. Among these, the 27 patients with Type I IFN diseases were further

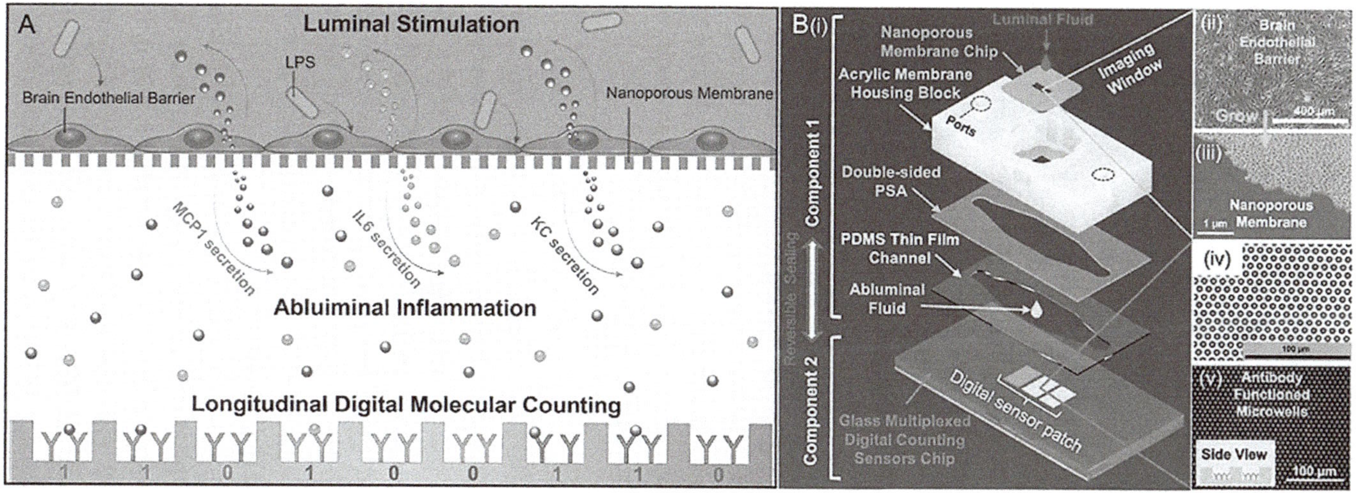

FIGURE 6.6 Monitoring cytokine concentrations using the integrated digital tissue-barrier-cytokine-counting-on-a-chip (DigiTACK): (A) schematic of the DigiTACK assay principle and (B) device characterization including (i) exploded view of a DigiTACK device, (ii) phase-contrast image of brain microvascular endothelial cells cultured within the device, (iii) SEM image of the NPN membrane, (iv) bright-field image of DGRIE-etched glass microwells, and (v) fluorescence image showing antibody functionalization within the microwells.

Source: Reprinted from *Biosens. Bioelectron.*, Vol. 224, Su S.H.; Song Y.J.; Stephens A. et al. A tissue chip with integrated digital immunosensors: in situ brain endothelial barrier cytokine secretion monitoring, pp. 115030 (ref 85). Copyright 2023, with permission from Elsevier.

grouped based on their genetic mutations. The IFN-α levels in patients with RVCL were not significantly different from those of healthy individuals (median value 1.6 fg/mL, interquartile range [IQR] 0.95–4.6 fg/mL) and thus served as a negative control. In contrast, patients with interferonopathies (median 310 fg/mL, IQR 71–2223 fg/mL), SLE (median 20 fg/mL, IQR 0.69–234 fg/mL), and JDM (median 56 fg/mL, IQR 14–120 fg/mL) exhibited significantly elevated IFN-α levels. Different cell types were isolated from patients' blood and subsequently lysed to measure IFN-α levels. The results indicated that STING-associated patients, a subgroup of type I interferonopathies, exhibited the highest IFN-α levels in CD14+ monocytes and plasmacytoid dendritic cells. In contrast, other cell types from STING-associated patients and those from patients with other diseases did not show significant differences in IFN-α levels compared to HCs.

Activation of the IFN-I pathway is a pathological hallmark of several systemic autoimmune diseases (SADs). However, the diversity of IFN-I subtypes and their low concentrations pose significant challenges in quantifying IFN-I proteins in biological samples using traditional methods such as ELISA. The development of the ultrasensitive Simoa has enabled the direct measurement of IFN proteins in the blood of patients with various immune-related diseases. This advancement offers valuable technical support for the diagnosis, classification, monitoring of disease progression, and prognosis of these conditions. The following section provides an overview of the application of DIs in detecting IFN levels in diseases such as SLE, tuberculosis, dengue virus infection, and Sjögren's syndrome.

SLE is a chronic autoimmune disorder of unknown etiology that can affect multiple organs and systems. Clinically, diagnosis is based on the SLE classification criteria established by the American College of Rheumatology; however, no quantitative diagnostic standards are currently available. It is widely believed that the dysregulation of IFNs plays a key role in the immune abnormalities observed in SLE. Elevated serum IFN-α levels have been linked to SLE activity and disease severity. Therefore, the strong association between IFN-α overexpression and SLE activity suggests that monitoring this cytokine may provide valuable insights for clinicians in assessing disease status. Mathian et al. employed digital ELISA to quantify IFN-α levels in the serum of 150 SLE patients.[87] Using IFN-α levels in HCs, they established an abnormal threshold for IFN-α at 136 fg/mL, and a threshold of 266 fg/mL distinguishing active from stable SLE. The AUC for differentiating active from stable SLE cases with this threshold was 0.83.

Tuberculosis (TB) is a chronic infectious disease caused by mycobacterium tuberculosis that poses a significant public health burden, particularly in high-incidence countries like China. The bacterium primarily enters the body through the respiratory tract and can lead to either active TB (ATB) or latent TB infection (LTBI). LTBI is characterized by the presence of tuberculosis-specific antigens in the body without clinical symptoms or radiological signs of disease, making individuals with LTBI a reservoir for potential future reactivation of TB. Approximately 5%–15% of individuals with LTBI may eventually develop active tuberculosis. As such, controlling and treating LTBI is critical to curbing the global TB epidemic. IFN-γ release assays are commonly used for TB screening, but certain test results can fall into an indeterminate range, making classification as either positive or negative difficult. Recent advances in highly sensitive DIs hold promise for providing more accurate and reliable diagnostic results, potentially improving the early detection and management of latent and active tuberculosis. Salah et al. conducted an analysis of 30 negative, 35 positive, 25 uncertain negative, 31 uncertain positive, and 30 uncertain samples using the Simoa.[88] The LOD of the assay was 0.002 IU/mL, and the batch-to-batch CV ranged from 3.7% to 8.2%. The results indicated that 83.3% of uncertain results could be reclassified as negative, 74% of uncertain positive results were confirmed as positive, and 72% of uncertain negative results were reclassified as either uncertain positive or positive. Llibre et al. measured the serum concentrations of IFN-α and IFN-β in patients with tuberculosis.[89] They found no significant difference in the levels of these proteins between active tuberculosis and LTBI patients, although the protein levels were slightly elevated compared to HCs. Interestingly, flu patients exhibited significantly higher levels of both proteins compared to HCs and tuberculosis patients. However, transcriptomic analyses revealed that the transcriptomic score of active tuberculosis patients was higher than that of LTBI patients, which contrasted with the protein level measurements in the blood. These findings offer new insights into the immune processes involved in mycobacterium TB infection.

Upasani et al. observed that the levels of Type I IFN in the blood of patients with dengue virus infection were significantly higher than those in healthy individuals, with IFN-α levels being more pronounced than IFN-β.[90] The LODs for IFN-α and IFN-β using the Simoa platform were 0.005 fg/mL and 0.05 pg/mL, respectively. A positive correlation was found between viral load and protein levels, with higher viral loads associated with elevated IFN levels in the blood. Additionally, mRNA expression of the IFN receptor subunit IFNAR2 and IFN-related IRF7 genes were upregulated in the peripheral blood mononuclear cells of the patients.

Sjögren's syndrome (SS) is an autoimmune disorder primarily affecting the exocrine glands, though it can also involve other organs, leading to immune-mediated damage. Huijser et al. investigated serum IFN-α2 levels, IFN-stimulated gene (ISG) expression in whole blood, and MxA (an IFN-induced antiviral protein) levels in whole blood among two independent cohorts of primary SS (pSS) patients (n = 85 and n = 110), a systemic sclerosis (SSc) cohort (n = 23), a SLE cohort (n = 24), and two HC groups (n = 40 and n = 28).[91] IFN-α2 levels were measured using the Simoa, while ISG expression was assessed by RT-PCR, and MxA levels were quantified via enzyme immunoassay. The results revealed that Simoa was able to detect IFN-α2 levels in 75.3% of pSS patients, 75% of SLE patients, 56.5% of SSc patients, and 45% of HC samples. Compared to HCs (median ≤5 fg/mL), IFN-α2 levels were significantly elevated in pSS (median 61.3 fg/mL), SSc (median 11.6 fg/mL), and SLE (median 313.5 fg/mL) patients. Notably, IFN-α2 levels in these diseases were positively correlated with ISG expression and MxA levels, with the strongest correlation observed in SLE patients. This study provides a direct assessment of the IFN activation pathway, offering valuable insights into its role in autoimmune diseases.

Trouillet-Assant et al. investigated the potential of using IFN-α as a biomarker to distinguish between bacterial and viral infections in febrile children.[92] The LOD for IFN-α using the Simoa was 30 fg/mL. Among the viral infection group, 2% (1/46) had IFN-α levels below the LOD, while 71% (39/55) of patients with bacterial infections had IFN-α levels below this threshold. The median IFN-α levels in viral infections (7,856 [3,096–62,395] fg/mL) were significantly higher than those observed in bacterial infections (406 [68–3,708] fg/mL). The IFN score, which was similar across both groups, showed a strong correlation with IFN-α levels.

Thanks to the ultrahigh sensitivity of DIs, numerous recent studies on SADs, viral infections, and associated drug treatments have employed the Simoa platform to directly measure IFN levels in the blood. In conjunction with transcriptomic profiling and IFN-induced protein expression technologies, these studies have explored the relationships between various types and subtypes of interferons, disease types, disease progression, and therapeutic interventions, offering new insights into the pathogenesis, evolution, and treatment of these diseases.[93–97]

C-Reactive Protein

CRP is synthesized by the liver in response to cytokines and was first identified in the serum of patients with acute inflammation. It was named for its ability to bind with the C-polysaccharide of Streptococcus pneumoniae cell walls. CRP levels are closely associated with the occurrence and progression of inflammation, and it also plays a role in various diseases, including cardiovascular disease and cancer. Current evidence suggests that low levels of CRP may be linked to an increased risk of cardiovascular disease. Structurally, CRP consists of five identical polypeptide subunits arranged in a spherical shape. These subunits covalently bind in a single-valent manner, resulting in a molecular weight of approximately 100 kDa. CRP is heat-sensitive, with a half-life of around 19 hours. In healthy individuals, the concentration of CRP in blood typically remains below 10 mg/L. During viral infections, CRP levels may rise to between 10 and 40 mg/L, while in active inflammation or bacterial infections, levels can increase to 40–200 mg/L. In severe bacterial infections, CRP concentrations may exceed 200 mg/L. CRP levels increase rapidly, often reaching up to 1000 times the baseline value, within 48 hours following trauma, tissue necrosis, or acute inflammation.[98] Conversely, consistently low CRP levels (below 5 mg/L) are associated with the development of conditions such as coronary heart disease, local ischemic stroke, and AMI. Furthermore, in patients with COVID-19, CRP levels are a useful marker of disease severity, with values exceeding 26.9 mg/L being indicative of a higher risk of progression to severe disease.

Altug et al. developed a DI based on optical abnormal transmission (EOT).[99] As illustrated in Figure 6.7, when GNPs were positioned within or near nanopores smaller than 100 nm, they induced a significant reduction in the EOT peak. This intensity change was leveraged to visualize individual GNPs. In this assay, antibodies were immobilized on the surface of a nanopore array, and an antigen was introduced along with GNP-labeled detection antibodies, forming a sandwich immunocomplex. The number of GNPs that resulted in a reduction of EOT was quantified under a bright-field microscope, providing a quantitative measure of the antigen. Background interference from nonspecifically adsorbed GNPs was subtracted to ensure accuracy. For CRP detection, the method demonstrated an LOD of 27 pg/mL in buffer and 69 pg/mL in cell culture medium. In 2020, they further developed a portable device weighing less than 1 kg and clinically tested the levels of C-reactive protein (CRP) and procalcitonin (PCT) in serum samples from sepsis patients, patients with

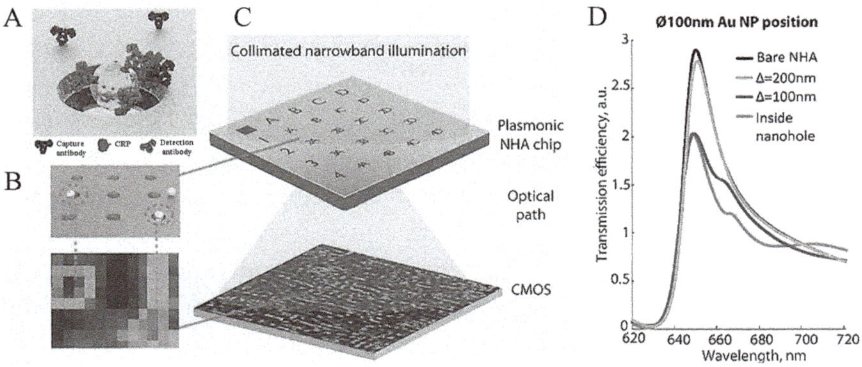

FIGURE 6.7 Nanoparticle-enhanced EOT-based DI. (A) schematic of the immunoreaction confined within a nanowell. (B, C) representative localized images of the nanowell array. (D) transmission spectra of EOT.

Source: Reproduced from Belushkin, A.; Yesilkoy, F.; Altug, H. et al. Nanoparticle-enhanced plasmonic biosensor for digital biomarker detection in a microarray. *ACS Nano*, 2018, 12, 4453–4461 (ref 99). Copyright 2018 American Chemical Society.

non-infectious systemic inflammatory response syndrome (SIRS), and healthy individuals.[100] The LODs were 36 pg/mL and 21.3 pg/mL for CRP and PCT, respectively, with a dynamic range of three orders of magnitude. The assay required only 20 μL of sample and could be completed within 15 minutes. The levels of both CRP and PCT in patients were significantly higher than those in healthy individuals, achieving an area under the curve (AUC) of 1 for distinguishing patients from HCs. The AUC for distinguishing sepsis from SIRS using PCT reached 0.995. Zhang et al. adopted a similar research approach but achieved an LOD of 2.36 ng/mL for CRP without the need for gold nanoparticle-labeled detection antibodies, instead directly immobilizing the immunocomplex into a nanopore.[101]

VIRUS DETECTION

Novel Coronavirus

The disease known as COVID-19 is caused by a novel coronavirus, SARS-CoV-2 (Severe Acute Respiratory Syndrome Coronavirus 2). SARS-CoV-2 is classified as a β-coronavirus and is an enveloped virus typically spherical or oval in shape, with a diameter ranging from 60 to 140 nm. The viral RNA genome is encased in a nucleocapsid, which is surrounded by a viral membrane that contains three key proteins: spike (S), membrane (M), and envelope (E) proteins. The S protein consists of two subunits, S1 and S2. The S1 subunit contains the N-terminal region and the receptor-binding domain (RBD) located at the C-terminal end. Upon entering the human respiratory tract, the RBD binds to the angiotensin-converting enzyme 2 (ACE2) receptor on the surface of host cells, initiating the infection process.[102] The nucleocapsid (N) protein is the primary antigen responsible for stimulating antibody production in infected individuals. COVID-19 is highly contagious, and severe cases can lead to multiorgan failure and death. Therefore, rapid and sensitive virus detection plays a crucial role in large-scale early screening, helping to control the spread of the virus and mitigate the epidemic. Common methods for detecting SARS-CoV-2 include nucleic acid testing, antigen testing, and serological testing. Among these, real-time reverse transcription quantitative PCR (RT-qPCR) is considered the gold standard, with sensitivity typically ranging from 10 to 150 copies per reaction. However, RT-qPCR has stringent requirements in terms of environmental conditions, specialized equipment, and skilled operators. The sample turnaround time for RT-PCR is generally between 4 to 6 hours, and in some cases, it may be longer. Lateral flow antigen tests serve as a complementary alternative to nucleic acid tests, offering rapid results and simplicity, making them suitable for at-home self-testing. However, lateral flow assays have lower sensitivity (ranging from 34% to 92%), with sensitivity decreasing as the viral load lowers. The agreement on positive test results between lateral flow assays and RT-PCR is only 22.9% to 71.4%.[103] From the perspective of interrupting transmission chains and preventing the spread of the virus, highly sensitive tests are crucial, as they minimize the likelihood of false negatives and ensure that more positive cases are identified. Serological testing, which detects COVID-19-specific antibodies (IgM/IgG) in the blood, is primarily useful for identifying past infections, as detectable antibodies typically appear about a week after infection. Consequently, serological tests are not suitable for rapid screening or detecting reinfections, but they can be valuable for determining whether an individual has been previously infected. Let us see the immune response during a COVID-19 infection,[104] highlighting the detection windows for various diagnostic methods. During the asymptomatic phase, both viral RNA and antibodies may be detectable; however, their concentrations are typically low, requiring highly sensitive testing methods. DIs may offer an effective alternative in such cases. RT-PCR, while highly sensitive, cannot provide information on disease progression, and serum tests may remain positive for weeks after seroconversion or recovery. Although serological tests can confirm whether an individual has mounted an immune response, they are not suitable for early detection or monitoring of active infections.

Nasopharyngeal swab sample collection is most commonly used to collect a sample, although they are uncomfortable for patients, require trained personnel, and carry a risk of transmission during the collection process. In contrast, saliva samples are gaining increasing attention due to their advantages, including ease and speed of collection, as well as the ability to be self-collected. However, a significant challenge with saliva samples is the lower concentration of antigens, which makes it more difficult to detect the virus using antigen-based tests. Breshears et al. utilized saline mouthwash as a sample for virus detection.[105] The sample was added to a paper-based chip, which was then evaporated to facilitate the attachment of the virus to the chip surface. Fluorescent particles, labeled with an anti-N protein antibody, were introduced to form fluorescent aggregates through an immunoreaction. These aggregates were imaged using a smartphone, and the fluorescence was quantified to determine the concentration of the novel coronavirus. The LOD was 10 ag/μL (approximately 1 fg per viral copy). In testing 27 samples, 13 of which were RT-qPCR positive, the method demonstrated a sensitivity of 100%, specificity of 86%, and accuracy of 93%. They also applied this assay to investigate the cross-reactivity of monoclonal antibodies against the S and N proteins of the Delta and Omicron variants of SARS-CoV-2.[106] The monoclonal antibody targeting the S protein (mAb-S) bound to both the S protein (O-S) and the N protein (O-N) of the Omicron variant. In contrast, the monoclonal antibody targeting the N protein (mAb-N) specifically bound only to the N protein (O-N) of Omicron. To address the bulky nature of existing DI devices (141 × 79 × 161 cm), Chiba et al. developed a compact desktop automatic DI device (32 × 60 × 57.5 cm) for detecting the N protein antigen of SARS-CoV-2.[107] The device featured three key innovations: first, it eliminated the need for flow channels or additional liquid reservoirs; instead, microhole arrays were created at the bottom of a cup-shaped device, with all reactions, dispersion, and detection occurring within the cup. Second, the magnetic bead loading process was simplified by adding an oil denser than the buffer solution to the cup. As the oil moved downward by gravity, it replaced the lower buffer, effectively dispersing the magnetic beads. Third, the cup was coated with black dye to reduce fluorescent background noise. This device achieved an LOD for the

N protein of 0.0043 pg/mL, with a CV of 14.2%, and a measurement time of 28 minutes. Using this method, they tested nasopharyngeal swab samples from 159 RT-PCR negative patients and 88 RT-PCR-positive patients. At a cutoff value of 0.029 pg/mL, the device demonstrated a sensitivity of 94.3% (83/88) for positive samples and a specificity of 98.1% (156/159) for negative samples.

Ogata et al. developed a viral antigen analysis and virus serology platform using the Simoa technology.[108] They quantitatively assessed three antigens (S1, S, N) and seven antibodies (total IgA, IgM, IgG, IgG1, IgG2, IgG3, IgG4) against SARS-CoV-2 in the plasma of COVID-19 patients. In the 3-plex analysis, the LODs for the S1, S, and N antigen tests were 5 pg/mL (0.07 pmol/L), 70 pg/mL (0.39 pmol/L), and 0.02 pg/mL (0.4 fmol/L), respectively. These variations in LODs were attributed to differences in the dissociation constants of the antigen-antibody interactions. Among 64 RT-PCR-positive patients, S1 and N antigens were detected in the plasma of 41 patients, while only 5 patients had detectable S protein. In the negative controls, a small number of individuals in both pre-epidemic healthy and pre-epidemic patient groups also had detectable levels of S1 and N proteins. The study further monitored changes in antigen and antibody levels in the plasma of hospitalized patients over time. Antigen levels were high during the early stages of infection but gradually decreased as the virus was cleared. In contrast, antibody levels increased progressively, reaching a steady state and indicating seroconversion, which typically occurred 7 ± 1 days after the first positive nasopharyngeal RT-PCR test. Virus clearance, defined as the point when both S1 and N concentrations became undetectable, occurred approximately 5 ± 1 days after seroconversion. The S1 protein concentration was found to be strongly correlated with the progression to severe disease. Patients were classified into three groups based on their S1 concentration: a zero concentration group, a low concentration group (6–50 pg/mL, 0.08–0.65 pmol/L), and a high-concentration group (>50 pg/mL, >0.65 pmol/L). The proportion of patients requiring ICU admission was 30% (7/23) in the zero concentration group, 52% (12/23) in the low concentration group, and 77% (14/18) in the high-concentration group. The time to intubation also correlated strongly with S1 concentration, with patients in the high-concentration group requiring intubation within one day of hospitalization.

Olsen et al. developed a platform for detecting the viral N protein using the Simoa technology.[109] They tested oropharyngeal swab samples from 148 RT-PCR-positive and 73 negative cases. With a cutoff value of 0.01 pg/mL, the method demonstrated a sensitivity of 95% and a specificity of

100%. Shan et al. employed the Simoa platform to detect the N protein in blood, dried blood, and saliva samples.[110] The LOD was 0.099 pg/mL (range: 0.046–0.204 pg/mL), with a lower LOQ of 0.313 pg/mL. Compared to RT-PCR, the accuracy of negative results was greater than 98% for symptomatic, asymptomatic, and PCR-positive patients before symptom onset across all sample matrices. Additionally, the accuracy of positive results exceeded 90%. Therefore, they concluded that the N protein could serve as a reliable biomarker for the early detection of both symptomatic and asymptomatic COVID-19 infections.

Zhao et al. employed photonic crystals modified with the S protein as a substrate and IgG secondary antibody-coated sea urchin-shaped GNPs as probes.[111] In the presence of specific antibodies against the virus's S protein, a sandwich-type immunocomplex formed on the surface of the photonic crystal. The GNPs significantly quenched the reflection intensity of the photonic crystal, allowing for the identification and counting of individual immunocomplexes, each corresponding to a single GNP. Using this method, they achieved quantitative detection of COVID-19 IgG in just 4 μL of serum within 15 minutes, with LOD and LOQ of 26.7 ± 7.7 pg/mL and 32.0 ± 8.9 pg/mL, respectively.

Ou et al. proposed that highly sensitive DIs could reduce the serum conversion time for detecting virus-specific antibodies, enabling earlier detection of viral infections.[112] They modified magnetic beads with the RBD of the S1 protein and the amino acid sequence (S2-78) from the S2 protein, respectively, to detect COVID-19-specific IgM and IgG antibodies in the serum of recovered patients using the Simoa platform. By monitoring antibody level changes in healthy individuals following vaccination, they found that Simoa detected a significant increase in IgG levels 7.5 days after the first dose and in IgM levels 8.6 days after vaccination. In contrast, flow cytometry required 21.5 and 24 days to detect similar increases, respectively.

Neutralizing antibodies are a class of antibodies that, produced by the body upon stimulation by viral surface antigens, bind to the virus and prevent its ability to infect host cells. Measuring antibody titers in serum is crucial for understanding the host's immune response to viral infection, evaluating the effectiveness of vaccines, and screening plasma for potential neutralizing antibody-based treatments. Gilboa et al. assessed neutralizing antibody levels in the plasma of COVID-19 patients using the Simoa platform in a competitive mode.[113] The underlying principle of the assay is as follows: magnetic beads coated with the S protein and ACE2 coupled with β-GaL formed a complex (magnetic beads-S protein-ACE2-enzyme). Upon the addition of neutralizing antibodies from the plasma of

COVID-19-recovered patients, these antibodies disrupted the binding of the S protein to ACE2, thereby reducing the complex formation. The degree of this reduction correlated with the concentration of neutralizing antibodies in the plasma. Compared to traditional methods, this approach was more sensitive and faster, with the entire process being completed in approximately two hours. Furthermore, it did not require live viruses or cells and can be conducted in a biosafety level 2 laboratory.

Norman et al. developed a multiplex Simoa assay to detect IgG, IgM, and IgA against four COVID-19 antigens: the S protein, RBD, S1 subunit, and N protein.[114] Using less than 1 μL of plasma, the assay successfully identified 12 distinct types of serum antibodies. The accuracy of the assay in distinguishing pre-epidemic HC samples from RT-PCR-positive patient samples was 99%.

Human Immunodeficiency Virus

HIV is the etiological agent of acquired immunodeficiency syndrome (AIDS). It is a chronic virus that progressively damages the immune system, rendering the body vulnerable to a wide array of bacterial, viral, and fungal infections. AIDS is associated with a high fatality rate. Despite the advent of combined antiretroviral therapy (cART), curing AIDS remains challenging due to the presence of viral reservoirs, which are formed when the virus integrates into the host genome. These reservoirs evade immune surveillance and clearance. HIV is spherical in shape, with a diameter ranging from 100 to 120 nm. Its outer envelope contains two glycoproteins, gp120 and gp41, which form spikes that protrude from the surface. Beneath the envelope lies the p17 matrix protein. The gp120 glycoprotein binds to host cell receptors, inducing the production of neutralizing antibodies. The primary receptor for HIV is the CD4, which is expressed on the surface of CD4+ T lymphocytes. The virus's core is encased by a capsid composed of the p24 protein, with approximately 2,000 copies of p24 present in each HIV virion. The core contains the p7 nucleocapsid protein, along with two strands of positive-sense RNA and the necessary enzymes for viral replication.

Detecting HIV is crucial for screening, diagnosis, and treatment guidance in AIDS management. Early diagnosis plays a vital role in preventing HIV transmission, as it enables timely initiation of antiretroviral therapy. Common methods for detecting HIV include antibody testing, viral antigen testing, and nucleic acid testing (NAT). Among these, NAT is the most sensitive, with a detection threshold of approximately 60 HIV RNA copies/mL (or 30 viral copies/mL) and is considered the gold standard for diagnosing HIV infection. Compared to antibody testing, which only detects antibodies after seroconversion, NAT significantly shortens the time between initial infection and detectable viral levels. However, in areas with limited medical resources and high disease prevalence, traditional immunoassays remain the primary screening tool. HIV is approximately 10 times more infectious during the acute phase than in the chronic phase, making highly sensitive diagnostic tests urgently needed for rapid screening during this period. The p24 antigen, which has a long half-life outside the virus particle, serves as a valuable marker for monitoring HIV infection, disease progression, and treatment efficacy. During the acute phase, HIV antibodies have not yet been produced, but both HIV RNA and p24 antigen levels are high in plasma. While the LOD of NAT is 30 viral copies/mL (equivalent to 6000 p24 antigen copies/mL or 3 fg/mL), this threshold is significantly lower than the LODs of traditional immunoassays (11–70 pg/mL). As a result, traditional immunoassays may fail to detect acute-phase infections, underscoring the need for more sensitive diagnostic tools in early HIV detection.

Chang et al. utilized the Simoa platform to detect the p24 antigen, achieving an LOD of 4.87 fg/mL for p24 and 90 RNA copies/mL (equivalent to 45 viral copies/mL) for cultured viruses.[115] The LODs for p24 and cultured viruses were comparable. The DI results showed a high correlation with those of NAT, and the sensitivity of the two methods was largely equivalent. Notably, the Simoa platform was able to detect abnormalities in p24 antigen levels 7–10 days earlier than conventional immunoassays. With the automation of the Simoa platform, an automated DI for p24 protein was also developed.[116] The LOD for spiked serum was 2.5 fg/mL, which corresponds to 63 RNA copies/mL, while the lower LOQ was 7.6 fg/mL. In a performance comparison of three assays for detecting viruses in acute HIV patients, the DI showed 100% agreement with NAT in terms of sensitivity, whereas conventional immunoassay methods demonstrated less than 40% agreement. Additionally, a combination of photonic crystals and GNPs was employed to detect HIV p24, with a serum LOQ of 1 pg/mL (42 fM, or 10,000 viral copies/mL).[117] Passaes et al. employed an acid-mediated dissociation method to break down immunocomplexes in plasma, followed by detection of p24 using DI.[118] They validated the screening accuracy for HIV carriers using a cutoff value based on the average p24 level (24 fg/mL) found in HIV-negative volunteer plasma. When the viral load exceeded 10^4 copies/mL, the detection accuracy for p24 was 18 out of 19. However, at lower viral loads, the accuracy significantly decreased to 1 out of 6. In plasma samples from 92 primary HIV patients, the accuracy was 72 out of 92, while in 137 chronic HIV infections, the accuracy increased to 98 out of 137. After 48 days of cART, 89 out of 108 samples showed no detectable p24.

The challenge of fully curing AIDS is largely attributed to the presence of HIV reservoirs. Consequently, the accurate and sensitive detection of these reservoirs is a critical application for DIs. Passaes et al. utilized the Simoa platform to detect p24 protein (ranging from 0.096 to 0.36 pg/mL) in cell culture supernatants produced by a single infected cell.[119] The detection range of the assay was found to be between 0.017 and 37.8 pg/mL. To monitor the reactivation of the HIV reservoir, the researchers isolated CD4+ T cells from the peripheral blood of cART-treated patients and activated them with an activator. Using flow cytometry, p24 was detectable after a median of 14 days post-activation. However, when employing DI, p24 was detectable after only 3.5 days (median), indicating that DI can detect reactivation 7–10 days earlier than flow cytometry. Furthermore, DI detected p24 in 43% of individuals who had reactivated after just 1 day. These findings provide a novel approach for detecting HIV reservoirs and potentially accelerating the identification of viral reactivation. Wu et al. further confirmed the potential of DI in measuring HIV reservoirs. They reactivated CD4+ T cells from ART-treated HIV-positive patients and employed the Simoa platform to measure p24 levels in CD4+ T-cell lysates.[120] The LOD of the assay was determined to be 5 fg/mL. Their study revealed that histone deacetylase inhibitors (HDACis) effectively stimulated the production of HIV p24 and virus-specific antibodies *in vitro*. Additionally, HDACis were found to induce the targeted killing of infected cells through the use of gp120/CD3 multispecific antibodies. This approach shows promise as a tool for evaluating the efficacy of HIV treatments.

Other Viruses

Dengue Virus

Simoa technology has been employed to detect antibodies, including IgG and IgM, against dengue virus type II.[121] Magnetic beads, coated with inactivated virus antigens, are used to capture virus-specific antibodies from plasma samples. Enzyme-labeled secondary antibodies specific for IgG or IgM are then introduced, forming a sandwich-like complex.

TABLE 6.6 Virus Dilution Folds

	IgG	IgM
ELISA	10,000	100
Simoa	10,000,000	1,000,000

The magnetic beads are subsequently collected and dispersed into 46 fL microchambers to facilitate an enzymatic reaction. Only those magnetic beads that bind to IgG or IgM can catalyze the substrate, resulting in fluorescence emission. The number of fluorescent microchambers is counted to quantify the levels of IgG or IgM. Both ELISA and Simoa techniques are utilized to detect plasma samples serially diluted with dengue virus antigens. A comparison of their sensitivities is shown in Table 6.6. The maximum dilution factor at which both methods can still detect positive results demonstrates that the sensitivity of Simoa is at least 1,000 times greater than that of ELISA.

Influenza Virus

Avian influenza virus (AIV) is a pathogen responsible for causing acute respiratory diseases in poultry and is characterized by multiple subtypes, some of which can infect humans, leading to influenza. Zhang Zhiling et al. proposed an electrochemical digital ELISA for detecting the H7N9 subtype of AIV.[122] In this assay, magnetic nanospheres were simultaneously conjugated with antibodies specific to the virus and ALP. Upon virus capture, the magnetic beads were placed onto the surface of a microelectrode array modified with monolayer GNPs (500 units). The ALP enzyme catalyzed the conversion of p-APP to p-AP, which subsequently reduced silver ions to metallic silver that deposits onto the electrode, generating an electrochemical signal. The virus was quantified by calculating the probability of electrochemical signals detected across the array. The LOD of this method was 7.8 fg/mL. In 2019, they achieved the simultaneous detection of three viruses—H9N2, H1N1, and H7N9—using a novel approach.[123] Magnetic nanoparticles were encapsulated with QDs and conjugated with antibodies specific to each virus, forming fluorescent magnetic multifunctional nanospheres with a diameter of 340–370 nm. Each virus was assigned a distinct fluorescence signature. A PDMS layer, containing a microhole array (500 holes), was placed over a glass substrate that had been simultaneously coated with three different antibodies. Following the immunoreaction, the fluorescent nanospheres were captured in the microholes, with the fluorescence color indicating the type of virus present. The number of fluorescent microholes was used to quantify the virus. The LODs for all three viruses were 0.02 pg·mL^{-1}.

Leiris et al. developed a microplate array DI for detecting the nucleoprotein of Influenza A.[124] After screening seven commercially available antibodies, they identified that the antibody with the highest dissociation constant yielded the best performance. In this assay, immunocomplexes formed on magnetic beads were dispersed into microplates, and the number of luminescent microplates was counted to quantify the nucleoprotein. The LODs in buffer solution and swab solution were found to be 4 ± 1 fM and 10 ± 2 fM, respectively. Assuming that each virus particle contains approximately 1,000 proteins, these limits correspond to 3.35 $\log_{10}$ and 4.79 $\log_{10}$ virus particles/mL (log values converted to virus particle numbers: $10^{3.35}$ = 2,239 and $10^{4.79}$ = 6.2 × 10^4). Noji et al. developed a neuraminidase (NA)-catalyzed approach for counting influenza virus particles,[125] eliminating the need for antibody capture. The method utilizes a mobile phone as an imaging device.[126] As illustrated in Figure 6.8, the transmembrane protein NA on the surface of the influenza virus catalyzed the hydrolysis of methylumbelliferyl N-acetylneuraminate (MUNANA), producing the fluorescent compound 4-methylumbelliferone (4-MU). The

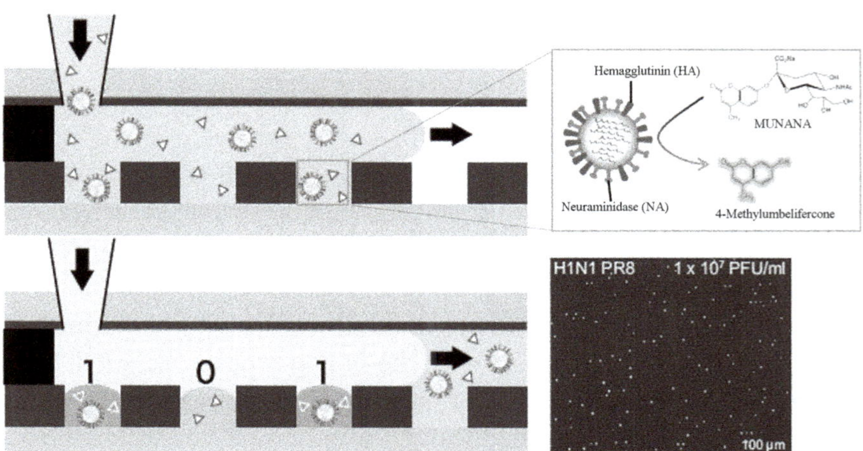

FIGURE 6.8 Virus-directly partitioned and detected DI.

Source: Reproduced from Tabata, K.V., Minagawa, Y., Kawaguchi, Y. et al. Antibody-free digital influenza virus counting based on neuraminidase activity, *Science Report*, 9, 1067 (2019). https://doi.org/10.1038/s41598-018-37994-6, licensed under CC BY 4.0.

H1N1 PR8 virus solution was directly mixed with the substrate and dispersed into the microplate array. The number of luminescent microplates was then counted to quantify the virus. The LOD of this method was 10^3 PFU/mL (10^5 particles/mL), which was significantly lower than the minimum detection threshold of 10^7 PFU/mL for common commercial rapid diagnostic tests, which were unable to detect viruses at 10^6 PFU/mL. This approach also offered the advantage of not requiring viral enrichment, as the sample was directly mixed with the substrate, enabling detection in just 15 minutes. These features make it highly suitable for bedside and portable testing applications.

Ashiba et al. employed 0.5 pL microplates (in contrast to tens of fL) for the discrete detection of Influenza A hemagglutinin (HA).[127] They compared the performance of NA-catalyzed detection with two other detections. The first method involved aggregation-induced emission (AIE) dye detection, where phenolamine, a peptide aptamer conjugated with a thiol derivative, emitted light upon binding to HA. This method provided results within 1 minute, with an LOD of 3×10^5copies/mL. The second method utilized high-density magnetic bead capture for virus detection via NA-catalysis. In this approach, HA antibodies were conjugated to 1 μm diameter magnetic beads. The high density of HA antibodies on the magnetic beads efficiently captured the virus. The virus was then reacted with MUNANA in the microplates for 10 minutes, followed by fluorescence imaging. The final concentration of magnetic beads was approximately 10^8 beads/mL, which was typical in DIs. Although the capture efficiency in conventional assays was low (2% with a 100-minute reaction, and 20% with shorter times), the high-density magnetic beads used in this method shortened the reaction time to 10 minutes and significantly increased the capture efficiency to 80%, resulting in a higher load of magnetic beads in each microplate. Consequently, detection sensitivity was enhanced, with an LOD of 1×10^2 copies/mL.

OTHER APPLICATIONS

Rhodopsin

Diabetic retinopathy (DR) is one of the most common and serious complications of diabetes, and it is a leading cause of vision impairment and blindness in adults. In addition to retinal disease, individuals with type II diabetes are also at an increased risk for other ocular conditions, including glaucoma, cataracts, and optic neuropathy. DR is particularly challenging because it often progresses without noticeable symptoms, meaning that it can only be detected once significant retinal damage has occurred. As such, early and accurate diagnosis is crucial to prevent irreversible retinal changes that can lead to vision loss and blindness. Currently, there are no identified blood biomarkers for detecting retinal damage or neuropathy in patients with DR. Petersen et al. based on the hypothesis that leakage from the blood-retinal barrier leads to the appearance of rhodopsin in the bloodstream, investigated the potential of rhodopsin as an early biomarker for retinal damage.[128] Using Simoa technology, they measured rhodopsin levels in the plasma of diabetic patients with DR. The LOD of their assay was 0.26 ng/L, with a lower LOQ of 3 ng/L and a CV of 19% at a concentration of 5 ng/L. However, their results showed no statistically significant difference in plasma rhodopsin concentrations between diabetic patients with DR (n = 466), diabetic patients without DR (n = 134), and HCs (n = 169). As such, their method did not successfully differentiate rhodopsin levels between diabetic patients with and without DR.

Dried Blood Spot

Dried blood spots (DBS) refer to blood (or other body fluids) collected in small amounts on filter paper and allowed to dry. Compared to plasma samples, DBS offers several advantages, including ease of sample collection, convenience in storage and transportation, enhanced sample stability, and reduced biological safety risks. These characteristics make DBS particularly valuable in resource-limited settings and in situations where blood collection is challenging. However, there are potential drawbacks to DBS testing, such as pre-analytical errors stemming from improper blood collection techniques, interference from blood cells, challenges in the effective extraction of analytes, and the potential for insufficient blood volume. With appropriate quality control and calibration strategies, single-molecule counting technologies can achieve accuracy and reproducibility comparable to traditional venous blood testing. Mukherjee et al. utilized Singulex's SMC technology to detect cTnI, pPSA, and CRP in fingertip dried DBS samples.[129] Their findings showed a high correlation with plasma samples measured by either single-molecule counting or commercial methods, with correlation slopes for cTnI, PSA, and CRP of 1.08, 1.04, and 0.99, respectively. These results demonstrated that single-molecule counting-DBS offered repeatable sensitivity, precision, and stability for quantifying low-abundance biomarkers, with minimal interference from hematocrit or normal variations in sample collection technique. In a subsequent study, they measured cTnI levels in athletes before and after completing a marathon using single-molecule counting-DBS.[130] The pre-race cTnI concentration in marathon runners was significantly higher than that in nonathletes (median 3.1 vs. 0.4 pg/mL), and 98% of athletes (41/42) exhibited a substantial increase in cTnI levels post-race (median 40.5 pg/mL).

Clostridium Difficile Toxins

As the normal human gut microbiota, Clostridioides difficile (C. difficile) is a Gram-positive, oxygen-sensitive, rod-shaped bacterium. However, it can become pathogenic when dysbiosis occurs due to prolonged use of certain antibiotics, which leads to the overgrowth of antibiotic-resistant microbes. These resistant organisms produce toxins, resulting in C. difficile infection (CDI). The clinical manifestations of CDI include diarrhea, abdominal pain, and systemic toxic symptoms, and in severe cases, the infection can be fatal. CDI is characterized by high recurrence, resistance, and mortality rates, as well as substantial healthcare costs. Early diagnosis of C. difficile toxins and identification of CDI are essential for effective treatment and prevention. C. difficile produces six toxins, of which toxins A (TcdA) and B (TcdB) are primarily responsible for diarrhea and colitis. In 2018, Singulex developed a single-molecule counting immunoassay for detecting TcdA and TcdB.[131] The LODs for TcdA and TcdB in buffer were found to be 0.8 and 0.3 pg/mL, respectively, while in fecal matter and urine, the corresponding values were 2.0 and 0.7 pg/mL. Compared to PCR, SMC demonstrated a sensitivity of 97.7% and a specificity of 100%. In 2019, Singulex conducted a prospective multicenter study to compare various assays for detecting CDI.[132] Using a cohort of 897 suspected CDI cases, a cutoff value of 12.0 pg/mL for the SMC immunoassay was established. A total of 1,005 suspected CDI samples were tested, and the results were validated using cell culture cytotoxicity neutralization assays (CCCNA). After correction based on CCCNA results, the SMC immunoassay showed a positive concordance rate of 96.3% and a negative concordance rate of 93.0%.

OUTLOOK

The concept of DIs was first proposed in 2010,[12] while the idea of single-molecule counting immunoassays dates back to the late 1990s and early 2000s.[133,134] Single-molecule counting is considered a form of digitization, with DIs representing binary encoding using single-molecule counting, or hexadecimal encoding through multi-molecule counting, and even statistical negative space without molecular counting. DIs are more inclusive, encompassing a broader range of applications. Over the past few decades, DIs have made substantial progress across various fields. This chapter highlights the areas most frequently discussed in the literature, including tumor biomarkers, neurodegenerative disease biomarkers, cytokines, and viruses. Among these, neurodegenerative disease biomarkers have been the most extensively studied using DIs. Consequently, this field has seen the most rapid progress and development. Two primary reasons explain this trend. First, ultrahigh sensitivity methods are required to detect neurodegenerative disease biomarkers in peripheral blood. Due to the blood–brain barrier, these biomarkers are found in relatively high concentrations in CSF but are rare in peripheral blood, making their detection challenging for conventional methods. DIs, however, meet the demand for convenient, accurate, and highly sensitive detection. Second, the Quanterix Simoa automated platform and related technologies have reached a relatively high level of maturity and are widely accepted within the neurodegenerative disease field. This has led to frequent reports of large cohort, long-term prospective studies. While the Simoa platform is not without its technical limitations, its high degree of automation and stability make it particularly user-friendly for nonexpert users, contributing to its widespread recognition and adoption.

Protein tumor biomarkers were among the earliest targets for DIs, with the aim of detecting abnormal biomarker levels at the very early stages of cancer. However, the commonly used tumor biomarkers often lack the sensitivity and specificity required to effectively distinguish patients at early stages of the disease. In many cases, tumor biomarkers have been used as model proteins for the development of DIs. However, achieving the ultimate goal of early cancer screening and diagnosis remains a significant challenge.

In population-based statistical analyses of disease biomarkers, it is common to observe statistically significant differences in biomarker levels between two groups, yet some overlap remains between the levels of these biomarkers. As a result, the reliability of accurately distinguishing early stage patients from control populations is often limited. This suggests that biomarkers, in such cases, serve primarily as a reference in individual diagnostics, and additional, more definitive parameters are needed for validation. In other words, this situation does not fully demonstrate the irreplaceability of DIs. Nevertheless, it is undeniable that DIs offer ultrahigh sensitivity. To maximize their potential, we believe that DIs should continue to evolve in the following directions:

1. Upgrade and Iteration of Instrumentation
   Under the premise of high stability, the sensitivity of DIs should be further enhanced, with a LOQ reaching the sub-zeptomolar (sub-zM) level. Additionally, greater emphasis should be placed on the development of homogeneous DIs. Ultimately, the DI platform should evolve into high-throughput, high-efficiency, multicomponent detection systems capable of automated point-of-care testing (POCT), with comprehensive sample-to-result capabilities.
2. Accelerating the Transition from Research to Clinical Practice
   The DI-based transition from research to clinical practice should be expedited. By accurately using biomarkers, DIs can achieve significant breakthroughs in clinical applications and demonstrate their unique value in improving patient outcomes.
3. Screening and Validation of Novel Biomarkers
   DIs should be integrated with proteomics and mass spectrometry technologies to enable large-scale searches, screenings, and validations of novel disease biomarkers, particularly in low-abundance plasma proteins.
4. Establishing Individualized Diagnosis and Treatment
   The population-based approach typically involves measuring the average baseline of biomarkers in healthy populations to establish diagnostic cutoffs, with deviations from this baseline serving as indicators for monitoring disease progression. However, numerous studies have shown that baselines are closely tied to the specific characteristics of the population being studied. In highly heterogeneous populations, each individual's baseline is unique. Establishing a personalized "3D barcode" of biomarkers—encompassing biomarker type, concentration, and temporal dynamics—will be fundamental to advancing individualized diagnosis and treatment in the future.

REFERENCES

1. Chen J, Zou GZ, Zhang XL, Jin WR. Ultrasensitive electrochemical immunoassay based on counting single magnetic nanobead by a combination of nanobead amplification and enzyme amplification, *Electrochemistry Communications*, 2009, 11, 1457–1459.
2. Poon CY, Wei L, Xu YL, et al. Quantification of cancer biomarkers in serum using scattering-based quantitative single particle intensity measurement with a dark-field microscope, *Analytical Chemistry*, 2016, 88, 8849–8856.

3. Ma J, Zhan L, Li RS, et al. Color-encoded assays for the simultaneous quantification of dual cancer biomarkers, *Analytical Chemistry*, 2017, 89, 8484–8489.
4. Wu X, Li T, Tao GY, et al. A universal and enzyme-free immunoassay platform for biomarker detection based on gold nanoparticle enumeration with a dark-field microscope, *Analyst*, 2017, 142, 4201–4205.
5. Chen YC, Tian YY, Yang Q, et al. Single-particle mobility analysis enables ratiometric detection of cancer markers under darkfield tracking microscopy, *Analytical Chemistry*, 2020, 92, 10233–10240.
6. Liu XJ, Huang CH, Dong XL, et al. Asynchrony of spectral blue-shifts of quantum dot based digital homogeneous immunoassay, *Chemical Communications*, 2018, 54, 13103–13106.
7. Zhang QQ, Zhang XB, Li JJ, Gai HW. Nonstochastic protein counting analysis for precision biomarker detection: suppressing poisson noise at ultralow concentration, *Analytical Chemistry*, 2020, 92, 654–658.
8. Tang HR, Wang H, Yang C, et al. Nanopore-based strategy for selective detection of single carcinoembryonic antigen (CEA) molecules, *Analytical Chemistry*, 2020, 92, 3042–3049.
9. Ahn S, Zhang P, Yu H, Lee S, Kang SH. Ultrasensitive detection of alpha-fetoprotein by total internal reflection scattering-based super-resolution microscopy for superlocalization of nano-immunoplasmonics, *Analytical Chemistry*, 2016, 88, 11070–11076.
10. Tian S, Zhang Z, Chen J, et al. Digital analysis with droplet-based microfluidic for the ultrasensitive detection of β-gal and AFP, *Talanta*, 2018, 186, 24–28.
11. Rissin DM, Kan CW, Campbell TG, et al. Single-molecule enzyme-linked immunosorbent assay detects serum proteins at subfemtomolar concentrations, *Nature Biotechnology*, 2010, 28, 595–599.
12. Rissin DM, Fournier DR, Piech T, et al. Simultaneous detection of single molecules and singulated ensembles of molecules enables immunoassays with broad dynamic range, *Analytical Chemistry*, 2011, 83, 2279–2285.
13. Kim SH, Iwai S, Araki S, et al. Large-scale femtoliter droplet array for digital counting of single biomolecules, *Lab on a Chip*, 2012, 12, 4986–4991.
14. Chen H, Li Z, Zhang L, et al. Quantitation of femtomolar-level protein biomarkers using a simple microbubbling digital assay and bright-field smartphone imaging, *Angewandte Chemie International Edition*, 2019, 58, 13922–13928.
15. Schubert SM, Walter SR, Manesse M, Walt DR. Protein counting in single cancer cells, *Analytical Chemistry*, 2016, 88, 2952–2957.
16. Shim JU, Ranasinghe RT, Smith CA, et al. Ultrarapid generation of femtoliter microfluidic droplets for single-molecule-counting immunoassays, *ACS Nano*, 2013, 7, 5955–5964.
17. Poon CY, Chan HM, Li HW. Direct detection of prostate specific antigen by darkfield microscopy using single immunotargeting silver nanoparticle, *Sensors and Actuators B: Chemical*, 2014, 190, 737–744.

18. Zhu L, Li GH, Sun SQ, Tan H, He Y. Digital immunoassay of a prostate-specific antigen using gold nanorods and magnetic nanoparticles, *RSC Advances*, 2017, 7, 27595–27602.
19. Farka Z, Mickert MJ, Hlaváček A, Skládal P, Gorris HH. Single molecule upconversion-linked immunosorbent assay with extended dynamic range for the sensitive detection of diagnostic biomarkers, *Analytical Chemistry*, 2017, 89, 11825–11830.
20. Mickert MJ, Farka Z, Kostiv U, et al. Measurement of sub-femtomolar concentrations of prostate-specific antigen through single-molecule counting with an upconversion-linked immunosorbent assay, *Analytical Chemistry*, 2019, 91, 9435–9441.
21. Li X, Wei L, Pan LL, et al. Homogeneous immunosorbent assay based on single-particle enumeration using upconversion nanoparticles for the sensitive detection of cancer biomarkers, *Analytical Chemistry*, 2018, 90, 4807–4814.
22. Liu XJ, Sun YY, Lin XY, et al. Digital duplex homogeneous immunoassay by counting immunocomplex labeled with quantum dots, *Analytical Chemistry*, 2021, 93, 3089–3095.
23. Liu XJ, Lin XY, Pan XY, Gai HW. Multiplexed homogeneous immunoassay based on counting single immunocomplexes together with dark-field and fluorescence microscope, *Analytical Chemistry*, 2022, 94, 5830–5837.
24. Liu CC, Xu XN, Li B, et al. Single-exosome-counting immunoassays for cancer diagnostics, *Nano Letters*, 2018, 18, 4226–4232.
25. Yang ZJ, Atiyas Y, Shen H, et al. Ultrasensitive single extracellular vesicle detection using high throughput droplet digital enzyme-linked immunosorbent assay, *Nano Letters*, 2022, 22, 4315–4324.
26. Morasso C, Ricciardi A, Sproviero D, et al. Fast quantification of extracellular vesicles levels in early breast cancer patients by single molecule detection array (SiMoA), *Breast Cancer Research and Treatment*, 2022, 192, 65–74.
27. Wei P, Wu F, Kang B, et al. Plasma extracellular vesicles detected by single molecule array technology as a liquid biopsy for colorectal cancer, *Journal of Extracellular Vesicles*, 2020, 9, 1809765.
28. Huang G, Zhu Y, Wen SH, et al. Single small extracellular vesicle (sEV) quantification by upconversion nanoparticles, *Nano Letters*, 2022, 22, 3761–3769.
29. Janelidze S, Stomrud E, Palmqvist S, et al. Plasma β-amyloid in Alzheimer's disease and vascular disease, *Scientific Reports*, 2016, 6, 26801.
30. Vergallo A, Mégret L, Lista S, et al. Plasma amyloid β 40/42 ratio predicts cerebral amyloidosis in cognitively normal individuals at risk for Alzheimer's disease, *Alzheimer's & Dementia*, 2019, 15, 764–775.
31. Pratishtha C, Mitra E, Kathryn G, et al. Ultrasensitive detection of plasma amyloid- as a biomarker for cognitively normal elderly individuals at risk of Alzheimer's Disease, *Journal of Alzheimer's Disease*, 2019, 71, 775–783.
32. Randall J, Mörtberg E, Provuncher GK, et al. Tau proteins in serum predict neurological outcome after hypoxic brain injury from cardiac arrest: results of a pilot study, *Resuscitation*, 2013, 84, 351–356.
33. Zetterberg H, Wilson D, Andreasson U, et al. Plasma tau levels in Alzheimer's disease, *Alzheimer's Research & Therapy*, 2013, 5, 9.

34. Ding XL, Zhang ST, Jiang LJ, Wang L, Li T, Lei P. Ultrasensitive assays for detection of plasma tau and phosphorylated tau 181 in Alzheimer's disease: a systematic review and meta-analysis, *Translational Neurodegeneration*, 2021, 10, 10.
35. Moscoso A, Grothe MJ, Ashton NJ, et al. Time course of phosphorylated-tau181 in blood across the Alzheimer's disease spectrum, *Brain*, 2021, 144, 325–339.
36. (a) Palmqvist S, Janelidze S, Quiroz YT, Zetterberg H, et al. Discriminative accuracy of plasma phospho-tau217 for alzheimer disease vs other neurodegenerative disorders, *JAMA-Journal of the American Medical Association*, 2020, 324, 772–781. (b) VandeVrede L, Cho H, Sanderson-Cimino M, et al. Detection of Alzheimer neuropathology in Alzheimer and non-Alzheimer clinical syndromes with blood-based biomarkers, *JAMA Neurology*, 2025, 82(4), 344–354.
37. Ashton NJ, Pascoal TA, Karikari TK, et al. Plasma p-tau231: a new biomarker for incipient Alzheimer's disease pathology, *Acta Neuropathologica*, 2021, 141, 709–724.
38. Kuhle J, Barro C, Andreasson U, et al. Comparison of three analytical platforms for quantification of the neurofilament light chain in blood samples: ELISA, electrochemiluminescence immunoassay and Simoa, *Clinical Chemistry Laboratory Medicine*, 2016, 54, 655–1661.
39. Gisslen M, Price RW, Andreasson U, et al. Plasma concentration of the neurofilament light protein (NFL) is a biomarker of CNS Injury in HIV infection: a cross-sectional study, *EBioMedicine*, 2016, 3, 135–140.
40. Rojas-Martinez J, Karydas A, Bang J, et al. Plasma neurofilament light chain predicts progression in progressive supranuclear palsy, *Annals of Clinical and Translational Neurology*, 2016, 3, 216–225.
41. Rohrer JD, Woollacott IOC, Dick KM, et al. Serum neurofilament light chain protein is a measure of disease intensity in frontotemporal dementia, *Neurology*, 2016, 87, 1329–1336.
42. Mattsson N, Andreasson U, Zetterberg H, Blennow K. Association of plasma neurofilament light with neurodegeneration in patients with alzheimer disease, *JAMA Neurology*, 2017, 74, 557–566.
43. Byrne LM, Rodrigues FB, Blennow K, et al. Neurofilament light protein in blood as a potential biomarker of neurodegeneration in Huntington's disease: a retrospective cohort analysis, *Lancet Neurol*, 2017, 16, 601–609.
44. Piehl F, Kockum I, Khademi M, et al. Plasma neurofilament light chain levels in patients with MS switching from injectable therapies to fingolimod, *Multiple Sclerosis Journal*, 2018, 24, 1046–1054.
45. Siller N, Kuhle J, Muthuraman M, et al. Serum neurofilament light chain is a biomarker of acute and chronic neuronal damage in early multiple sclerosis, *Multiple Sclerosis Journal*, 2019, 25, 678–686.
46. Oskar H, Shorena J, Sara H, et al. Blood-based NfL: a biomarker for differential diagnosis of parkinsonian disorder, *Neurology*, 2017, 88, 930–937.

47. Ashton JN, Janelidze S, Ahmad ALK, et al. A multicentre validation study of the diagnostic value of plasma neurofilament light, *Nature Communications*, 2021, 12, 3400–3412.
48. Gleerup HS, Sanna F, Hogh P, et al. Saliva neurofilament light chain is not a diagnostic biomarker for neurodegeneration in a mixed memory clinic population, *Frontiers in Aging Neuroscience*, 2021, 13, 659898.
49. Oeckl P, Halbgebauer S, Anderl-Straub S, et al. Glial fibrillary acidic protein in serum is increased in Alzheimer's disease and correlates with cognitive impairment, *Journal of Alzheimer's Disease*, 2019, 67, 481–488.
50. Elahi FM, Casaletto KB, La Joiel R, et al. Plasma biomarkers of astrocytic and neuronal dysfunction in early- and late-onset Alzheimer's disease, *Alzheimer's & Dementia*, 2020, 16, 681–695.
51. (a) Verberk IMW, Thijssen E, Koelewijn J, et al. Combination of plasma amyloid beta (1-42/1-40) and glial fibrillary acidic protein strongly associates with cerebral amyloid pathology, *Alzheimer's Research & Therapy*, 2020, 12, 118. (b) Verberk IMW, Laarhuis MB, van den Bosch K, et al. Serum markers glial fibrillary acidic protein and neurofilament light for prognosis and monitoring in cognitively normal older people: a prospective memory clinic-based cohort study. *Lancet Healthy Longevity*, 2021, 2: e87–e95.
52. Asken BM, Elahi FM, La Joie R, et al. Plasma glial fibrillary acidic protein levels differ along the spectra of amyloid burden and clinical disease stage, *Journal of Alzheimer's Disease*, 2020, 78, 265–276.
53. Benussi A, Ashton NJ, Karikari TK, et al. Serum glial fibrillary acidic protein (GFAP) is a marker of disease severity in frontotemporal lobar degeneration, *Journal of Alzheimer's Disease*, 2020, 77, 1129–1141.
54. Chatterjee P, Pedrini S, Stoops E, et al. Plasma glial fibrillary acidic protein is elevated in cognitively normal older adults at risk of Alzheimer's disease, *Translational Psychiatry*, 2021, 11, 27.
55. Cicognola C, Janelidze S, Hertze J, et al. Plasma glial fibrillary acidic protein detects Alzheimer pathology and predicts future conversion to Alzheimer dementia in patients with mild cognitive impairment, *Alzheimer's Research & Therapy*, 2021, 13, 68.
56. Ng ASL, Tan YJ, Lu Z, et al. Plasma alpha-synuclein detected by single molecule array is increased in PD, *Annals of Clinical and Translational Neurology*, 2019, 6, 615–619.
57. Cariulo C, Martufi P, Verani M, et al. Phospho-S129 alpha-synuclein is present in human plasma but not in cerebrospinal fluid as determined by an ultrasensitive immunoassay, *Frontiers in Neuroscience*, 2019, 13, 889.
58. Youssef P, Kim WS, Halliday GM, Lewis SJG, Dzamko N. Comparison of different platform immunoassays for the measurement of plasma alpha-synuclein in Parkinson's disease patients, *Journal of Parkinson's Disease*, 2021, 11, 1761–1772.
59. Sugarman MA, Zetterberg H, Blennow K, et al. A longitudinal examination of plasma neurofilament light and total tau for the clinical detection and monitoring of Alzheimer's disease, *Neurobiology of Aging*, 2020, 94, 60–70.

60. Wu X, Xiao ZX, Yi JW, et al. Development of a plasma biomarker diagnostic model incorporating ultrasensitive digital immunoassay as a screening strategy for Alzheimer disease in a chinese population, *Clinical Chemistry*, 2021, 67, 1628–1639.
61. Huang YL, Li YH, Xie F, Guo QH. Associations of plasma phosphorylated tau181 and neurofilament light chain with brain amyloid burden and cognition in objectively defined subtle cognitive decline patients, *CNS Neurosci Therapeutics*, 2022, 28, 2195–2205.
62. de Wolf F, Ghanbari M, Licher S, et al. Plasma tau, neurofilament light chain and amyloid-β levels and risk of dementia; a population-based cohort study, *Brain*, 2020, 143, 1220–1232.
63. Simren J, Leuzy A, Karikari TK, et al. The diagnostic and prognostic capabilities of plasma biomarkers in Alzheimer's disease, *Alzheimer's Dement*, 2021, 17, 1145–1156.
64. Li Q, Li Z, Han XX, et al. A panel of plasma biomarkers for differential diagnosis of parkinsonian syndromes, *Frontier in Neuroscience*, 2022, 16, 805953.
65. Wu AHB, Fukushima N, Puskas R, Todd J, Goix P. Development and preliminary clinical validation of a high sensitivity assay for cardiac troponin using a capillary flow (single molecule) fluorescence detector, *Clinical Chemistry*, 2006, 52, 2157–2159.
66. Neumann JT, Havulinna AS, Zeller T, et al. Comparison of three troponins as predictors of future cardiovascular events-prospective results from the FINRISK and BiomaCarRE studies, *PLoS One*, 2014, 9, e90063.
67. Neumann JT, Sörensen NA, Rübsamen N, et al. Evaluation of a new ultra-sensitivity troponin I assay in patients with suspected myocardial infarction, *International Journal of Cardiology*, 2019, 283, 35–40.
68. Wang Y, Yang JZ, Chen C, et al. One-step digital immunoassay for rapid and sensitive detection of cardiac troponin, *ACS Sensors*, 2020, 5, 1126–1131.
69. Jing WW, Wang Y, Chen C, et al. Gradient-based rapid digital immunoassay for high-sensitivity cardiac troponin T (hs-cTnT) detection in 1 uL plasma, *ACS Sensors*, 2021, 6, 399–407.
70. Jarolim P, Patel PP, Conrad MJ, Chang L, et al. Fully automated ultrasensitive digital immunoassay for cardiac troponin I based on single molecule array technology, *Clinical Chemistry*, 2015, 61, 1283–1291.
71. Empana JP, Lerner I, Perier MC, et al. Ultrasensitive troponin I and incident cardiovascular disease, *Arteriosclerosis, Thrombosis, and Vascular Biology*, 2022, 42, 1471–1481.
72. Linkov F, Ferris RL, Yurkovetsky Z, et al. Multiplex analysis of cytokines as biomarkers that differentiate benign and malignant thyroid diseases, *Proteomics Clinical Applications*, 2008, 2, 1575–1585.
73. Dai S, Feng C, Li W, et al. Quantitative detection of tumor necrosis factor-alpha by single molecule counting based on a hybridization chain reaction, *Biosensors and Bioelectronics*, 2014, 60, 180–184.
74. Wu DL, Milutinovic MD, Walt DR, et al. Single molecule array (Simoa) assay with optimal antibody pairs for cytokine detection in human serum samples, *Analyst*, 2015, 140, 6277–6282.

75. Wu D, Katilius E, Olivas E, et al. Incorporation of slow off-rate modified aptamers reagents in single molecule array assays for cytokine detection with ultrahigh sensitivity, *Analytical Chemistry*, 2016, 88, 8385–8389.
76. Wu DL, Dinh TL, Bausk PB, Walt DR. Long-term measurements of human inflammatory cytokines reveal complex baseline variations between individuals, *The American Journal of Pathology*, 2017, 187, 2620–2626.
77. Akdis M, Burgler S, Crameri R, et al. Interleukins, from 1 to 37, and interferon-g: receptors, functions, and roles in diseases, *The Journal of Allergy and Clinical Immunology*, 2011, 127, 701–721.
78. Song L, Hanlon DW, Chang L, et al. Single molecule measurements of tumor necrosis factor α and interleukin-6 in the plasma of patients with Crohn's disease, *Journal of Immunological Methods*, 2011, 372, 177–186.
79. Rivnak AJ, Rissin DM, Kan CW, et al. A fully-automated, six-plex single molecule immunoassay for measuring cytokines in blood, *Journal of Immunological Methods*, 2015, 424, 20–27.
80. Gupta V, Kalia N, Yadav M, et al. Optimization and qualification of the single molecule array digital immunoassay for IL-12p70 in plasma of cancer patients, *Bioanalysis*, 2018, 10, 1413–1425.
81. Song Y, Ye Y, Su SH, et al. A digital protein microarray for COVID-19 cytokine storm monitoring, *Lab on a Chip*, 2021, 21, 331–343.
82. Flora C, Singer BH, Ghosh M, et al. Machine learning-based cytokine microarray digital immunoassay analysis, *Biosensors and Bioelectronics*, 2021, 180, 113088.
83. Gao Z, Song Y, Hsiao TY, et al. Machine-learning-assisted microfluidic nanoplasmonic digital immunoassay for cytokine storm profiling in COVID-19 patients, *ACS Nano*, 2021, 15, 18023–18036.
84. Wu C, Dougan TJ, Walt DR. High-throughput, high-multiplex digital protein detection with attomolar sensitivity, *ACS Nano*, 2022, 16, 1025–1035.
85. Su SH, Song YJ, Stephens A, et al. A tissue chip with integrated digital immunosensors: in situ brain endothelial barrier cytokine secretion monitoring, *Biosensors and Bioelectronics*, 2023, 224, 115030.
86. Rodero MP, Decalf J, Bondet V, et al. Detection of interferon alpha protein reveals differential levels and cellular sources in disease, *Journal of Experimental Medicine*, 2017, 214, 1547–1555.
87. Mathian A, Mouries-Martin S, Dorgham K, et al. Monitoring disease activity in systemic lupus erythematosus with single-molecule array digital enzyme-linked immunosorbent assay quantification of serum interferon-α, *Arthritis & Rheumatology*, 2019, 71, 756–765.
88. Ben Salah, E, Dorgham, K, Lesenechal, M, et al. Assessment of an ultra-sensitive IFNI gamma immunoassay prototype for latent tuberculosis diagnosis, *European Cytokine Network*, 2018, 29, 136–145.
89. Llibre A, Bilek N, Bondet V, et al. Plasma type I IFN protein concentrations in human tuberculosis, *Frontiers in Cellular and Infection Microbiology*, 2019, 9, 296.
90. Upasani V, Scagnolari C, Frasca F, et al. Decreased type I interferon production by plasmacytoid dendritic cells contributes to severe dengue, *Frontiers in Immunology*, 2020, 11, 605087.

91. Huijser E, Gopfert J, Brkic Z, et al. Serum IFN-alpha2 measured by single-molecule array associates with systemic disease manifestations in Sjogren's syndrome, *Rheumatology*, 2021, 61, 2156–2166.
92. Trouillet-Assant S, Viel S, Ouziel A, et al. Type I interferon in children with viral or bacterial infections, *Clinical Chemistry*, 2020, 66, 802–808.
93. Vieira M, Régnier P, Maciejewski-Duval A, et al. Interferon signature in giant cell arteritis aortitis, *Journal of Autoimmunity*, 2022, 127, 102796.
94. Mähönen K, Hau A, Bondet V, et al. Activation of NLRP3 inflammasome in the skin of patients with systemic and cutaneous lupus erythematosus, *Acta Dermato-Venereologica*, 2022, 102, 00708.
95. Trutschel D, Bost P, Mariette X, et al. Variability of primary Sjögren's syndrome is driven by interferon-α and interferon-α blood levels are associated with the class II HLA-DQ locus, *Arthritis & Rheumatology*, 2022, 74, 1991–2002.
96. Devi-Marulkar P, Moraes-Cabe C, Campagne P, et al. Altered immune phenotypes and HLA-DQB1 gene variation in multiple sclerosis patients failing interferon β treatment, *Frontiers in Immunology*, 2021, 12, 628375.
97. Casey KA, Smith MA, Sinibaldi D, et al. Modulation of cardiometabolic disease markers by Type I interferon inhibition in systemic lupus erythematosus, *Arthritis & Rheumatology*, 2021, 73, 459–471.
98. Nagy-Simon T, Hada A-M, Suarasan S, et al. Recent advances on the development of plasmon-assisted biosensors for detection of C-reactive protein, *Journal of Molecular Structure*, 2021, 1246, 131178.
99. Belushkin A, Yesilkoy F, Altug H, et al. Nanoparticle-enhanced plasmonic biosensor for digital biomarker detection in a microarray, *ACS Nano*, 2018, 12, 4453–4461.
100. Belushkin A, Yesilkoy F, González-López JJ, et al. Rapid and digital detection of inflammatory biomarkers enabled by a novel portable nanoplasmonic imager, *Small*, 2020, 16, 1906108.
101. Zhang W, Dang T, Li Y, et al. Digital plasmonic immunosorbent assay for dynamic imaging detection of protein binding, *Sensors and Actuators B: Chemical*, 2021, 348, 130711.
102. National Health Commission of the People's Republic of China, "Diagnosis and Treatment Protocol for COVID-19 (Trial Version 10)" [2023-1-5]. http://www.nhc.gov.cn/.
103. Scheiblauer H, Filomena A, Nitsche A, et al. Comparative sensitivity evaluation for 122 CE-marked rapid diagnostic tests for SARS-CoV-2 antigen, Germany, September 2020 to April 2021, *Eurosurveillance*, 2021, 26, 2100441.
104. Azkur AK, Akdis M, Azkur D, et al. Immune response to SARS-CoV-2 and mechanisms of immunopathological changes in COVID-19, *Allergy*, 2020, 75, 1564–1581.
105. Breshears LE, Nguyen BT, Akarapipad P, et al. Sensitive, smartphone-based SARS-CoV-2 detection from clinical saline gargle samples, *PNAS Nexus*, 2022, 1, 1–9.

106. Kim S, Eades C, Yoon JY. COVID-19 variants' cross-reactivity on the paper microfluidic particle counting immunoassay, *Analytical and Bioanalytical Chemistry*, 2022, 414, 7957–7965.
107. Chiba R, Miyakawa K, Aoki K, et al. Development of a fully automated desktop analyzer and ultrahigh sensitivity digital immunoassay for SARS-CoV-2 nucleocapsid antigen detection, *Biomedicines*, 2022, 10, 2291.
108. Ogata AF, Maley AM, Wu C, et al. Ultra-sensitive serial profiling of SARS-CoV-2 antigens and antibodies in plasma to understand disease progression in COVID-19 patients with severe disease, *Clinical Chemistry*, 2020, 66, 1562–1572.
109. Olsen DA, Brasen CL, Kahns S, et al. Quantifying SARS-CoV-2 nucleocapsid antigen in oropharyngeal swabs using single molecule array technology, *Scientific Reports*, 2021, 11, 20323.
110. Shan D, Johnson JM, Fernandes SC, et al. N-protein presents early in blood, dried blood and saliva during asymptomatic and symptomatic SARS-CoV-2 infection, *Nature Communications*, 2021, 12, 1931.
111. Zhao B, Che C, Wang W, et al. Single-step, wash-free digital immunoassay for rapid quantitative analysis of serological antibody against SARS-CoV-2 by photonic resonator absorption microscopy, *Talanta*, 2021, 225, 122004.
112. Ou F, Lai D, Kuang X, et al. Ultrasensitive monitoring of SARS-CoV-2-specific antibody responses based on a digital approach reveals one week of IgG seroconversion, *Biosensors and Bioelectronics*, 2022, 217, 114710.
113. Gilboa T, Cohen L, Cheng CA, et al. A SARS-CoV-2 neutralization assay using single molecule arrays, *Angewandte Chemie International Edition*, 2021, 60, 25966–25972.
114. Norman M, Gilboa T, Ogata AF, et al. Ultrasensitive high-resolution profiling of early seroconversion in patients with COVID-19, *Nature Biomedical Engineering*, 2020, 4, 1180–1187.
115. Chang L, Song L, Fournier DR, et al. Simple diffusion-constrained immunoassay for p 24 protein with the sensitivity of nucleic acid amplification for detecting acute HIV infection, *Journal of Virological Methods*, 2013, 188, 153–160.
116. Chang L, Stone M, Busch M, et al. Rapid, fully automated digital immunoassay for p 24 protein with the sensitivity of nucleic acid amplification for detecting acute HIV infection, *Clinical Chemistry*, 2015, 61, 1372–1380.
117. Che C, Li N, Long KD, et al. Activate capture and digital counting (AC + DC) assay for protein biomarker detection integrated with a self-powered microfluidic cartridge, *Lab on a Chip*, 2019, 19, 3943–3953.
118. Passaes C, Delagreverie HM, Avettand-Fenoel V, et al. Ultrasensitive detection of p24 in plasma samples from people with primary and chronic HIV-1 infection, *Journal of Virology*, 2021, 95, e00016–21.
119. Passaes CPB, Bruel T, Decalf J, et al. Ultrasensitive HIV-1 p24 assay detects single infected cells and differences in reservoir induction by latency reversal agents, *Journal of Virology*, 2017, 91, e02296–16.

120. Wu G, Swanson M, Talla A, et al. HDAC inhibition induces HIV-1 protein and enables immune-based clearance following latency reversal, *JCI Insight*, 2017, 2, e92901.
121. Gaylord ST, Abdul-Aziz S, Walt DR. Single-molecule arrays for ultrasensitive detection of host immune response to dengue virus infection, *Journal of Clinical Microbiology*, 2015, 53, 1722–1724.
122. Wu Z, Guo WJ, Bai YY, et al. Digital single virus electrochemical enzyme-linked immunoassay for ultrasensitive H7N9 avian influenza virus counting, *Analytical Chemistry*, 2018, 90, 1683–1690.
123. Wu Z, Zeng T, Guo WJ, et al. Digital single virus immunoassay for ultrasensitive multiplex avian influenza virus detection based on fluorescent magnetic multifunctional nanospheres, *ACS Applied Materials & Interfaces*, 2019, 11, 5762–5770.
124. Leirs K, Tewari KP, Decrop D, et al. Bioassay development for ultrasensitive detection of influenza a nucleoprotein using digital ELISA, *Analytical Chemistry*, 2016, 88, 8450–8458.
125. Tabata KV, Minagawa Y, Kawaguchi Y, et al. Antibody-free digital influenza virus counting based on neuraminidase activity, *Scientific Reports*, 2019, 9, 1067.
126. Minagawa Y, Ueno H, Tabata KV, et al. Mobile imaging platform for digital influenza virus counting, *Lab on a Chip*, 2019, 19, 2678–2687.
127. Yasuura M, Fukuda T, Hatano K, et al. Quick and ultra-sensitive digital assay of influenza virus using sub-picoliter microwells, *Analytica Chimica Acta*, 2022, 1213, 339926.
128. Petersen ERB, Olsen DA, Christensen H, et al. Rhodopsin in plasma from patients with diabetic retinopathy - development and validation of digital ELISA by Single Molecule Array (Simoa) technology, *Journal Immunological Methods*, 2017, 446, 60–69.
129. Mukherjee A, Dang T, Morrell H, et al. Expanding the utility of high-sensitivity dried blood spot immunoassay testing with single molecule counting, *Journal of Applied Laboratory Medicine*, 2018, 2, 674–686.
130. Dang T, Morrell H, Estis J, et al. Effect of health and training on ultrasensitive cardiac troponin in marathon runners, *Journal of Applied Laboratory Medicine*, 2019, 3, 775–787.
131. Sandlund J, Bartolome A, Biscocho S, et al. Ultrasensitive detection of Clostridioides difficile toxins a and b by use of automated single-molecule counting technology, *Journal of Clinical Microbiology*, 2018, 56, e00908–18.
132. Hansen G, Young S. Ultrasensitive detection of Clostridioides difficile toxins in stool by use of single-molecule counting technology: comparison with detection of free toxin by cell culture cytotoxicity neutralization assay, *Journal of Clinical Microbiology*, 2019, 57, e00719–19.
133. Rosenzweig Z, Yeung ES. Laser-based particle-counting microimmunoassay for the analysis of single human erythrocytes, *Analytical Chemistry*, 1994, 66, 10, 1771–1776.
134. Schultz S, Smith DR, Mock JJ, Schultz DA. Single-target molecule detection with nonbleaching multicolor optical immunolabels, *Proceedings of the National Academy of Sciences of the United States of America*, 2000, 97, 996–1001.

CHAPTER 7

Outlook of Digital Immunoassays

Digital immunoassay (DI) has emerged as the most sensitive immunoassay technology to date. Commercial instruments have been available for over a decade, playing a significant role in the early detection of biomarkers for neurodegenerative diseases. This has laid a solid foundation for applications such as early detection and screening of major diseases, biomarker discovery, and personalized diagnostics and treatment. However, the clinical application of DI is still not fully realized. Its potential to address key challenges in early cancer screening has not been fully demonstrated, and its irreplaceable role in the medical field needs to be further established. From a technological standpoint, there remains room for improvement and refinement; from an application standpoint, there is a need for more efforts toward biomarker discovery and validation; and from an instrumentation standpoint, further advancements in automation, portability, and multiplexing are required.

TECHNOLOGICAL ASPECTS

Although the LOD of DI has reached the aM level, it has not yet attained its ultimate, potential sensitivity. The fundamental reason is that detection efficiency does not reach 100%. Major sources of detection efficiency loss include losses during the wash and separation steps in the immunoassay process, losses during sample transfer, incomplete dissociation

DOI: 10.1201/9781003729211-7

of immunocomplexes, and failure to detect all discrete units. The first-generation Simoa microwell array enzyme immunoassay had an overall detection efficiency of 4.8%, with 500,000 magnetic beads participating in the reaction but only 23,000 beads ultimately detected. The upgraded version of the instrument improved the efficiency of magnetic bead entry into microwells using magnetic and capillary forces, achieving a detection efficiency of 47.2%. The LOD improved from 313 aM to 0.72 aM.[1] Soft discretization does not need a physical compartment, so the efficiency of immunocomplex deposition onto the substrate largely determines detection efficiency. At pH 7, protein adsorption efficiency on a glass substrate is typically around 1%. When using a hydrophobic substrate (PDMS), the deposition efficiency can reach 20%.[2] In 2020, both our group and the Walt group proposed enhancing immunocomplex deposition rates using evaporation to dryness.[3,4] Although evaporation increased the deposition efficiency to nearly 100%, losses still occur during operations before and after deposition, which are difficult to avoid. For example, separation and washing processes result in a 12% loss, pipette transfer causes an 8.3% loss, and imaging analysis accounts for a 10% loss.[5] Minimizing sample transfer, reducing operational steps, and improving discretization and detection efficiency are the main technological challenges of DI. Based on current technologies, we believe homogeneous soft discretization-based DI holds promise to solve these issues. Homogeneous assays require one-step mixing and direct detection, eliminating the need for separation and washing, thus minimizing operation and transfer steps for both reaction and sample loading. Soft discretization does not require physical compartments, avoiding issues such as poor bead loading in microwells or droplet encapsulation failures. In theory, discretization efficiency can reach 100%. Moreover, homogeneous immunoassays also have the advantages of reduced cumulative error and easier implementation of fully automated and portable instruments. Developing novel signal recognition and amplification techniques is a key direction for advancing homogeneous DI technology.

APPLICATION ASPECTS

In its early application stages, DI primarily focused on detecting tumor biomarkers in blood, but it has yet to make a breakthrough. The main reason lies in the insufficient specificity of existing tumor biomarkers. There are too few clinically recognized specific tumor biomarkers. Early diagnosis based solely on tumor biomarker concentrations in blood is not yet widely accepted, and pathological diagnosis remains the gold standard for cancer confirmation. While biomarker concentrations rise in cancer patients, healthy individuals also have measurable amounts of these biomarkers, and the concentration boundaries between disease and non-disease states are not sharp enough to necessitate DI. In other words, conventional immunoassay technologies are sufficient to detect existing biomarker concentration ranges. Therefore, from the application perspective, rather than continuing to use existing biomarkers, it is more promising to leverage DI's high sensitivity to screen and validate new biomarkers present at low concentrations in blood for developing early cancer diagnostic technologies. Ideal biomarkers should be either extremely low in concentration (below 10^{-15} mol/L) or entirely absent in healthy individuals and have strong associations and causal links with the disease. Such biomarkers likely exist in blood but have not been discovered due to insufficient detection sensitivity. Fully utilizing the sensitivity advantage of DI to build precise and reliable biomarker panels is an important application area.

Another potential application of DI lies in personalized monitoring of tumor biomarkers. Currently, diagnostic standards using biomarkers are based on the average values from "normal" populations. Test samples are compared to these standards, and values above the upper limit are considered positive, while those below are considered negative. These thresholds and standards are based on population averages, but this diagnostic and screening strategy lacks precision. Individual expression of tumor biomarkers varies significantly—by as much as 1,000-fold.[6] For individuals with extremely low biomarker expression, even if levels rise 1,000-fold and indicate disease, the concentration may still fall within the "normal" range, missing opportunities for early diagnosis and screening. Conversely, individuals with high biomarker expression may have concentrations exceeding the threshold while still healthy, leading to false positives and overtreatment. Biomarker levels in healthy individuals do not follow a normal distribution, so using population averages as thresholds is inappropriate. The ideal screening and diagnostic thresholds should vary by individual. By tracking changes in an individual's biomarker levels over time, one can establish correlations between biomarker levels and tumor progression, enabling assessment of that individual's cancer risk — i.e., personalized tumor biomarker monitoring. DI's high sensitivity can also be leveraged in the early detection of cardiovascular diseases and viral infections.

INSTRUMENTATION ASPECTS

The Simoa system by Quanterix is the flagship representative of DI instruments. However, because it employs heterogeneous hard discretization and sample pretreatment requires many steps, resulting in significant sample loss and limiting further improvements in detection sensitivity. Additionally, to automate sample pretreatment, the instrument is relatively large, leaving little room for miniaturization, which makes point-of-care testing difficult. The development trend for DI instruments should be toward achieving detection limits down to 10^{18} mol/L, a sufficiently wide dynamic range, portability, multiplexing capabilities, and fully automated, all-in-one systems capable of "sample-in, answer-out" workflows.

In summary, the future of DI instrumentation lies in developing fully automated analyzers based on novel homogeneous DI technology. Looking back at the completion of the Human Genome Project at the end of the 20th century, the emergence of high-throughput sequencers shortened sequencing time and accelerated the project's completion. Similarly, the successful development of efficient DI analyzers could mark the right moment to propose a "biomarker panel project", offering opportunities for large-scale biomarker screening and validation, and in turn driving the clinical application of individualized early screening and early diagnosis of major diseases.

REFERENCES

1. Mickert MJ, Farka Z, Kostiv U, Hlavacek A, Horak D, Skladal P, Gorris HH. Measurement of sub-femtomolar concentrations of prostate-specific antigen through single-molecule counting with an upconversion-linked immunosorbent assay, *Analytical Chemistry*, 2019, 91, 9435–9441.
2. Li N, Tang H, Gai HW, Dong XL, Wang Q, Yeung ES. Determination of protein surface excess on a liquid/solid interface by single-molecule counting, *Analytical and Bioanalytical Chemistry*, 2009, 394, 1879–1885.
3. Zhang QQ, Zhang XB, Li JJ, Gai HW. Nonstochastic Protein counting analysis for precision biomarker detection: suppressing Poisson noise at ultralow concentration, *Analytical Chemistry*, 2020, 92, 654–658.
4. Wu C, Garden PM, Walt DR. Ultrasensitive detection of attomolar protein concentrations by dropcast single molecule assays, *Journal of the American Chemical Society*, 2020, 142, 12314–12323.

5. Kan CW, Tobos CI, Rissin DM, et al. Digital enzyme-linked immunosorbent assays with sub-attomolar detection limits based on low numbers of capture beads combined with high efficiency bead analysis, *Lab on a Chip*, 2020, 20, 2122–2135.
6. Wu D, Dinh TL, Bausk BP, Walt DR. Long-term measurements of human inflammatory cytokines reveal complex baseline variations between individuals, *American Journal of Pathology*, 2017, 187, 2620–2626.

For Product Safety Concerns and Information please contact our EU
representative GPSR@taylorandfrancis.com
Taylor & Francis Verlag GmbH, Kaufingerstraße 24, 80331 München, Germany

www.ingramcontent.com/pod-product-compliance
Ingram Content Group UK Ltd.
Pitfield, Milton Keynes, MK11 3LW, UK
UKHW022216190926
475702UK00003B/378
* 9 7 8 1 0 4 1 2 1 3 6 2 8 *